Lecture Notes in Electrical Engineering

Volume 787

The book series *Lecture Notes in Electrical Engineering* (LNEE) publishes the latest developments in Electrical Engineering - quickly, informally and in high quality. While original research reported in proceedings and monographs has traditionally formed the core of LNEE, we also encourage authors to submit books devoted to supporting student education and professional training in the various fields and applications areas of electrical engineering. The series cover classical and emerging topics concerning:

- Communication Engineering, Information Theory and Networks
- Electronics Engineering and Microelectronics
- Signal, Image and Speech Processing
- Wireless and Mobile Communication
- Circuits and Systems
- Energy Systems, Power Electronics and Electrical Machines
- Electro-optical Engineering
- Instrumentation Engineering
- Avionics Engineering
- Control Systems
- Internet-of-Things and Cybersecurity
- Biomedical Devices, MEMS and NEMS

For general information about this book series, comments or suggestions, please contact leontina.dicecco@springer.com.

To submit a proposal or request further information, please contact the Publishing Editor in your country:

China

Jasmine Dou, Editor (jasmine.dou@springer.com)

India, Japan, Rest of Asia

Swati Meherishi, Editorial Director (Swati.Meherishi@springer.com)

Southeast Asia, Australia, New Zealand

Ramesh Nath Premnath, Editor (ramesh.premnath@springernature.com)

USA, Canada:

Michael Luby, Senior Editor (michael.luby@springer.com)

All other Countries:

Leontina Di Cecco, Senior Editor (leontina.dicecco@springer.com)

**** This series is indexed by EI Compendex and Scopus databases. ****

More information about this series at http://www.springer.com/series/7818

Shiban Kishen Koul · Richa Bharadwaj

Wearable Antennas and Body Centric Communication

Present and Future

Springer

Shiban Kishen Koul
Centre for Applied Research in Electronics
Indian Institute of Technology Delhi
New Delhi, India

Richa Bharadwaj
Centre for Applied Research in Electronics
Indian Institute of Technology Delhi
New Delhi, India

ISSN 1876-1100 ISSN 1876-1119 (electronic)
Lecture Notes in Electrical Engineering
ISBN 978-981-16-3975-3 ISBN 978-981-16-3973-9 (eBook)
https://doi.org/10.1007/978-981-16-3973-9

This Springer imprint is published by the registered company Springer Nature Singapore Pte Ltd.
The registered company address is: 152 Beach Road, #21-01/04 Gateway East, Singapore 189721, Singapore

We would like to dedicate our efforts to our families who inspired and supported us through everything.

We would also like to acknowledge our students and colleagues for their support throughout.

Preface

Wireless body area network (WBAN) technology is providing attractive new possibilities in wearable communication considering increase in demand of connectivity, information-centric users and the ever-evolving wireless world. From applications in day-to-day activities and general well-being to specific domains such as healthcare, telemedicine, defence, sports, entertainment, search, and rescue emergency operations, WBANs form an integral part in enhancing quality of life. The rising popularity of commercial wearable gadgets, such as fitness trackers and smart watches which provide real-time information regarding various health stats, enhanced detection, and sensing capabilities, has paved way for several research advancements in the domain of wearable sensing technologies. The upcoming era of Internet of Things (IoT) is revolutionizing the way gadgets connect with wearable devices being the key focus, operating in variable and dynamic environments over short and long range. The future generation of wearable devices will be compact, low cost, lightweight, efficient, low power, portable, and accessible, provide flexibility of integration, and work with high data rates and high-quality wireless connectivity.

Antenna is one of the key components of the WBAN which is integrated with wearable gadgets and clothing to provide robust wireless connectivity between the wearable devices suitable for a wide range of applications. Research and development in the field of antennas and propagation for body-centric communication is an upcoming area due to the miniaturization of devices, new fabrication technologies, advancement in material science, and availability of wide range of the electromagnetic spectrum for operation of the wearable devices. They must efficiently support various channels ranging from on-body communications to off-body/body-to-body and even in-body implantable communications.

Antennas and radio wave propagation constitute the basic elements of the wireless channel which determine the quality and the reliability of the wireless link and hence have a great impact on the quality of service offered by a whole system. Ultra-wideband (UWB) (3.6–10 GHz) and 60 GHz millimeter-wave (mmWave) (57–64) GHz frequency bands are considered as attractive solutions for future WBANs due to the high data rates, compact devices, and availability of wide bandwidth. Many research activities have been focused on the design and development of wearable antennas and characterization of the body-centric propagation channel which need to consider various challenges of working in proximity with the human body, dynamic scenarios, and variable environments.

This book highlights the recent progress and state-of-the-art techniques in the field of antennas for body-centric communication at UWB and 60 GHz mmWaves frequencies. Work related to current trends, research aspects in wearable antenna design, optimization of antenna parameters, and characterization and modelling of the channel are reported for various types of body-centric links. Various applications have also been discussed such as localization and tracking of the human subject, monitoring of physical activities using wearable technology. Radar-based applications such as monitoring of vital sign parameters, tracking of human subject, and medical imaging have also been reported. Finally, IoT applications and machine learning approach have been described which aim to enhance the overall performance in various domains such as healthcare systems, smart home, and smart cities. This book will serve as a comprehensive resource for graduate students, researchers, and professionals in academia as well as industry in the field of antennas and propagation, microwave engineering, and wireless communication.

Chapter 1 introduces the wireless body area networks and gives an overview of the applications, current and future technologies, and an outline of the antenna and propagation aspects for wireless body-centric communication. The scope of the book is also provided with a summary of the content of the chapters.

Chapter 2 describes various aspects of the on-body propagation from antenna design to channel modelling in the UWB and mmWave frequency range. Key features related to on-body antenna design and requirements, simulation and phantom-based study, and body-centric channel characterization for static and dynamic scenarios are reported.

Chapter 3 focuses on modelling and characterization of the off-body and body-to-body propagation channels for UWB and mmWave frequency range. Theoretical, numerical, simulation-based, and experimental investigations are reported to understand the channel behaviour in the presence of the human subject.

Chapter 4 presents latest trends in wearable flexible antenna design covering aspects such as fabrication techniques, substrate material selection, and novel designs suitable for UWB and mmWave communication. Electromagnetic and mechanical properties of such antennas are discussed for free space and on-body scenarios.

Chapter 5 gives an overview of in-body and on-body antenna design and channel characteristics of implantable UWB communication systems suitable for several medical applications such as capsule endoscopy and vital body parameter monitoring.

Chapter 6 gives an insight on the factors affecting the localization accuracy while tracking a human subject in an indoor environment using simple and effective techniques based on channel information and time of arrival localization techniques.

Chapter 7 presents work related to monitoring and assessment of physical activities using channel information, gait movement, and joint angle estimation during flexion/extension of limbs using wearable UWB technology.

Chapter 8 presents recent advances and state-of-the-art techniques related to IR-UWB and mmWave radar system design, vital sign monitoring, detection, daily activity monitoring, fall detection, sleep monitoring, gait analysis, and gesture recognition.

Chapter 9 discusses various research studies based on UWB radar imaging for medical and through-wall detection applications. Several antenna designs, development of image reconstruction algorithms, and providing set-up details of the complete microwave imaging systems are reported.

Chapter 10 presents an overview of the Internet of Things (IoT) and explores the role of machine learning in enhancing overall performance with the focus on healthcare applications. Various technologies and state-of-the-art techniques related to compact antenna design for body-centric communication for IoT applications are discussed in this chapter.

New Delhi, India

Shiban Kishen Koul
Richa Bharadwaj

Contents

About the Authors

Shiban Kishen Koul (Life Fellow, IEEE) received the B.E. in electrical engineering from the Regional Engineering College, Srinagar, in 1977, and the M.Tech. and Ph.D. degrees in microwave engineering from IIT Delhi, New Delhi, India, in 1979 and 1983, respectively. He is Emeritus Professor in the Indian Institute of Technology Delhi since 2019. He served as Deputy Director (Strategy and Planning) in IIT Delhi from 2012 to 2016 and Mentor Deputy Director (Strategy and Planning, International Affairs) in IIT Jammu, J&K, India, from 2018 to 2020. He also served as Chairman of Astra Microwave Products Limited, Hyderabad, from 2009 to 2019 and Dr R. P. Shenoy Astra Microwave Chair Professor at IIT Delhi from 2014 to 2019. He has successfully completed 38 major sponsored projects, 52 consultancy projects, and 61 technology development projects. He has authored or co-authored 506 research articles, 13 state-of-the-art books, 4 chapters, and 2 e-books. He holds 16 patents, 6 copyrights, and one trademark. He has guided 25 Ph.D. theses and more than 100 master's theses. His current research interests include RF MEMS, nonlinear device modelling, microwave and millimetre wave active and passive circuit design, and reconfigurable microwave circuits including antennas. He is Fellow of the Indian National Academy of Engineering, India, and the Institution of Electronics and Telecommunication Engineers (IETE), India. He served as Distinguished Microwave Lecturer of IEEE MTT-S from 2012 to 2014. He was a recipient of numerous awards including the Indian National Science Academy (INSA) Young Scientist Award, in 1986; the Top Invention Award of the National Research Development Council for his contributions to the indigenous development of ferrite phase shifter technology, in 1991; the VASVIK Award for the development of Ka-band components and phase shifters, in 1994; Ram Lal Wadhwa Gold Medal from the Institution of Electronics and Communication Engineers (IETE), in 1995; the Academic Excellence Award from the Indian government for his pioneering contributions to phase control modules for Rajendra Radar, in 1998; the Shri Om Prakash Bhasin Award in the field of electronics and information technology, in 2009; the VASVIK Award for the contributions made to the area of information and communication technology (ICT), in 2012; the Teaching Excellence Award from IIT Delhi, in 2012; the M. N. Saha Memorial

Award from IETE, in 2013; and the IEEE MTT Society Distinguished Educator Award, in 2014. He is Chief Editor of IETE Journal of Research and Associate Editor of the International Journal of Microwave and Wireless Technologies, Cambridge University Press.

Richa Bharadwaj (Member, IEEE) received the Bachelors of Engineering degree (Hons.) in electronics and communication from Panjab University Chandigarh, India, in 2008, the M.S. degree in micro- and nanotechnologies for integrated systems from Politecnico di Torino, Turin, Italy; INPG Grenoble, Grenoble, France; and EPFL Lausanne, Lausanne, Switzerland, in 2010, and the Ph.D. degree in electronic engineering with the specialization in ultra-wideband technology from the School of Electronics and Computer Science, Antennas and Electromagnetics Research Group, Queen Mary University of London, London, UK, in 2015. She is currently Postdoctoral Fellow at the Centre for Applied Research in Electronics, Indian Institute of Technology Delhi, New Delhi, India. She has authored or co-authored two chapters and several research publications in leading international journals and peer-reviewed conferences. Her current research interests include ultra-wideband communication, 3D localization, wireless sensor networks, body-centric communication, radio propagation characterization and modelling, miniaturized antenna design, and flexible and wearable communication. She was awarded the C. J. Reddy Best Paper Award for Young Professionals at The Indian Conference on Antennas and Propagation (INCAP 2019) held at Ahmedabad, India. She is a reviewer for several leading transactions and journals in the fields of antennas and propagation, wireless communication, sensors, and vehicular technology.

Abbreviations

1D	One-Dimensional
2D	Two-Dimensional
3D	Three-Dimensional
5G	Fifth Generation
6G	Sixth Generation
A	Received Signal Amplitude
AAV	Absolute Acceleration Variation
AB_R	Abdomen Right
ABS	Acrylonitrile Butadiene Styrene
AD	Arctangent Demodulation
ADC	Analog-To-Digital Converter
ADL	Activities of Daily Living
AFD	Average Fade Duration
AI	Artificial Intelligence
AM	Additive Manufacture
AN	Ankle
AOA	Angle of Arrival
ATA-FGP	All-Textile Antenna with Full Ground Plane
AVA	Antipodal Vivaldi Antenna
B	Back
BANs	Body Area Networks
BAVA	Balanced Antipodal Vivaldi Antenna
BCNs	Body-Centric Networks
BCWN	Body-Centric Wireless Networks
BCWS	Body-Centric Wireless Sensor
BDT	Boosted Decision Tree
BLE	Bluetooth Low Energy
BMI	Body Mass Index
B-MI	Brain–Machine Interface
BOS	Base of Support

B-P	Back Projection
BP	Blood Pressure
BR	Breath Rate
BSF	Body Shadowing Factor
BSs	Base Stations
CAD	Computer-Aided Design
CDF	Cumulative Distribution Function
CF	Coherence Factor
CFAR	Constant False Alarm Rate
CF-DAS	Coherence Factor Delay and Sum
CFR	Channel Frequency Response
CIR	Channel Impulse Response
CNN	Convolutional Neural Network
CP	Circular Polarization
CPI	Coherent Processing Interval
CPW	Coplanar Waveguide
CS	Compressed Sensing
CSAR	Circular Synthetic Aperture Radar
CSD	Complex Signal Demodulation
CSF	Cerebrospinal Fluid
CT	Computed Tomography
CTBV	Continuous Time Binary Valued
CW	Continuous Wave
CWT	Continuous Wavelet Transform
DAQ	Multifunction Data Acquisition
DAS	Delay And Sum
DCNN	Deep Convolutional Neural Network
DCT	Discrete Cosine Transform
DL	Deep Learning
DMAS	Delay-Multiply-And-Sum
DNNs	Deep Neural Networks
DOP	Dilution of Precision
DRA	Dielectric Resonance Antenna
D-S	Displacement Signal
DS	Doppler Spectrogram
DSN	Noise Threshold
EBG	Electromagnetic Band Gap
ECG	Electrocardiograph
ECTSRLS	Equality Constrained Taylor Series Robust Least Squares
ECU	Electronic Control Unit
EEG	Electroencephalogram
EEMD	Ensemble Empirical Mode Decomposition
EFIR	Extended Finite Impulse Response
EHR	Electronic Health Record
EKF	Extended Kalman Filter

EM	Electromagnetic
EMD	Empirical Mode Decomposition
EMG	Electromyograph
ESD	Ensemble Subspace Discriminant
ETSA	Exponentially Tapered Slot Antenna
F	Face
FA	Frequency Accumulation
FCC	Federal Communications Commission
FDM	Fused Deposition Modelling
FDTD	Finite-Difference Time-Domain
FED	Feature Embedding Dimension
FEM	Finite Element Method
FFF	Fuse Filament Fabrication
FFT	Fast Fourier Transform
FIR	Finite Impulse Response
FIT	Finite Integration Technique
FMCW	Frequency-Modulated Continuous Wave
FN	False Negative
FP	False Positive
FPCB	Flexible Printed Circuit Board
FPGA	Field-Programmable Gate Array
FPS	Frames Per Second
FR	Front
FS	Free Space
FTI	Feature Time Index
FVPIEF	First Valley-Peak of IMF Energy Function
GAF	Graphene-Assembled Film
GBP	Global Back Projection
GDOP	Geometric Dilution of Precision
GI	Gastrointestinal
GO	Geometrical Optics
GPIB	General-Purpose Interface Bus
GPS	Global Positioning System
GSM	Global System for Mobile Communications
H	Horizontal
HAPA	Harmonic Path
HDOP	Horizontal Dilution of Precision
HEDL	Half Elliptical-Shaped Dielectric Lens
HFSS	High-Frequency Structure Simulator
HMLD	Harmonic Multiple Loop Detection
HOC	Higher-Order Cumulant
HR	Heart Rate
I	In-Phase
IAA	Iterative Adaptive Approach
IB2IB	In-Body to In-Body

IB2OB	In-Body to On-Body
IC	Integrated Circuit
ICT	Information and Communication Technology
IFFT	Inverse Fast Fourier Transform
IMF	Intrinsic Mode Function
IMU	Inertial Measurement Unit
IN	Inner
IoT	Internet of Things
IR	Infrared
IR-UWB	Impulse Radio-Ultra-Wideband
ISI	Intersymbol Interference
ISM	Industrial–Scientific–Medical
ITU	International Telecommunication Union
KMC	K-Means Clustering
k-NN	K-Nearest Neighbour
L.	Left
L. AK	Left Ankle
LAN	Local Area Network
LCP	Liquid Crystal Polymer
LCR	Level Crossing Rate
LHM	Left-Handed Metamaterial
LO	Local Oscillator
LOS	Line of Sight
LS-SVM	Least Squares Support Vector Machine
LSTM	Long Short-Term Memory
LTCC	Low-Temperature Co-Fired Ceramic
LWA	Leaky-Wave Antenna
MARG sensors	Magnetic, Angular Rate, And Gravity
MAVA	Modified Antipodal Vivaldi Antenna
MC-SVM	Multi-Class Support Vector Machine
MCU	Microcontroller Unit
MDS	Micro-Doppler Signatures
MEMS	Microelectromechanical Systems
MHT	Multi-Hypothesis Tracking
MI	Microwave Imaging
MICS	Medical Implant Communications Service
MIMO	Multiple Input Multiple Output
MIS	Microwave Imaging System
ML	Machine Learning
MLDS	Millimetre-Wave Life Detection System
MLE	Maximum Likelihood Estimation
MLP	Multi-Layer Perceptron
mmWave	Millimetre Wave (mmW)
MoM	Method of Moments
MPA	Microstrip Patch Antenna

MPCs	Multipath Components
MPOC	Modified-Phase-Only-Correlator
MRC	Maximum Ratio Combining
MRI	Magnetic Resonance Imaging
MS	Mobile Station
MStrip	Microstrip
MTMs	Metamaterials
MWCNTs	Multi-Walled Carbon Nanotubes
MWDAS	Modified Weighted-Delay And Sum
MWI	Microwave Imaging
NB	Naive Bayes
NCA	Neighbourhood Component Analysis
NLOS	Non-Line-of-Sight
NN	Neural Network
OSUA	Octagonally Shaped UWB Antenna
OUT	Outer
PA	Power Amplifier
PANI	Polyaniline
PANs	Personal Area Networks
PBDEEMD	Pseudo-Bi-Dimensional Ensemble Empirical Mode Decomposition
PC	Personal Computer
PCA	Principal Component Analysis
PCB	Printed Circuit Board
PD	Power Detector
PDF	Probability Distribution Function
PDMS	Polydimethylsiloxane
PDP	Power Delay Profile
PEDOT:PSS	Poly3,4-ethylenedioxythiophene Polystyrene Sulfonate
PEN	Polyethylene Naphthalate
PET	Polyethylene Terephthalate
PH	Personalized Healthcare
PIFAs	Planar Inverted-F Antennas
PIIC	Position-Information-Indexed Classifier
PL	Path Loss
PL_0	PL at reference distance
PLA	Polylactic Acid
PMA	Printed Monopole Antenna
PNLOS	Partial NLOS
PPG	Photoplethysmogram or Photoplethysmography
PR	Pattern Recognition
P_r	Received Signal Power
PRF	Pulse Repetition Frequency
PSADEA	Parallel Surrogate Model-Assisted Hybrid Differential Evolution For Antenna Optimization

PSG	Polysomnography
PSSPs	Parasitic Surrounding Stacked Patches
P_t	Transmit Power
PTFE	Polytetrafluoroethylene
Q	Quadrature
QC	Quasi-Circulator
QL	Quadruple Loop
R.	Right
R. AK	Right Ankle
R. SH	Right Shoulder
R. SL	Right Step Length
R. SW	Right Stride Width
R.TH	Thigh Region
RCSRR	Rectangular Complementary Split-Ring Resonator
RD	Range-Doppler
RF	Radio Frequency
RF	Random Forest
RFID	Radio Frequency Identification
RGW	Ridge Gap Waveguide Feed
rms	Root Mean Square
RMSE	Root Mean Squared Estimator
ROI	Region of Interest
RP	Range Point
RPM	Range-Point Migration
RR	Respiration Rate
RSNR	Relative Signal-to-Noise Ratio
RSRR	Rectangular Split-Ring Resonator
RSS	Received Signal Strength
RT	Ray-Tracing
RT-TOF	Round Trip-Time of Flight
RVSM	Remote Vital Sign Monitoring
Rx	Receiver
S_{11}	Reflection Coefficient
S_{21}	Transmission Response
SAGE	Space Alternating Generalized Expectation Maximization
SAR	Specific Absorption Rate
SCADA	Supervisory Control and Data Acquisition
SCNR	Signal-to-Clutter Noise Ratio
SDLA	Successive Detection Logarithmic Amplifier
SFCW	Stepped-Frequency Continuous Wave
SFF	System Fidelity Factor
SHAPA	Spectrum-Averaged Harmonic Path
SIW	Substrate Integrated Waveguide
SL	Side Left
SLE	Step Length Estimation

SMA	Sub-miniature Version A
SMOTE	Synthetic Minority Oversampling Technique
SMR	Signal-to-Mean Ratio
SNCR	Signal-to-Noise Clutter Ratio
SNR	Signal-to-Noise Ratio
SoC	System-on-Chip
SP	Strongest Path
SPGP	Sparse Pseudo-Input Gaussian Process
SPN	SleepPoseNet
SPO_2	Saturation of Peripheral Oxygen
SPT	Sleep Postural Transition
SR	Side Right
SRR	Split-Ring Resonator
SSM	State-Space Method
SSRR	Square Split-Ring Resonator
ST	S Transform
STDEV	Standard Deviation
STFT	Short-Time Fourier Transform
SUS	Scene Under Surveillance
S-V	Saleh-Valenzuela
SVD	Singular Value Decomposition
SVM	Support Vector Machine
$tan\delta$	Loss Tangent
TDOA	Time Difference of Arrival
TE	Transverse Electric
T–F	Time–Frequency
THz	Terahertz
TLs	Transmission Lines
TM	Transverse Magnetic
TMMs	Tissue Mimicking Materials
TN	True Negative
TO_L	Upper Torso Left
TOA	Time of Arrival
TOF	Time of Flight
TP	True Positive
TROI	Time Region of Interest
TSA	Tapered Slot Antenna
TTW	Through-the-Wall
TW	Through Wall
TWDP	Two-Wave Diffuse Power
TWI	Through-Wall Imaging
TWIR	Through-Wall Imaging Radar
TWR	Through-the-Wall Radar
Tx	Transmitter
UAV	Unmanned Aerial Vehicle

US	Ultrasound
UTD	Uniform Theory of Diffraction
UWB	Ultra-Wideband
UWB-SP	Ultra-Wideband Short Pulse Radar
V	Vertical
VDOP	Vertical Dilution of Precision
VMD	Variational Mode Decomposition
VNA	Vector Network Analyser
WBANs	Wireless Body Area Networks
WCE	Wireless Capsule Endoscopy
WiMAX	Worldwide Interoperability for Microwave Access
WLAN	Wireless Local Area Network
WMTS	Wireless Medical Telemetry System
WPAN	Wireless Personal Area Network
WR	Wrist
WRTFT	Weighted Range-Time–Frequency Transform
WSNs	Wireless Sensor Networks
XETS	Exponentially Tapered Slot-Based Antenna
γ	PL Exponent
ε_r	Relative Permittivity
κ	Kurtosis
μD	Micro-Doppler
σ_τ	RMS Delay Spread
τ_m	Mean Excess Delay

Chapter 1
Introduction to Body Centric Wireless Communication

1.1 Body Centric Wireless Communication

Body centric wireless networks (BCWN) communication systems are one of the most attractive venues for the next generation wireless technologies due to a huge range of applications offering wide variety of services. BCWN will be part of the forthcoming convergence and personalization across the various domains, which include personal area networks, (PANs), and body area networks, (BANs). Advancements in the miniaturisation of wearable hardware, embedded software, digital signal processing, and biomedical engineering have made human to human networking possible incorporating wearable sensors and communications. Body-centric wireless communications are attracting considerable attention due to their potential applications in several domains such as healthcare, remote monitoring, assisted living, localization and tracking, entertainment, security and defence services and general wellbeing [1, 2]. Therefore, a lot of emphasis has been given on the design aspects of the wearable antenna as well as body centric propagation analysis and modelling [3].

Though personal health care is the dominant field, its applications cover navigation and tracking, detection and localization, entertainment, security, wearable computer technology, sports and fitness, infotainment and gaming, augmented reality, and smart watches [1, 4]. For example, a wearable smart watch can monitor various health parameters, provide feedback to user, healthcare personnel and aid in remote monitoring [5]. They are also expected to play a very important role in 5G, 6G and Internet of Things (IoT) [6–8]. Figure 1.1 presents a wide variety of wearable sensors that can provide important information regarding the physical and health status of an individual. Ambient assisted living system with applications related to improved drug management, remote patient monitoring, telehealth, behaviour modification etc. are important applications of WBANs [8].

Body centric communication takes its place within the sphere of personal area networks and body area networks (PANs and BANs). Various body-centric links

S. K. Koul and R. Bharadwaj, *Wearable Antennas and Body Centric Communication*, Lecture Notes in Electrical Engineering 787,
https://doi.org/10.1007/978-981-16-3973-9_1

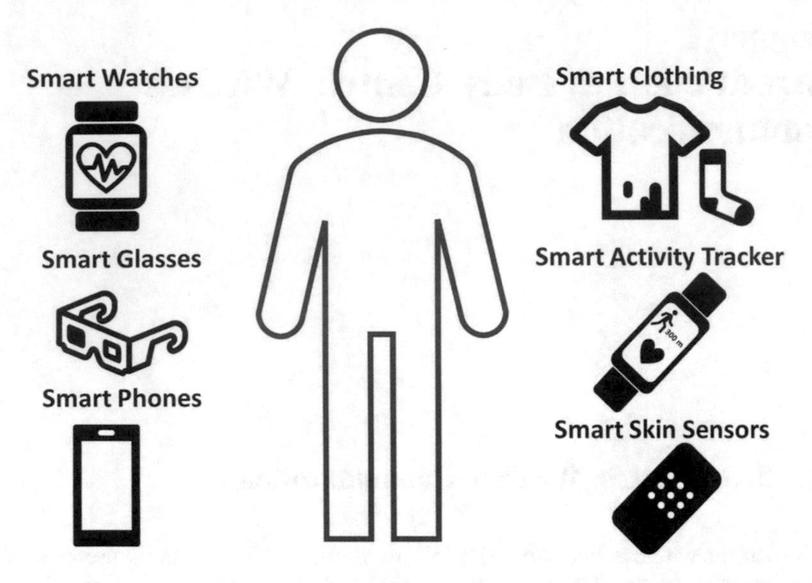

Fig. 1.1 Different types of wearable technology

can be formed in a realistic wireless network scenario such as on-body, off-body, body-to-body, and in-body [1–3, 9, 10] depending on the application and communication requirements. Figure 1.2 shows the schematic for various body centric links.

On-body: On-body communication refers to the link between the body mounted devices communicating wirelessly. In this scenario, the transmitter (Tx) and receiver (Rx) antennas are mounted on the body and communication takes place between these antennas in which most of the channel is on the surface of the body.
Off-body: Off-body communication defines the radio link between body worn devices and base units or mobile devices located in surrounding environment. The off-body communication link consists of one of the antennas placed on the human subject and another antenna placed at short-distance away from the human subject. In this scenario only of the antenna of the communication link is on the body and the propagation takes placed in the surrounding space between the wearable and base station antennas.

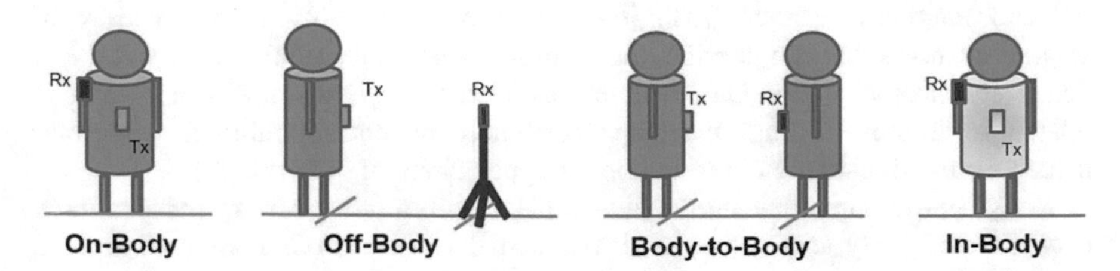

Fig. 1.2 Body centric communication links: on-body, off-body, body-to-body and in-body

Body-to-body: In body-to-body scenario, each antenna is placed on the body of the human subject leading to presence of two or more human subjects and communication taking place between the wearable antennas placed on each of the subject.
In-body: In-body communication refers to the communication between wireless medical implants and on-body nodes. For in-body communication links, a significant part of the channel is inside the body and implanted transceivers are used. This type of link has several applications in the medical domain for diagnosis of underlying diseases and monitoring body parameters.

Wearable devices/sensors are an important part of the wireless body-centric communication system which can be worn by the human subject and have the potential to communicate with other devices. They connect with other devices through their integrated wireless modules, which interface with other elements such as batteries, sensors, and antenna. Wearable devices include wrist watches, exercise shoes, virtual reality glasses, wearable sensors for medical applications, and integration of compact devices with clothing/shoes etc. [11]. Wearable devices/sensors should be unobtrusive, versatile, and they work close to the human body with minimal deterioration [4–6]. Smart wearable sensors/devices will revolutionize technology leading to enhancement in the quality of life, social interaction and activities providing social and economic upliftment.

In recent times there has been a surge of usages of wearable sensors, especially in the medical sciences, where there are a lot of different applications in monitoring physiological activities. In the medical field, it is possible to monitor patients' body temperature, heart rate, brain activity, muscle motion and other critical data (Fig. 1.3). It is important to have very light sensors that could be worn on the body to perform standard medical monitoring. All the physiological signals as well as physical activities of the patient are possible to be monitored with the help of wearable sensors [12]. During the rehabilitation stage the wearable sensors may provide audio feedback, virtual reality images and other rehabilitative services. The system can be tuned to the requirement of individual patient. The whole activity can be monitored remotely by doctors, nurses, or caregivers [4–6].

1.2 The Wireless Body Area Networks

The ever-growing miniaturization of electronic hardware and embedded systems, combined with recent developments in wearable communication technology, are leading to the creation of body-centric wireless communication systems which refer to human-self and human-to-human networking with the use of wearable and implanted wireless sensors. A wireless body area network (WBAN) typically consists of a collection of low power, miniaturised, invasive or non-invasive, lightweight devices with wireless communication capabilities that operate in the proximity of a human body. These devices can be placed in, on, or around the body,

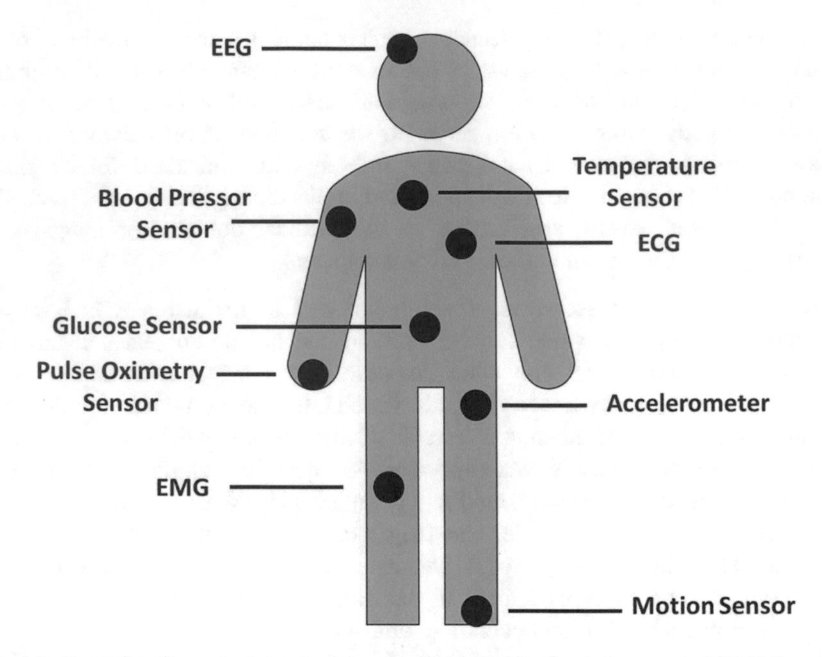

Fig. 1.3 Example of a patient monitoring using a wireless body area network (WBAN)

and are often wireless sensor nodes that can monitor the human body functions and characteristics from the surrounding environment [13, 14].

It has abundant envisioned applications in personal healthcare, medical field, smart home, personal entertainment, gaming, ambient intelligent areas and identification systems, space exploration and military. This leads to various system requirements with a wide variation in terms of frequency, performance metrics, throughput, flexible architectures, and protocols. The main communication standard solutions considered as reference are IEEE 802.15.4, IEEE 802.15.6, and Bluetooth Low Energy. The IEEE 802.15.6 was specifically designed for wireless communications in the vicinity of, or inside, a human body [13–15].

In recent years, body-centric wireless communication has become an integral part of human lives leading to increase in the antennas and propagation aspect for body-centric communication systems. The IEEE 802.15 standardization group has been established to standardize applications intended for on-body, off-body, or in-body communication. Recently, there has been extensive research into on-body antennas and propagation at 2.45 GHz, ultra-wideband (UWB) and 60 GHz millimeter wave (mmWave) bands [16–18]. Studies on many related topics have been carried out, including wearable antenna design, on-body channel characterization, the effect of human body presence on the link performance etc.

WBAN consists of a group of tiny nodes that are equipped with biomedical sensors, motion detectors, and wireless communication. These nodes are situated in the clothing of a person, body, or skin and distributed either on the human body

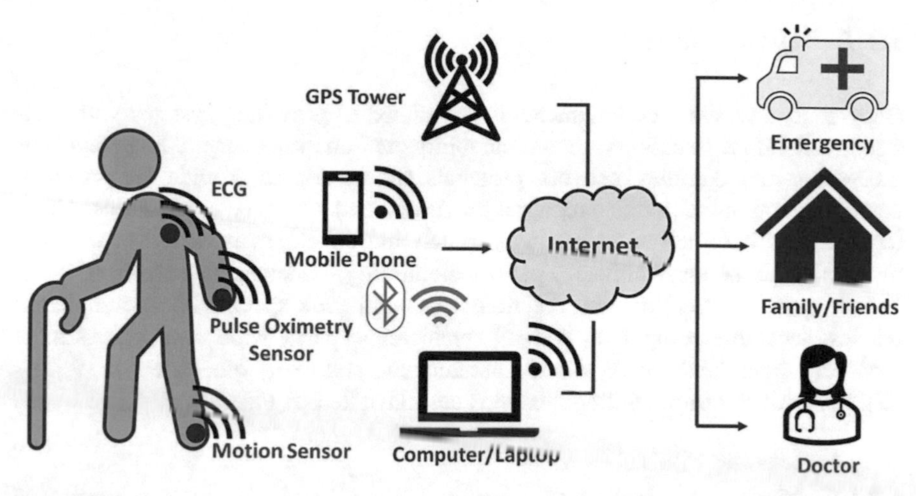

Fig. 1.4 Typical WBAN architecture. Illustration of a remote health monitoring system based on wearable sensors

(i.e., wearable sensor nodes) or implanted inside the body and they are under the control of a coordinator node [14, 18]. These sensors/devices monitors' physiological signals and record body health indicators such as blood pressure, glucose level, heart rate, body temperature etc. while a person is performing day-to-day or specific activities. These nodes form a network between the sensors and a control device. Figure 1.4 shows a generic WBAN application scenario. Basically, a WBAN system consists of several tiny sensor nodes and a gateway node used to connect them to remote locations such as hospital, doctor, family and emergency services [18].

Although body-centric communications have advanced in recent years, there are several challenges that need to be considered while designing efficient, accurate, low power wearable communication devices. Domains in which research activities are carried out range from optimization of antenna design taking into account the proximity of the human subject, compact light weight structure, minimisation of interference with other equipment and susceptibility to observation and jamming; channel characterization and modelling for better understanding of the body-centric propagation phenomenon; compact transceiver design using advance miniaturization techniques; application of signal processing algorithms and machine learning approaches to enhance the performance of the WBAN wearable devices and communication links.

1.2.1 Applications

Body-centric wireless communications are aimed at providing systems with constant availability, scalability, reconfigurability, and unobtrusiveness. High levels of processing and complex network protocols are needed to provide the powerful computational functionalities required for current and advanced applications. These requirements have led to increasing research and development activities with the main interests being healthcare, patient monitoring, personal identification, navigation, personal multimedia entertainment, and task-specific/fully compatible wireless wearable computers. WBAN applications span a wide area such as military, ubiquitous health care, sport, entertainment, and many other areas [1, 9, 18–22]. Some of the important applications are illustrated in Fig. 1.5.

1.2.1.1 Medical Applications

WBANs have a huge potential to revolutionize the future of health care monitoring by enhancing diagnosis and detection of many life-threatening diseases, providing real time patient monitoring, activity monitoring, assisted living and general

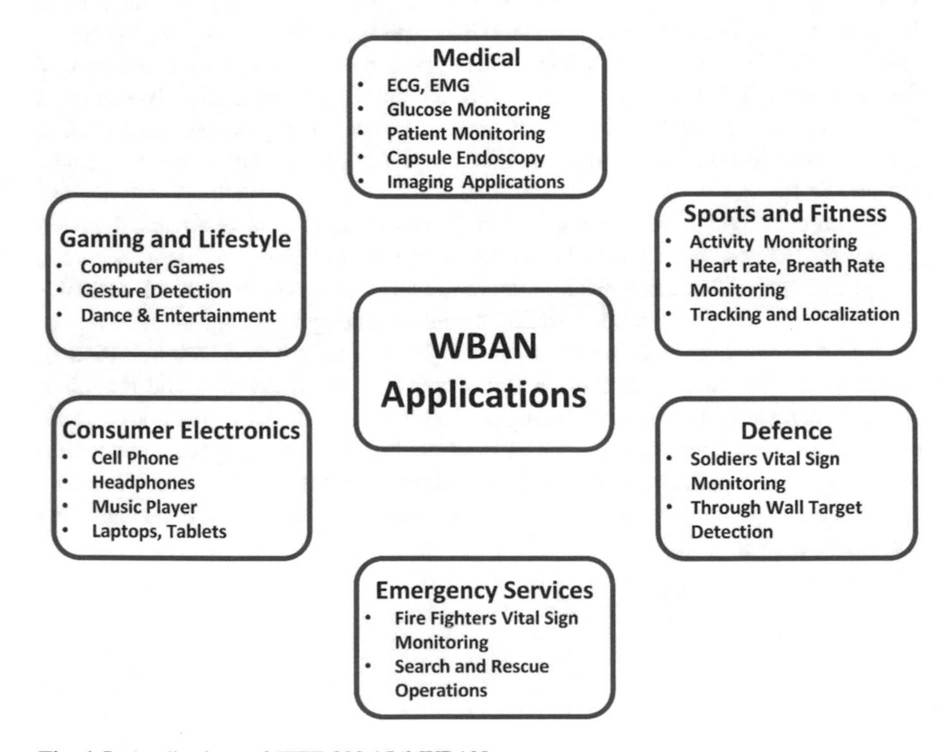

Fig. 1.5 Applications of IEEE 802.15.6 WBAN

wellness. Using WBANs in medical applications allows for continuous monitoring of one's physiological parameters such as blood pressure, respiration rate, heartbeat, and body temperature. In cases where abnormal conditions are detected, data being collected by the sensors can be sent to a gateway such as a cell phone [1, 20, 23]. The gateway then delivers its data via a cellular network or the Internet to a remote location such as an emergency centre or a doctor's room based on which an action can be taken.

Remote health monitoring allows continuous and real assessment of the progress of the patients health, reduces cost of the rehabilitation process and avoidance of the unnecessary frequent movement to the health care facility. Health monitoring leverages for early disease detection, reduces mortality rate, and consequently extends people's lifespan. The emergence of WBANs opens new possibilities in patient care and monitoring using e-health, telehealth, in-body telemetry systems which will find catering to the aged population providing them health and assistance related services, leading to enhancement in the quality of life [24, 25].

1.2.1.2 Defense, Security and Military

The activity of soldiers in the battlefield can be monitored more closely by WBANs, hence accessing soldier fatigue and battle readiness. This can be achieved through a WBAN consisting of cameras, biometric sensors, GPS (Global Positioning System) and wireless networking combined with an aggregation device for communication with other soldiers and centralized monitoring [1, 20, 21]. To increase overall performance of soldiers on the battlefield, works are undertaken to integrate wireless communications systems to all equipment, such as weapons, sighting systems, helmet. However, to prevent ambushes, a secure communication channel should exist among the soldiers. WBANs can also be used by policemen and fire-fighters to enhance the capabilities during search, rescue and emergency situations. The use of WBANs in harsh environments can be instrumental in reducing the probability of injury while providing improved monitoring and care in case of injury.

1.2.1.3 Sports

A WBAN-based monitoring system can also be extended to monitor athletes' performance to assist them in their training activities. Using sensor nodes, a trainer can get both on and off the track performance data [1, 20, 21]. For example, for a cricketer or a tennis player, movement of arms and body postures are very important for their success. In this case, a trainer can obtain data from a player via a WBAN and store those data in computers for further analysis. The location of the athlete can be tracked using wearable sensors placed on the body and base station nodes placed in the field and training ground. In the area of sport and training there is an increasing trend of using various wearable sensors such as measurement of heart rate, breathing rate and sweat rate using wearable sensors.

1.2.1.4 Lifestyle and Entertainment

WBAN can also play a vital role in our daily life. It enables some basic services like navigation support while walking, driving, exploring a new city etc. Infant monitoring, wireless wearable music system, making video call using big screen TV, playback of audio and videos from portable devices to TV or Audio system are some examples of WBAN application. Entertainment is also a very promising field for WBAN [1, 26]. The film industry, for example, benefits from motion capturing and postproduction mechanisms to produce movies in which actors perform the objects roles. Using the on-body accelerometers and gyroscopes for capturing motions facilitates the possibility of tracking the different positions of body parts. WBAN can be used in different spheres of entertainment applications such as real time video and audio streaming, consumer electronics like microphones, MP3-players, cameras and for gaming, virtual reality, ambient intelligence areas, personal item tracking and social networking [26].

1.2.1.5 Miscellaneous Applications

In future, deployment of WBAN may not only be limited to above mentioned applications but may also be extended to many other applications, including training, industrial, security/safety, smart homes/vehicles, entertainment, and other logistic applications. Appliances such as microphones, MP3-players, cameras, head-mounted displays, and advanced computer appliances can be used as devices integrated in WBANs for virtual reality and gaming purposes, inventory tracking, exchanging digital profile and consumer electronics. Sensors placed in homes/offices can detect a non-medical emergency such as fire in the home or flammable/toxic gas in the house and must urgently communicate this information to body-worn devices to warn the occupants of the emergency condition [4, 5]. The emotional status of a human being can be monitored anywhere and anytime through monitoring emotion-related physiological signals like Electrocardiograph (ECG), Electromyograph (EMG) through wearable sensors worn on the human body. This can be achieved through wearable biosensors that can be integrated in blood pressure sensors, earrings or watches, respiration sensors in T-shirts, conductivity sensors deployed in shoes and more. Secure authentication is possible by utilizing both physiological and behavioural biometrics such as iris recognition, fingerprints and facial patterns keeping in view the unique behavioural/physical characteristics of the human subject.

1.3 State-of-the-Art Technologies

Most of the wearable wireless sensors used so far in body-area network (BAN) applications like health monitoring, ambient assisted living, military applications, entertainment, and sport applications are operating mainly in the industrial–scientific–medical (ISM) bands, and many BAN-related studies have been carried out at 2.45 and 5.8 GHz, in the ultra-wideband (UWB) spectrum (3.1–10.6 GHz) and mmWave 60 GHz frequency range.

WBANs can be categorised according to the wireless communication technology employed. In this section an overview of these technologies considering Radio Frequency (RF) solutions is presented. Most of the works that can be found in the literature are about WBANs based on RF techniques, which can be classified according to the frequency band they operate in. Worldwide communication authorities regulate the use of the frequency spectrum, however, and allocate various frequency bands for body centric applications. The IEEE 802.15 Task Group 6 delivered a report that provides an overview of frequency regulations for medical applications in different countries and regions [1, 13, 14].

1.3.1 Wireless Medical Telemetry System (WMTS) and Medical Implant Communications Service (MICS) Bands

Wireless Medical Telemetry System (WMTS) and Medical Implant Communications Service (MICS) bands were allocated exclusively for body-worn and implanted medical applications, which require simple point-to-point communication. The Medical Implant Communication System (MICS) band is 402–405 MHz, and the Wireless Medical Telemetry Services (WMTS) operates on various bands between 420 and 1430 MHz [1, 9]. Within the MICS band it is possible to achieve a bit rate up to 400 kbps and a communication range around 2 m to satisfy the requirements of application such as cardiac pacemakers, implanted defibrillator, and neuro stimulator [27]. Application like the swallowable camera pill may require bit rate in the order of 1 Mbps, which are achievable in the WMTS band. An example of WBAN using a combination of devices operating in the MICS band for short range intra-BAN communication, and in the WMTS band for medium range communication with a central data collector is reported in [27].

1.3.2 Industrial, Scientific, and Medical (ISM) Band

The unlicensed Industrial, Scientific, and Medical (ISM) bands are defined by the International Telecommunication Union (ITU) with some of them being subject to specific country's radio regulations. The FCC allocates different frequencies for different purposes. The 900, 2400 and 5000 MHz frequency bands are set aside by the FCC for ISM applications. Being unlicensed, the ISM bands are prone to coexistence issues that must be taken seriously into account by the WBAN designer. The band between 2.4 and 2.5 GHz is often preferred among the others because of its worldwide availability [1, 9].

These bands are used for consumer and commercial Wi-Fi and WLAN applications as well as for commercial Radio Frequency Identification (RFID) Supervisory Control and Data Acquisition (SCADA) applications. The 900 MHz ISM band is very narrow, and this limits the maximum data rates. Typically, applications such as SCADA and RFID use the 900 MHz band as the data rate requirements are lower than applications found in the 2.4–5 GHz frequency bands [28].

For the home user and commercial business 2.4 GHz is the primary band one uses for Wi-Fi, Bluetooth, cordless phone, printer, keyboard, mouse, and gaming controller applications. Voice, video, and data communications typically use 2.4 GHz systems requiring higher data rates (up to 300 Mbps for 802.11n applications). The 5 GHz frequency is often used in commercial Wi-Fi applications. 5 GHz is also the frequency used for the emerging standard 802.11ac which provides up to 1.3 Gbps of wireless data throughput [28, 29].

1.3.3 Ultra-Wideband Technology (UWB)

In recent years there has been an increasing interest in using ultra-wideband (UWB) technology for short-range wireless communication. The IEEE 802.15.4a standard has adopted UWB as one of the technologies for robust transmission in WBANs and WPANs [16, 17]. UWB signals operating between 3.1 to 10.6 GHz have an inherent noise-like behavior due to their extremely low maximum effective isotropic radiated power (EIRP) spectral density of −41.3 dBm/MHz [1]. The bandwidth of such systems can also be defined as more than 25% of the center frequency or more than 1.5 GHz. This makes them difficult to detect and robust against jamming, potentially rescinding the need for complex encryption algorithms in tiny transceivers.

The high-rate throughput capability is a great support to WBAN considering the number of wearables and sensors that are used for remote monitoring these days. IR-UWB radios are very suitable for WBAN applications as they are generally of low-complexity by design and consume less power and increase battery life due to the use of very short duration pulses for transmission. Impulse radio (IR) transceivers have simple structure and very low power consumption, which facilitates their miniaturization [30]. UWB technology can offer broad capacity, high data rate, efficient communication links, short-range communications at a relatively low level of energy usage, reduced effect of multipath fading which is very desirable for WBANs. In UWB-BANs, the human tissues are not affected by closely placed WBAN devices and conforms to the size and cost constraints of BAN devices [3].

Wireless connectivity is provided through the deployment of lightweight and compact UWB antennas, unlike its narrowband counterpart, the design of UWB antenna is determined not only by its return loss characteristics but also by its ability to preserve the pulse shape as it employs the unorthodox carrier-free modulation in impulse radio systems. Antennas for UWB systems are required to have very broad impedance bandwidth, stable and constant channel transfer response and high efficiency [30, 31]. Owing to such characteristics, UWB has emerged as a solution for the radio communication with several applications in medical and healthcare domain. Applications include vital-sign information such as heart rate (HR), respiration motion, blood pressure (BP) giving estimate of patient's health, remote monitoring of patient, medical imaging applications using IR-UWB radar technology, activity classification, capsule endoscopy to detect diseases, localization/tracking of patient movement and activities monitoring etc. [31–35]. UWB technology also caters to applications in domain of sports, general fitness, security, military, and entertainment sectors.

1.3.4 mmWave 60 GHz Technology

Since the lower frequency spectrum-based devices are becoming more over-crowded, increasing the operational frequency to 60 GHz millimeter wave (mmWave) (57–64 GHz) is an attractive complementary option [36]. Recent advances in millimetre-wave technologies increased the interest to the 60-GHz band for body-centric short-range communications and 5G technology applications. In this unlicensed frequency range, a 7–9 GHz bandwidth is typical depending on the country norms. Along with providing large bandwidth, short wavelengths of 10–1 mm enable mmWave integration of the whole transceiver inside a small chip. Compared to the lower part of the microwave spectrum, the unlicensed 57–64 GHz range provides significant advantages such as high data rates (7 Gb/s), enhanced security, low interference with adjacent networks and comparatively smaller on-body devices in comparison to the systems operating at lower frequencies [36, 37]. Furthermore, the penetration depth at 60 GHz is around 0.5 mm, the penetration is mainly limited to the superficial layers of human skin.

At 60 GHz, path loss is more severe than at 2 or 5 GHz and implementing path loss in an integrated transceiver is a challenging assignment [36, 37]. As the path loss is very high at 60 GHz, a high gain directional antenna is required to compensate for channel loss, and radiation must focus on the receiver [36, 38]. The use of millimetre wave systems for BANs will have a high impact, in the defence sector where communications emanating from a dismounted soldier leads to detection, location and vulnerability to enemy attack. The high atmospheric attenuation in the 60 GHz band will lead to much higher levels of security against detection, interception, and jamming. In addition, millimetre wave BANs will also benefit in sectors such as healthcare, personal entertainment, sports training, and emergency services.

1.3.5 THz Technology

The rising market for wireless applications has provided researchers with a continuous need for low cost, effective, and small size system design [39]. Wireless communication technology has undergone unprecedented increase in demand for high data rates with high spectral efficiency and strong wide band fading reduction over the last few years. To fulfil the demands for high speed communication and miniaturised devices, higher frequencies are considered, hence, the unallocated terahertz (THz) spectrum i.e. 0.1–10 THz has been explored to provide wireless communication.

Terahertz band involves the electromagnetic waves with frequencies from 0.1 to 10 THz providing very high data rates over distance of 10 m, low interference, and miniaturized devices. The term normally applies to electromagnetic radiation with frequencies between 300 and 3000 GHz [39, 40]. Since Terahertz radiation has lower frequency as compared to X-rays, thus making it much safer and preferable for use in medical imaging for diagnostic as well as screening. The terahertz frequency band has been verified to be a reliable connection for communication between short range devices or point-to-point connectivity which also incorporates WBAN. The THz frequency is finding vast applications related to body-centric communication due to its non-ionizing properties and compactness [39].

The THz antenna can have very small dimensions scaling down to nanometre range. The antenna could be wired, log-periodic, aperture, microstrip or reflector, lens. Microstrip antenna is smaller in size, low weight, thin profile, easy to fabricate and may be used for body-centric applications [41]. The terahertz spectrum has several applications relating to sensing, communication and ranging in various sectors such as healthcare, industry, sports and military. Application of THz in emerging healthcare is evident, with various notable applications in, e.g., security, health monitoring, disease diagnosis, medical imaging, biology, material spectroscopy and sensing, medicine, pharmaceuticals, and communications.

1.4 Wearable Antenna and Body-Centric Propagation Aspects

Wearable antenna designs have been proposed at frequencies such as the 2.4 GHz ISM band, UWB systems and 60 GHz frequency range. The BCN antenna design is a very challenging task due to multiple requirements to ensure mobility, reliable link, compact design, and robustness being in proximity of the human body. Broad bandwidth, multiband operation and high gain are some of the requirements for future 5G/IoT applications to ensure interconnectivity between body-worn sensors, body-worn access points, and remote processing units.

The topic of antennas and propagation for body centric wireless communications continues to be of significance and serves many applications in the field of healthcare, entertainment, security, and defence. Wearable antennas must be compact, lightweight, low profile and preferably flexible, conformal to the body surface and unobtrusive, to provide freedom of movement while the human subject performs activities. Keeping the above requirements into consideration, planar printed antennas realized on thin and flexible substrates are preferred and several configurations have been presented in the open literature operating in various frequency bands [1–3, 31, 42–44]. They include patch antennas, planar inverted-F antennas (PIFAs), printed monopoles/dipoles, loops, slot antennas. Wearable multi-band antennas, circularly polarized antennas, tunable antennas, as well as dual-polarized and dual-mode [42] antennas have been designed and used to improve the performance of the body-centric communication systems. Antenna arrays and multiple antenna radiating systems have also been proposed to implement (multiple input multiple output) MIMO systems, diversity schemes and electrical beam steering [31, 42–46].

Various types of antennas operating in narrow band, wideband, implantable, fabric and multiple antennas are reported in open literature catering to a wide range of applications. The sensitivity of antenna performance to body proximity and the study of the effect of various antenna characteristics such as return loss, impedance matching, gain, efficiency, and the radiation pattern are of immense importance. Wearable antennas are designed keeping in view the proximity with the human body and the random actions that take place due to change in body postures and natural movements having high degree of variability. The specific absorption rate (SAR) is an essential factor for evaluation when an antenna operates on or near the human body. To protect the human body from radio wave exposure, antennas used in WBAN's must have a low specific absorption rate [1].

Textile antennas have gained a lot of attention over the past few years since they can be integrated directly in the clothing providing higher degree of flexibility and comfort. Textile fabrics have low dielectric constant leading to reduction in surface wave losses and improvement in the bandwidth [45–47]. Planar antenna layouts have been implemented by using electro textiles such as embroidered conductive polymer fibers, Shieldit®, Flectron® and non-conductive textile substrates such as fleece fabric, cotton/polyester, felt, foam [46, 47]. Single-band, dual-band, multi-band and ultra-wideband textile antennas have been presented with good performance in free space and on-body. Unlike antennas fabricated on rigid PCB substrates, textile antennas are affected by the environment, humidity, and may undergo bending/stretching deformations and mechanical compression which in turn effects the antenna performance [31, 44].

The antennas operating in the vicinity of the human body are subjected to electromagnetic distortions due to absorptions in the lossy human body tissues and reflections/scattering from the body surface [1–3]. The human body effect on the antenna performance is studied and investigated for various wearable antennas using simulations, modelling, realistic phantoms (liquid, semi-solid, solid) and human volunteers. It is important to focus on the selection of the type of antenna based on the desired application and properties of the material such as permittivity, thickness and loss tangent which will influence the performance of the wearable device [1, 9].

The design of reliable and efficient BANs, operating at various frequency bands, requires detailed analysis of the communication channel [48–51]. The wearable antenna performance should be evaluated through proper statistical parameters related to body-centric propagation analysis and modelling. Numerical full-wave simulations and measurements in different environments such as anechoic chamber, indoor/outdoor provide reliable and accurate results giving an estimate of the propagation channel in the vicinity of the human subject. The availability of accurate numerical human phantoms, obtained by means of advanced imaging techniques, make simulations a powerful tool for analysing the body-centric radio channel in great detail suitable for such purposes. For realistic measurements solid, semisolid, or liquid phantoms are manufactured to mimic the electrical properties of the actual human body [1, 2].

1.5 Scope of the Book

Body centric communication is emerging as an important domain to enhance quality of life, enriching user experience and increasing over all connectivity. There is considerable ongoing research related to the antennas and propagation for body centric communications systems. The aim of this book is to provide a review of recent progress and current trends in the field of antennas for body-centric communication at UWB and 60 GHz millimetre waves frequencies.

Work related to current trends, research aspects in wearable antenna design, optimization of antenna parameters, characterisation and modelling of the channel are reported for various types of body-centric links such as on-body, off-body, body-to-body and in-body communication. Various applications have also been discussed such as localization and tracking of the human subject, monitoring of physical activities using wearable technology. Radar based applications such as monitoring of vital-sign parameters, tracking of human subject and medical imaging have also been reported. Finally, IoT applications and machine learning approach have been described which aim to enhance the overall performance in various domains such as healthcare systems, smart home, and smart cities.

Chapter 2 describes various aspects of the on-body propagation from antenna design to channel modelling in the UWB and mmWave frequency range. Key features related to on-body antenna design and requirements for the frequencies are studied. Further, body centric channel study is presented considering simulation-based analysis, phantom based study, and realistic measurements for static and dynamic scenarios.

Chapter 3 focuses on modelling and characterization of off-body and body-to-body propagation channels for UWB and mmWave frequency range. Theoretical, numerical, simulation-based, and experimental investigations are reported to understand the channel behaviour in the presence of the human subject in different environments and node locations on the body for static and dynamic scenarios.

Chapter 4 presents an overview of different flexible substrate materials ranging from polymer based to textile for wearable antenna design. Various fabrication techniques such as lithography, screen printing, ink-jet printing is being reported. Simulation, design, and fabrication aspects of flexible wearable antenna suitable for UWB and mmWave range have been discussed. Electromagnetic and mechanical properties of such antennas have been discussed for free space and on-body scenarios.

Chapter 5 gives an overview of in-body and on-body antenna design and channel characteristics of implantable UWB communication systems suitable for several medical applications like capsule endoscopy and vital body parameters monitoring. The path loss and shadow fading models are presented and analysed based on the numerical techniques using human and animal anatomical models, phantom-based measurements and in-vivo living animal experiments.

Chapter 6 gives an insight into propagation characteristics when antennas are placed on various body locations and the factors affecting the localization accuracy while tracking a human subject in an indoor environment. Simple and effective techniques based on channel information and time of arrival localization techniques are presented in this chapter for classification/mitigation of non-line-of-sight (NLOS) links and localization accuracy enhancement.

Chapter 7 presents work related to monitoring and assessment of physical activities, gait movement, joint angles estimation during flexion/extension of limbs using wearable UWB technology. Detailed investigation of monitoring progress of physical exercises for upper and lower limb movements is reported using time of arrival and channel information. Algorithms and methods used for classification and monitoring of daily physical activities and gait movement using UWB wearable technology is also presented in this chapter.

Chapter 8 presents recent advances in IR-UWB and mmWave radar system design for healthcare, such as vital signs measurements, through-wall vitals measurement and detection, daily activity monitoring, fall detection, sleep monitoring, gait analysis and gesture recognition. The aim of this study is to present various state-of-the-art techniques and algorithms for non-invasive detection and monitoring applications such as vital-sign monitoring, classification and activity recognition, localization and tracking of human subjects.

Chapter 9 discusses various research studies reported in simulation and experimental form related to UWB imaging technology for various medical applications. Several antenna designs, development of image reconstruction algorithms, and providing set up details of the complete microwave imaging systems are reported. Through wall UWB radar description for detection and localization of the human subject in complex and cluttered environments is also presented.

Chapter 10 presents an overview of the Internet of Things (IoT) and discusses various applications and potentials of the IoT technology with focus on healthcare applications. Various technologies and state-of-the art techniques related to compact antenna design for body-centric communication for IoT applications are discussed. This chapter also explores the role of artificial intelligence (AI) and machine learning (ML) for healthcare IoT applications which would aid in overall performance of the healthcare system.

References

1. Hall PS, Hao Y (2012) Antennas and propagation for body-centric wireless communications. Artech House
2. Hall PS, Hao Y, Cotton SL (2010) Advances in antennas and propagation for body centric wireless communications. In: Proceedings of the fourth European conference on antennas and propagation, Barcelona, Spain, pp 1–7
3. Bharadwaj R, Koul SK (2019) Experimental analysis of ultra wideband body-to-body communication channel characterization in an indoor environment. IEEE Trans Ant Propag 67(3):1779–1789
4. Mukhopadhyay SC (2015) Wearable sensors for human activity monitoring: a review. IEEE Sens J 15(3):1321–1330
5. Wearable devices with mHealth apps: integration and implementation. https://mobisoftinfotech.com/resources/blog/mhealth-wearable-devices/
6. Dian FJ, Vahidnia R, Rahmati A (2020) Wearables and the Internet of Things (IoT), applications, opportunities, and challenges: a survey. IEEE Access 8:69200–69211
7. Akyildiz F, Kak A, Nie S (2020) 6G and beyond: the future of wireless communications systems. IEEE Access 8:133995–134030
8. Rodrigues JJPC et al (2018) Enabling technologies for the internet of health things. IEEE Access 6:13129–13141
9. Abbasi QH, Rehman MU, Qaraqe K, Alomainy A (2016) Advances in body-centric wireless communication: applications and state-of-the-art. The Institution of Engineering and Technology (IET), London, U.K., Jul 2016
10. Arbia D, Alam M, Moullec Y, Hamida E (2017) Communication challenges in on-body and body-to-body wearable wireless networks—a connectivity perspective. Technologies 5(3):43
11. Yetisen AK, Martinez-Hurtado JL, Ünal B, Khademhosseini A, Butt H (2018) Wearables in medicine. Adv Mater (Deerfield Beach, Fla.) 30(33):1706910
12. Meharouech A, Elias J, Mehaoua A (2019) Moving towards body-to-body sensor networks for ubiquitous applications: a survey. J Sens Actuator Netw 8(2):27
13. Cavallari R, Martelli F, Rosini R, Buratti C, Verdone R (2014) A survey on wireless body area networks: technologies and design challenges. IEEE Commun Surv Tutorials 16 (3):1635–1657
14. Movassaghi S, Abolhasan M, Lipman J, Smith D, Jamalipour A (2014) Wireless body area networks: a survey. IEEE Commun Surv Tutorials 16(3):1658–1686
15. Gravina R, Fortino G (2020) Wearable body sensor networks: state-of-the-art and research directions. IEEE Sens J 21(11):12511–12522
16. Čuljak I, Vasić ŽL, Mihaldinec H, Džapo H (2020) Wireless body sensor communication systems based on UWB and IBC technologies: state-of-the-art and open challenges. Sensors 20(12):3587
17. Viittala H, Hamalainen M, Iinatti J, Taparugssanagorn A (2009) Different experimental WBAN channel models and IEEE 802.15.6 models: comparison and effects. In: 2nd International symposium on applied sciences in biomedical and communication technologies, Bratislava, Slovakia, pp 1–5
18. Wang Q, Chen W, Markopoulos P (2014) Literature review on wearable systems in upper extremity rehabilitation. In: IEEE-EMBS international conference on biomedical and health informatics (BHI), Valencia, Spain, pp 551–555

19. Saboor A, Ahmad R, Ahmed W, A K.Kiani, Moullec YL, Alam MM (2019) On research challenges in hybrid medium-access control protocols for IEEE 802.15.6 WBANs. IEEE Sens J 19(19):8543–8555
20. Yan H et al (2015) An emerging technology—wearable wireless sensor networks with applications in human health condition monitoring. J Manag Anal 2(2):1–137
21. Arefin M, Ali M, Haque A (2017) Wireless body area network: an overview and various applications. J Comput Commun 5:53–64
22. Negra R, Jemili I, Belghith A (2016) Wireless body area networks: applications and technologies. Procedia Comput Sci 83:1274–1281
23. Leu FY et al (2017) A smartphone-based wearable sensors for monitoring real-time physiological data. Comput Electr Eng 65:376–392
24. Fan Y, Xu P, Jin H, Ma J, Qin L (2019) Vital sign measurement in telemedicine rehabilitation based on intelligent wearable medical devices. IEEE Access 7:54819–54823
25. Kiourti A, Nikita KS (2017) A review of in-body biotelemetry devices: implantables, ingestibles, and injectables. IEEE Trans Biomed Eng 64(7):1422–1430
26. Salayma M, Al-Dubai A, Romdhani I, Nasser Y (2017) Wireless body area network (WBAN): a survey on reliability fault tolerance and technologies coexistence. ACM Comput Surveys 50(1):1–38
27. Islam MN, Yuce MR (2016) Review of medical implant communication system (MICS) band and network. ICT Express 2(4):188–194
28. ISM band of frequencies and allocation. https://www.data-alliance.net/blog/ism-band-of-frequencies-and-allocation/
29. What are the advantages and disadvantages of ISM band frequencies? https://www.l-com.com/frequently-asked-questions/advantages-and-disadvantages-of-ism-band-frequencies.
30. Yan S, Soh PJ, Vandenbosch GAE (2018) Wearable ultrawideband technology—a review of ultrawideband antennas, propagation channels, and applications in wireless body area networks. IEEE Access 6:42177–42185
31. Alani S et al (2020) A review on UWB antenna sensor for wireless body area networks. In: 2020 4th International symposium on multidisciplinary studies and innovative technologies (ISMSIT), Istanbul, Turkey, pp 1–10
32. Bharadwaj R, Swaisaenyakorn S, Parini CG, Batchelor JC, Alomainy A (2017) Impulse radio ultra-wideband communications for localization and tracking of human body and limbs movement for healthcare applications. IEEE Trans Ant Propag 65(12):7298–7309
33. Bharadwaj R, Koul SK (2021) Assessment of limb movement activities using wearable ultra-wideband technology. IEEE Trans Ant Propag 69(4):2316–2325
34. Danko Š, Stopjaková V (2016) Wireless ultra-wide band communication for implantable sensors. In: International conference on emerging e-learning technologies and applications (ICETA), Stary Smokovec, Slovakia, pp 43–48
35. Fioranelli F, Kernec JL, Shah SA (2019) Radar for health care: recognizing human activities and monitoring vital signs. IEEE Potentials 38(4):16–23
36. Pellegrini A et al (2013) Antennas and propagation for body-centric wireless communications at millimeter-wave frequencies: a review. IEEE Ant Propag Mag 55(4):262–287
37. Ur-Rehman M, Malik NA, Yang X, Abbasi QH, Zhang Z, Zhao N (2017) A low profile antenna for millimeter-wave body-centric applications. IEEE Trans Ant Propag 65(12):6329–6337
38. Zhadobov M (2018) Millimeter-wave technologies for body-centric applications. In: 43rd International conference on infrared, millimeter, and terahertz waves (IRMMW-THz), Nagoya, 2018, p 1
39. Rubani Q, Gupta SH, Pani S, Kumar A (2019) Design and analysis of a terahertz antenna for wireless body area networks. Optik 179:684–690
40. Akyildiz IF, Jornet JM, Han C (2014) Terahertz band: next frontier for wireless communications. Phys Commun 12:16–23
41. Xu R et al (2020) A review of broadband low-cost and high-gain low-terahertz antennas for wireless communications applications. IEEE Access 8:57615–57629

42. Mahmood SN, Ismail A, Soh AC, Zakaria Z, Alani S (2020) Recent wearable antenna technologies. Int J Adv Sci Technol 29(04):14
43. Mahmood SN et al (2020) ON-OFF body ultra-wideband (UWB) antenna for wireless body area networks (WBAN): a review. IEEE Access 8:150844–150863
44. Alharbi S, Shubair R, Kiourti A (2018) Flexible antennas for wearable applications: recent advances and design challenges. In: Proceedings in European conference on antennas and propagation, pp 1–2, Apr 2018
45. Kirtania SG, Elger AW, Hasan MR, Wisniewska A, Sekhar K, Karacolak T, Sekhar PK (2020) Flexible antennas: a review. Micromachines 11:847
46. Guraliuc AR, Zhadobov M, Valerio G, Sauleau R (2014) Enhancement of on-body propagation at 60 GHz using electro textiles. IEEE Antennas Wirel Propag Lett 13:603–606
47. Yadav A, Kumar Singh V, Kumar Bhoi A, Marques G, Garcia-Zapirain B, de la Torre DI (2020) Wireless body area networks: UWB wearable textile antenna for telemedicine and mobile health systems. Micromachines 11:558
48. Cotton SL, D'Errico R, Oestges C (2014) A review of radio channel models for body centric communications. Radio Sci 49(6):371–388
49. Bharadwaj R, Koul SK (2019) UWB channel analysis using hybrid antenna configuration for BAN localization applications. In: 2nd Indian conference on antennas and propagation (InCAP 2019), 19–22 Dec 2019, Ahmedabad, India
50. Alomainy A, Sani A, Rahman A, Santas JG, Hao Y (2009) Transient characteristics of wearable antennas and radio propagation channels for ultrawideband body-centric wireless communications. IEEE Trans Antennas Propag 57(4):875–884
51. Bharadwaj R, Koul SK (2017) Study and analysis of channel characteristics of ultra-wideband communication links using wearable antennas. In: Asia Pacific microwave conference, APMC 2017, Kuala Lumpur, Malaysia, 13–16 Nov 2017

Chapter 2
On-Body Radio Propagation: UWB and mmW Technologies

2.1 Introduction

Ultra-wideband (UWB) and 60 GHz mmWave communication have been considered as potential technologies for wireless body area networks (WBANs). Ultra-wideband (3.1–10.6 GHz) is a short-range communication technology with attractive characteristics for body-centric communication. Some of the advantages include high bandwidth, high data rate transmission capabilities (typically 100 Mbps), low power spectral densities (41.3 dBm/MHz), ensuring low interference with other narrow-band wireless devices [1]. It works at low-power which is good for human body exposure and for longer life of the system, provides high data rate, nano second pulse width, that provides immunity to multipath interference, enabling the compatibility of such technology for wearable applications [1–3].

Millimetre-wave (mmWave) technologies have gained interest in recent times with the rise of the fifth generation (5G) communication providing high data rates, network capacity and low latency leading to advancement in the communication service [1]. mmWave technologies can also resolve the problem of spectrum congestion due to the operating frequency bands which range from 28 to 73 GHz suitable for both indoor and outdoor environments. It provides low interference, co-existence with other WBAN technologies and communication systems making it suitable for various applications. The 60 GHz communication is an ideal candidate for future wearable devices due to advantages such as compact RF components and antennas, visibility and susceptibility to interference, wide frequency range of unlicensed spectrum (57–64 GHz) and high data rates [4, 5].

The on-body communication takes place between the two or more nodes, placed on the human body. It can be, for instance, a sensor collecting monitoring data, which is then sent to the on-body device usually located on the wrist or waist. In on-body communications, the influence of the human body on the antenna and propagation characteristics is significant which needs in-depth analysis to design robust wearable communication system [1, 6]. On-body communication links are

S. K. Koul and R. Bharadwaj, *Wearable Antennas and Body Centric Communication*, Lecture Notes in Electrical Engineering 787,
https://doi.org/10.1007/978-981-16-3973-9_2

one of the most researched topics with works carried out in domains such as antenna design, analytical or electromagnetic simulation, channel modelling and various aspects of the body centric communication system performance.

This chapter describes various aspects of the on-body propagation from antenna design to channel modelling in the UWB and mmWave frequency range. The chapter starts with key aspects of the on-body antenna design and requirements for the frequencies studied. Various state-of the-art designs and techniques for enhancing the performance of the on-body antennas is described. Further body centric channel study is presented considering simulation-based analysis, phantom based study, and realistic measurements for static and dynamic scenarios. Finally, key points regarding some of important aspects of on-body communication is provided.

2.2 Wearable Antenna Requirements

Antenna is a key component of the body-centric wireless communication system and one of the most interesting topics for research and development. Design and fabrication of compact antennas for Body Area Networks (BAN) plays an important role in the performance of the wireless communication systems to enable communication between the wearable devices. Such systems are of great interest for various applications including home/office, sports, multimedia, health care, biomedical, defence and security.

2.2.1 Design Strategy

Miniaturized antennas for body centric communication should be designed to provide ease and comfort to the user without compromising on the antenna performance and desired parameters. The antenna designer should aim at low profile, compact size, robust, conformal to the body, light weight antenna keeping in view the frequency of operation and specific application [1–4]. For example, UWB antenna design requires broad bandwidth and mmWave antennas have the requirement of compact high gain antennas.

The antenna designers should also focus on low cost and fabrication technologies leading to more commercially viable products. The performance and related antenna parameters such as antenna matching, gain and efficiency should have minimum degradation when placed in the proximity of the human body. The design should also meet the requirements for human radiation exposure such as specific absorption rate (SAR) which is the measure of the rate at which energy is absorbed per unit mass by a human body when exposed to a radio frequency (RF) electromagnetic field [2].

Key requirements for the UWB antennas are wide bandwidth, gain flatness and phase linearity, i.e., constant group delay. Ultra-wideband antennas are specifically designed to transmit/receive very short time durations of the electromagnetic energy [1, 2]. For body centric applications high fidelity of the received signal is desired to preserve the pulse shape so that a robust communication link can be formed between the transmitter and receiver antenna. Recent UWB antenna development tends to focus on small planar antennas as they are more feasible in terms of manufacturing and integrating with the system board. Planar antennas also provide large impedance bandwidth and a vast range of the achievable radiation properties. Design strategies such as coplanar waveguide (CPW)-fed bowtie/triangular patch antenna, and UWB antenna with partial ground plane are commonly used for body-centric communication with patches of various geometrical shapes and sizes. Different approaches such as band notch antennas to avoid interference with narrowband frequencies, slots in patches for enhanced bandwidth, use of flexible substrates, low-temperature co-fired ceramic (LTCC) technologies are also reported in open literature [2]

For 60 GHz communication, the antennas are generally based on planar substrate technology. Additional requirements of a high gain and directional radiation pattern are desired keeping in view the characteristics of the mmWave frequency band. This is because of the high attenuation at 60 GHz, high antenna gains are needed which can boost the signal and lead to a robust communication link. In most designs, the radiating elements are shielded from the body by introducing a ground plane or an electromagnetic bandgap (EBG) backing layer leading to higher antenna efficiency [4, 5].

2.2.2 *Simulation Based Approach: Performance Analysis*

Simulation tools such as CST microwave studio and Ansoft High Frequency Structure Simulator (HFSS) are used to design antennas for free space/on-body applications and to investigate the antenna performance when in close proximity with the human phantom [2]. Various models of the human phantom are available with the commercially available simulation tools which are close to the realistic human body. Simplified models are also reported which can be homogenous or layered human model with cylindrical and rectangular shape being the most preferable form of depicting the human body. Compact, efficient and cost-effective antennas such as various microstrip patch antennas, array antennas, advance designs such as substrate integrated waveguide (SIW), electromagnetic bandgap structures (EBG), metamaterials, and Vivaldi antenna suitable for wearable devices and base stations for short range communication are designed with the aid of simulation tools [2]. Study regarding identification of the most suitable bandwidth of the antenna for a particular application and desired frequency range keeping in view various constraints such as power requirements, chipset design and manufacturing, overall performance of the wearable device has to be taken into consideration.

2.2.3 UWB Antennas Design for On-Body Communication

Several kinds of the planar UWB antennas are proposed in the literature such as monopole antennas, slot type UWB antennas, tapered slot antennas, fractal/slot UWB antennas leading to bandwidth enhancement and good radiation characteristics. UWB antennas with band notch properties have also been proposed in open literature to avoid interference with the narrowband.

Figure 2.1a illustrates the designed and fabricated very compact UWB coplanar waveguide (CPW) fed circular monopole antenna with dimensions (14×18 mm^2). The return loss over the band of operation is well below −10 dB and overall good performance for free space and on-body scenarios [7]. The antenna has omni-directional radiation characteristics in the x–y plane. The designed and fabricated UWB octagonal antenna shown in Fig. 2.1b has dimensions of (40×36 mm^2) and its radiation pattern is omni-directional in nature [8]. The antenna performs well in the frequency range with −10 dB return loss in free space, on-body and indoor environments. A low-profile UWB antenna based on 2D monocone antenna structure is fabricated on FR4 substrate is proposed for wireless body area networks [9]. The antenna is low-weight can be produced by printed circuit board manufacturing technique. The antenna has an enhanced impedance bandwidth of about 162% in the range of 2.5 to 24 GHz.

A directional antenna is designed using a bevelled y-shape monopole backed with fork slotted electromagnetic band gap (EBG) structure with a size of 8×8 mm^2 which acts as a reflector (Fig. 2.1c) [10]. It also acts as a protective shield against the electromagnetic radiation emerging from the monopole antenna towards the body tissue. The proposed antenna is developed on a 0.2 mm thin RO4003 substrate with a dielectric constant (ε_r) of 3.55 and loss tangent ($\tan\delta$) of 0.0027. The proposed monopole antenna has a small footprint with low profile of 20×25 mm^2. The antenna provides a peak gain of 6.25 dBi with high radiation and total efficiency. The calculated SAR value is 0.695 W/Kg which is well below the desired 1.6 W/kg. The antenna is developed on a 0.2 mm thin RO4003 substrate with a ε_r of 3.55 and $\tan\delta$ of 0.0027 with dimensions 20×25 mm^2. In [11] a design based on etching a Q-slot on a rectangular radiator, optimized to produce wide bandwidth in free space and close to the human body is proposed. The antenna is fabricated on FR4 substrate (with $\varepsilon_r = 3$, $\tan\delta = 0.01$) of 1.6 mm thickness and has dimensions of 36.6×39 mm^2. A thin Q-slot is etched on the radiator to generate additional resonance in the structure and the back of the substrate has a partial rectangular ground plane as shown in Fig. 2.1c.

A triple-notch band planar ultra-wideband (UWB) antenna is proposed for wireless body area networks (WBANs) to suppress unwanted signals of conventional narrowband communication technologies (Fig. 2.1d) [12]. The notch bands are Worldwide Interoperability for Microwave Access (WiMAX) (3.3–3.8 GHz), Wireless Local Area Network (WLAN) (5.1–5.825 GHz), and *X*-band downlink satellite communication systems (7.25–7.75 GHz). The overall antenna structure is fabricated on FR4 substrate with a size of 12×19 mm^2. A complementary split ring resonator

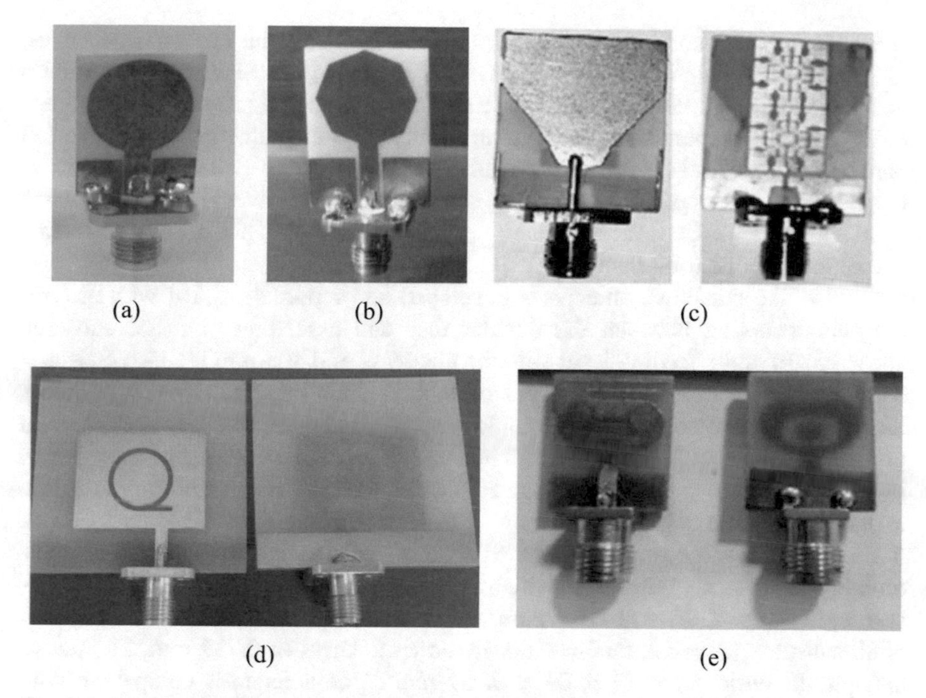

Fig. 2.1 UWB antennas: **a** Compact circular monopole (Reprinted with permission from IEEE [7]). **b** Compact octagonal monopole (Reprinted with permission from Elsevier [8]). **c** Bevelled y-shape monopole backed with fork slotted electromagnetic band gap (EBG) structure ([10], Reprinted with permission from IEEE). **d** Q-slot etched antenna (Reprinted with permission from IEEE [11]). **e** Triple-notch band planar antenna (Reprinted with permission from IEEE [12])

slot and two L-shaped stubs are introduced on an elliptical-shaped radiating patch to obtain UWB coverage from 2.9 to 12 GHz with three notch bands. A cactus shaped UWB monopole antenna fabricated on liquid crystal polymer (LCP) with dimensions of 20×28 mm^2 operating in the 2.85–11.85 GHz range is reported in [13].

2.2.4 60 GHz On-Body Antenna Design and Analysis

Due to very small size and high precision requirement during the fabrication process, antennas are particularly challenging aspect of the mmWave technology. Several technologies have been proposed in the literature such as low temperature co-fired ceramics (LTCC), substrate integrated waveguide (SIW) and array antennas for enhanced gain.

Initial designs reported are conventional Yagi array and SIW Yagi antenna. Each individual conventional Yagi array is designed with a driven dipole, 18 directors, and one reflector, printed on both sides of a 0.127 mm thick RT/Duroid 5880 substrate (ε_r = 2.2 and tanδ = 0.0009). Impedance bandwidth of −10 dB from 56 to 62 GHz with gain of 15 dBi at 60 GHz is reported [14]. SIW structures offer a compact, low loss, flexible, and cost-effective solution for integrating active circuits, passive components and radiating elements on the same substrate. The left part of the antenna is a planar waveguide structure, where two periodic rows of metallic vias are applied to form the sidewalls of the waveguide. The antenna operates in the 58–61 GHz frequency band with measured gains of 12.5 dBi (free space) and 10 dBi (on-phantom) [15].

A disc-like mmWave antenna is developed and reported in [16] with electromagnetic coupling between the circular disc and a feeding pin. The substrate Duroid 5870 of the feeding layer with the thickness of 0.508 mm has also dielectric constant ε_r = 2.33 and loss tangent of 0.0036. This concept enables relatively broadband characteristics with omnidirectional radiation pattern and vertical polarization suitable for on-body communication. Fabricated prototype is presented in Fig. 2.2a and is well matched from 59.3 to 63.4 GHz. An antenna analysed close to the body tissue exhibits efficiency of at least 25% and the maximum gain of 5.2 dB. A planar end-fire H-plane SIW horn antenna is designed for 60 GHz on-body channel measurements which operates with a bandwidth 6.1 GHz bandwidth [17] (Fig. 2.2b). The SIW horn antenna has been designed with RT Duroid 5880 substrate (ε_r = 2.2, tanδ = 0.003) with a thickness of 0.787 mm. The planar antenna dimensions are 17 × 24 × 0.787 mm^3. The antenna is compatible with BAN applications. The radiation is directive with a 6.6 dB end-fire gain.

A compact wideband aperture-coupled patch antenna array (8 × 8) with 64 radiating elements based on ridge gap waveguide feed (RGW) layer for 60-GHz applications has been proposed by Zarifi et al. [18] allowing to achieve a wideband patch antenna array with high gain and radiation efficiency (Fig. 2.2c). The structure has dimension of 28 × 28 × 7 mm^3. More than 75% efficiency and higher than 21.5-dBi gain, a sidelobe level below −13 dB in both the E- and H-planes is reported. LTCC technology has been applied to 60-GHz antenna design due to its advantages such as light weight, compactness, and high fabrication accuracy. A 2 × 2 microstrip patch antenna (MPA) array with the parasitic surrounding

(a) (b) (c)

Fig. 2.2 Antennas for mmWave communication, **a** disk shaped antenna (Reprinted with permission from IEEE [16]) and **b** end fire antenna (Reprinted with permission from IEEE [17]), **c** wideband patch antenna array (Reprinted with permission from IEEE [18])

stacked patches (PSSPs) based on LTCC multilayer technology has been proposed for bandwidth and gain enhancement in [19]. The proposed antenna array can achieve a gain of 10.5 dBi and a wide bandwidth of 27.3% at 60 GHz. An LTCC based 60-GHz differential-fed 4 × 4 patch antenna array with the soft-surface structure is proposed in [20]. Wideband patches with L-shaped feeding scheme are used as antenna elements. The differential SIW feeding network with low insertion loss is applied for the integration of antenna array. Such techniques enhance the performance in terms of high gain (18.62 dBi at 61.5 GHz), low cross-polarization level (<−25 dB), and wide bandwidth (11.7% at −10 dB).

Cavity-backed patch antenna arrays with full corporate SIW feed networks have been presented and investigated at 60-GHz band by Li and Luk [21]. Two large arrays of 64 and 256 radiating elements are designed, fabricated, and measured to demonstrate the good performance of the proposed arrays. A 60 GHz Yagi-uda circular array antenna with omni-directional pattern with 2.86 dB gain for millimetre wave WBAN applications is proposed in [22]. Center fed octagonal plates are used for feeding 8 Yagi-uda antennas located at the vertices of top and bottom octagonal plates.

2.2.5 *Effect of Feeding Structures*

2.2.5.1 UWB Antenna

A study of the influence of the feeding structure on the performance of two printed UWB monopole antennas near a human arm has been carried out by Koohestani et al. [23]. Two identical versions of the same UWB monopole antenna were design with different feeding structures, coplanar waveguide (CPW)- and microstrip (MStrip)-fed, as shown in (Fig. 2.3a, b respectively) [24]. Table 2.1 presents the dimensions of the parameters of both structures. In both versions, the radiator patch comprises two semicircles with different radii. The backside of the CPW–fed antenna substrate is devoid of any metallization while the backside of the MStrip–fed antenna supports a finite ground plane. The difference in the designs is with respect to the feeding line width (w for CPW and w′ for MStrip) that was calculated to have an input impedance of 50 Ω. The used substrate was RT/Duroid™ 5880 with thickness of 1.57 mm, relative permittivity of 2.2 and loss tangent of 0.0009.

The effects on UWB monopoles of a dielectric loading sandwich technique, consisting of gluing two commercial substrate pads on both sides of an antenna, have been studied. It was found that this technique increases the antenna electrical size and improves matching resilience without significant efficiency decrease [24]. The dielectric loading sandwich technique was applied to both antennas adequately choosing pad size and permittivity. Two identical rectangular slices of de–metalized commercial substrate (RO3003™) with a permittivity of 3.0, loss tangent of 0.001

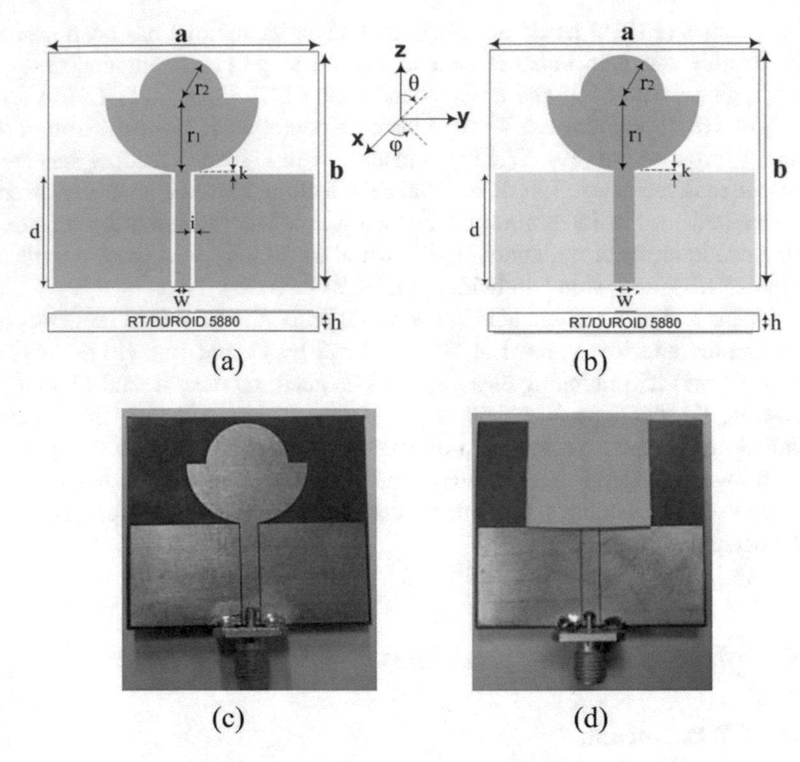

Fig. 2.3 Printed UWB monopole antennas: **a** CPW-fed, **b** MStrip-fed, **c** CPW fed prototype antenna: unloaded and **d** sandwich loaded UWB antenna (Reprinted with permission from IEEE [24])

Table 2.1 Optimal parameter values for CPW and MStrip-fed antennas [24]

A	b	r_1	r_2	d	w	k	I	h	w'
44	38	11	7.5	18	3.2	0.2	0.15	1.57	4.9

and thickness of 1.52 mm were padded on each side of the two structures. Prototype of the CPW-fed UWB antenna and the dielectric loaded antenna can be seen in Fig. 2.3c, d respectively. The width and length of the dielectric pads are 19 and 22 mm, respectively. The input reflection coefficient (S_{11}) and system fidelity factor (SFF) of both antennas were evaluated in free space and close to an arm at 3 mm distance. In the proximity of the arm, the CPW-fed antenna detunes $\sim$39% less than the MStrip-fed antenna. Furthermore, the feeding structure has minimal effect on pulse distortion as there is a small system fidelity factor value difference between CPW- and MStrip-fed antennas; the average deviation from CPW- to MStrip-fed is 2% in free space and 0.4% in presence of the arm.

2.2.5.2 60 GHz Antenna

Quantitative assessment of electromagnetic exposure and resulting heating induced by mmWave on-body mounted antennas has been explored in [25]. Four-patch antenna arrays with three feeding topologies have been considered, and the impact of the presence of a skin-equivalent phantom on antenna performances has been investigated. The different feeding techniques for 60 GHz on-body antennas are compared which are: A1; patch antenna array fed by microstrip lines; A2; aperture-coupled patch antenna array excited by microstrip lines; and A3; aperture-coupled patch antenna array excited by stripline lines [25]. Figure 2.4a–e shows the schematic and dimensions of the proposed configurations. All structures are printed on RT Duroid 5880 substrate ($\varepsilon_r = 2.2$ and $\tan\delta = 0.003$ at 60 GHz) with a 17-μm-thick copper ground plane. Near-field interaction between representative antenna arrays for off-body communications with three feeding topologies and human body is compared in terms of matching, radiation and user exposure. The presence of a ground plane results in an exposure reduction by more than 70 and 8 times in terms of peak and averaged levels, respectively. The sensitivity of the antenna reflection coefficient to the human body presence is also reduced. Computed gain of the fabricated antenna arrays are presented in Table 2.2.

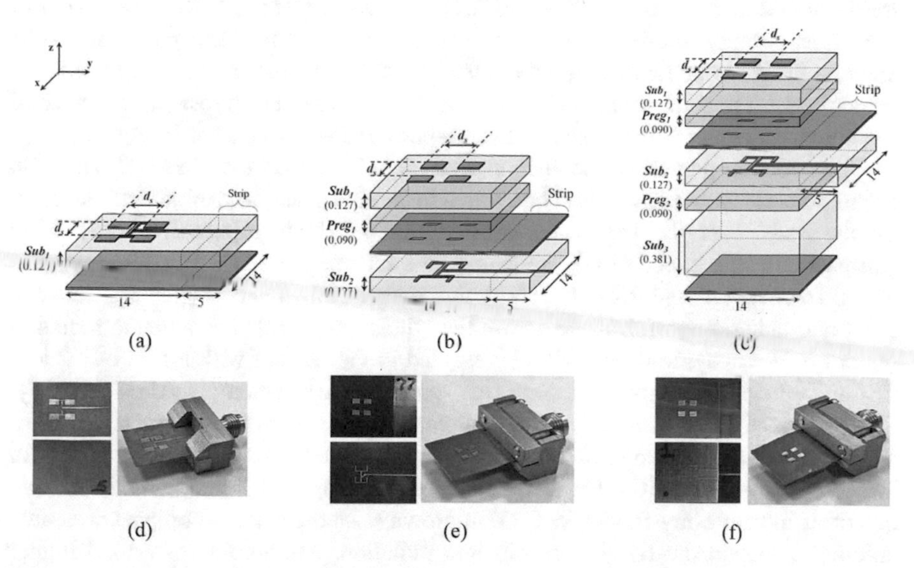

Fig. 2.4 Schematic (**a–c**) and manufactured antenna array prototypes (**d–f**) [A1–A3] with V-band connectors: dimensions in mm (Reprinted with permission from IEEE [25])

Table 2.2 Computed gain at 60 GHz for antenna arrays A1, A2 and A3 (Reprinted with permission from IEEE [25])

Peak gain [dBi]	Gain (dBi)		
	A1	A2	A3
Free space	9.4	8.8	9.6
On phantom	9.8	9.5	10

2.3 Influence of Wearable Antenna Location on the Radiation Pattern

2.3.1 Variation of Antenna Radiation Pattern with Body Location and Limb Movements

Numerical analysis was carried out using CST microwave simulations to obtain the radiation patterns when the antenna is placed on different locations of the human body model in free space and for different limb movements for the 3–9 GHz bandwidth (Fig. 2.5a) [26]. For the human body model, a homogeneous tissue that comprises bone, fat, muscle, and skin was considered. Weights assigned to calculate weighted average of the dielectric permittivity and conductivity are: 10% skin, 30% fat, 40% muscle, and 20% bone (average of bone cancellous, bone cortical and bone marrow), which resulted in a weighted averaged relative permittivity of 25.87 and conductivity of 3.14 S/m at 6.5 GHz [27, 28]. The height of the human body model of average build was 1.72 m. Tapered slot UWB antenna (TSA) schematic and fabricated prototype is presented in Fig. 2.5b, c respectively. The total antenna size is $27 \times 6\ \text{mm}^2$. The return loss performance of the antenna is good for free space and on-body scenarios as seen in Fig. 2.5d.

The radiation pattern of the antenna and its influence on the observed path gain are important issues for wireless body area networks. The effect of antenna location on the body over the frequency band of 3–10 GHz is reported in [26] using compact and cost effective UWB tapered slot antennas (TSAs). The TSA operates in the frequency range 2.2–11 GHz with excellent impedance matching, constant gain, and radiation performance across the whole band [27]. The antenna radiation pattern in free space and on-body is presented in Fig. 2.5e, f with a gain of −2 to 2 dBi and efficiency of around 80% in free space and 63% when placed on the body.

Analysis of the human body limb movement influence on the radiation pattern of a wearable antenna during different activities are carried out. The analysis is done at 3, 6, 9 GHz of the 3–10 GHz UWB range of frequencies. Simulations are carried out on a human body model in CST microwave studio with a compact wearable antenna to obtain the body-worn antenna radiation patterns for lower and higher frequencies. This study gives an insight into the variation of the radiation patterns of a compact UWB antenna depending upon the position of the wearable antenna on the body. Results conclude that the radiation pattern of the wearable antenna changes significantly in terms of shape, size, level of distortion and direction of maximum radiation with different limb movement activities and depends upon the

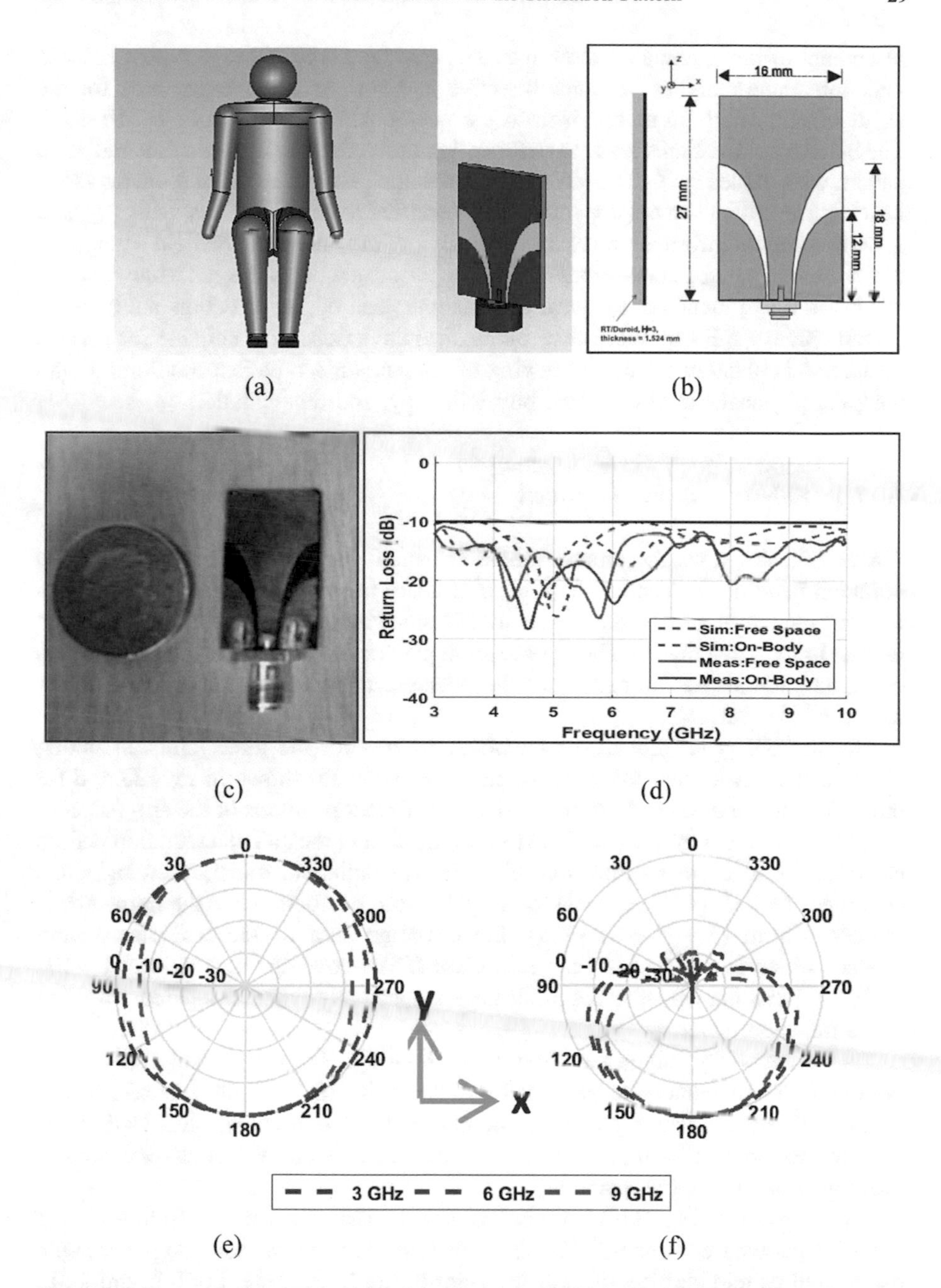

Fig. 2.5 **a** Human body model schematic with average built and height in sitting position and compact tapered slot UWB antenna shown in inset. **b** TSA antenna schematic and dimensions. **c** Fabricated TSA antenna. **d** Simulated and Measured return loss for free space and on-body scenario. The TSA antenna azimuth radiation pattern for 3, 6, 9 GHz when in **e** free space scenario and **f** when placed on the human body model (Reproduced courtesy of The Electromagnetics Academy [26])

placement of the antenna on the limbs. The coverage area of the wearable antenna radiation pattern highly becomes directive and shrinks in coverage area for the shoulder/thigh node in comparison to the wrist/ankle wearable node by 10–15%. The bending of the limbs leads to deformation and reduction in area of the radiation pattern with values as high as 30–40% when compared to free space scenario as the bending angle between the upper and lower arm/leg reduces. The analysis presented gives directional information regarding maximum radiation and the field strength of the radiation pattern for various activities performed. The present study reports results on the influence of the wearable antenna position, on detection and tracking performance of RF and microwave biomedical devices/sensors suitable for various healthcare applications such as tracking of human subject, patient monitoring, gait analysis, physical exercises, yoga, physiotherapy, and rehabilitation.

2.3.1.1 Sideways Arm Movement

Figure 2.6a, b shows the human model schematic in sitting posture and the 3D radiation pattern (X–Z and X–Y plane at 3 GHz) for 90° position of the right arm for the antenna placed on the shoulder. The sideways arm angle is measured with respect to the torso region. The 3D radiation pattern shows the direction and region of maximum radiation strength when the wearable antenna is placed on the shoulder region of the arm.

Detailed 2D polar plot of the radiation patterns for the three joints: shoulder, elbow and wrist for various positions and frequency are shown in Fig. 2.6c, d for the X–Z plane and X–Y plane, respectively. Different positions of the arm (0°, 30°, 60°, 90°, 120°) and frequencies lead to modification in the antenna radiation pattern performance in terms of direction of maximum radiation, radiation strength and coverage area (2D polar plot). For each body-worn position, when the frequency is increased from (3 $\rightarrow$ 6 $\rightarrow$ 9 GHz), the coverage area of the radiation pattern reduces which is depicted in Fig. 2.6c, Case I. The coverage area reduces by 10–20%, 25–30% and 30–40% for 3/6/9 GHz respectively when compared with free space radiation patterns.

As the frequency is increased from (3 $\rightarrow$ 6 $\rightarrow$ 9 GHz), the coverage area of the radiation pattern reduces as seen in Fig. 2.6c, Case II. With the increase of frequency, the electric length gets reduced, and the antenna becomes electrically large to some extent at the higher frequencies and more edge reflections are created leading to more directive patterns.

As observed in Fig. 2.6d Case I, the arm position 0° and 30° follow similar radiation pattern trend and 90°/120° arm position have similar radiation pattern with the overall pattern shifting towards the right in the X–Y plane. For 90° and 120°, there is more distortion of the radiation pattern due to the position of the arm leading to deviation in radiation pattern trend followed by the lower angle arm position. There is more variation in the radiation pattern plot for 9 GHz frequency in comparison to 3 or 6 GHz frequency which can be observed in Fig. 2.6d Case I. As the coverage area for the 9 GHz radiation pattern is least, the variation for

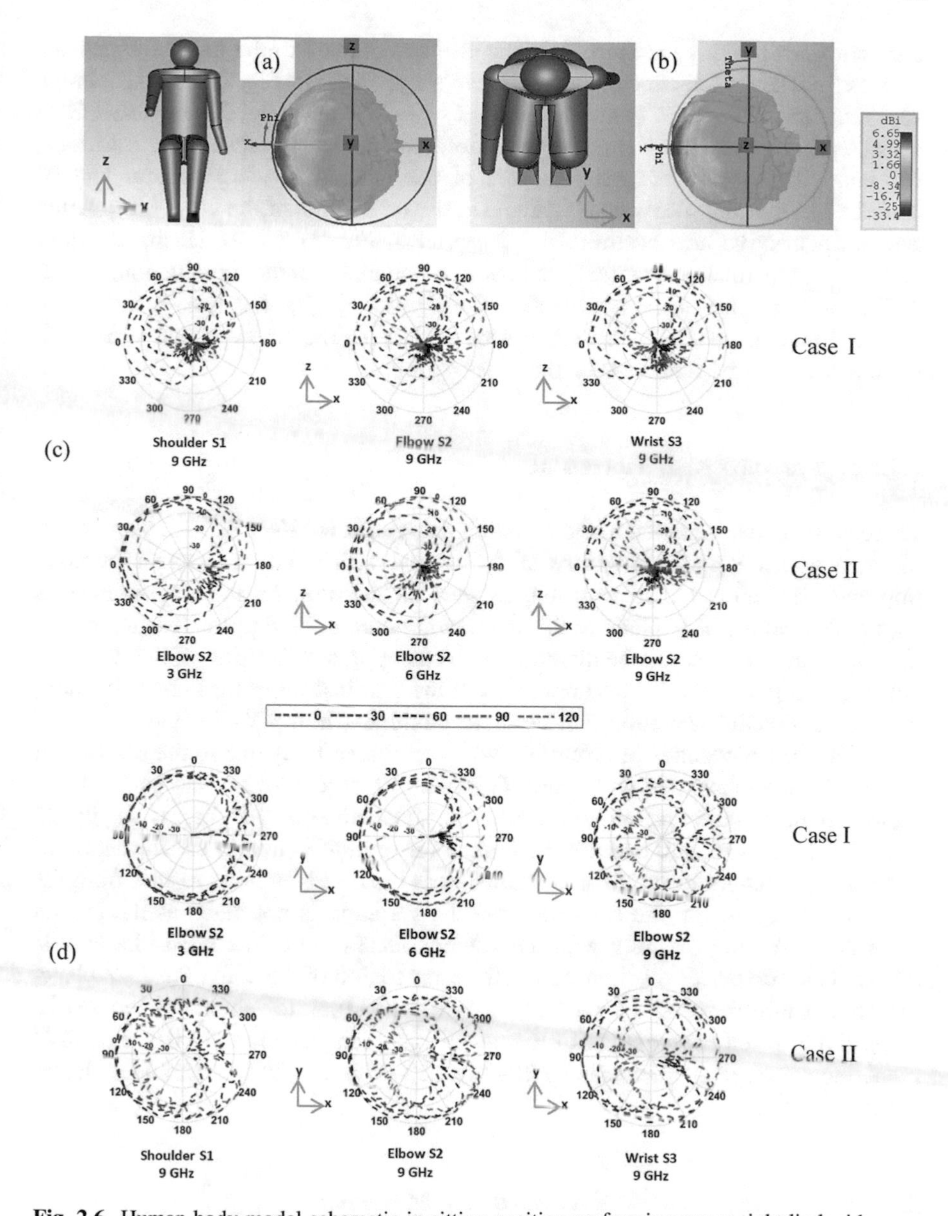

Fig. 2.6 Human body model schematic in sitting position performing upper right limb sideways movement: 90° (shoulder: 3 GHz) and 3D radiation pattern image **a** X–Z plane, **b** X–Y plane, **c** X–Z and **d** X-Y plane normalized radiation plots for Case I: different frequencies (3/6/9 GHz): elbow; Case II: same frequency (9 GHz): shoulder/elbow/wrist/ for 0° to 120° (Reproduced courtesy of The Electromagnetics Academy [26])

different limb positions is more prominent in comparison to the lower frequencies such as 3 GHz chosen which have wider coverage area. For Fig. 2.6d, the coverage area reduces in the X–Y plane by 10–20%, 10–25% and 10–30% for 3/6/9 GHz respectively keeping the free space radiation patterns as reference. Arm positions, 60° and 90°, has higher coverage area of the radiation pattern contour and 0° position has minimum coverage area due to the orientation of the wearable antenna and the respective arm position. As observed in Fig. 2.6d Case II, the shoulder region leads to smaller radiation patterns in comparison to the wrist region, due to the variation in size of the arm. Hence, the shoulder region has less spread of the radiation pattern contour, which increases further for the elbow region and maximum coverage area obtained for the wrist region.

2.3.1.2 Forward Arm Movement

There is not much variation in the radiation patterns in the X–Y or X–Z plane when the human arm is performing forward movement which is dependent on the position and orientation of the wearable antenna on the arm. As the displacement is taking place in the Y–Z plane of the simulation volume, X–Z plane does not depict the variation in the main lobe direction with changing arm position. If the wearable antenna was placed over the front region of the arm instead of the side of the arm, main lobe direction variation will be more prominent in the Y–Z plane.

Till 30° the antenna is in proximity with the human body due to the position of the arm for the elbow and wrist joint. This leads to larger effect of the human torso region on the radiation pattern of the wearable antenna. After 30° there is significant role of the arm structure and not the human torso/thigh region in the formation of the radiation pattern of the wearable antenna as the arm is moving further from the torso/thigh region. For the shoulder joint the variation is not much as the human torso is always in proximity with the antenna location. For the wrist, the case is different as it displaces maximum from the torso region of the body. For X–Z plane, the coverage area reduces by 10–25%, 5–10%, 5–10% for shoulder/elbow/wrist respectively when compared with free space radiation pattern at 3 GHz. For X–Y plane the coverage area reduces by 25–30%, 20–25%, 15–20% or shoulder/elbow/ wrist respectively.

2.3.1.3 Sideway/Forward/Backward Leg Movement

The forward and backward leg movement shows similar trend to that of the forward arm movement as the placement of the wearable antennas is same as that of the arms which is facing side outwards. For the leg movement, the lower torso/thigh region is considered as reference to measure the angles. The sideways movement of the leg with antennas located at the similar position as in the case of forward/ backward leg movement, will lead to variation in the direction of the radiation pattern in the X–Z plane. The leg is moved in forward, backward, and sideways direction with an interval of 30° in each direction. The main variation in the radiation pattern of the wearable antenna is the coverage area and maximum lobe direction in the radiation pattern. In comparison to the shoulder/wrist, there is significant reduction (10–15%) in the radiation pattern area for the antenna placed on the thigh/ankle region.

2.3.1.4 Limb Bending

Radiation patterns have also been analysed for upper and lower limbs bending which significantly vary with antenna orientation and bending angle. Limb bending is another common activity performed during physical exercises, rehabilitation, and physiotherapy. It gives indication regarding the flexibility and range of movement of the elbow and knee joint. By placing antennas/wearable devices on the various joints, one can monitor the limb bending progress and performance. The antenna is placed on the wrist/ankle and elbow/knee for studying the sideways arm and backward leg bending scenario. For the limb bending activity, the arm at 180° vertical position (straight) is considered as reference and the leg at 180° vertical position is considered as reference. The angle is formed between the upper and lower region of the arm/leg is the bending angle.

The 3D radiation pattern for X–Z and X–Y plane view showing 30° arm bend is presented in Fig. 2.7a, b. It can be observed that the 3D radiation pattern is more distorted for the limb bending activity in comparison to simple limb forward/sideways/backward movements.

As the arm is bent from 180° to 30° (Fig. 2.7 c, d), distortion of the radiation pattern starts to take place near 120° bend of the forearm for the antenna placed on the wrist. For the antenna placed on the elbow distortion of the pattern takes place near 90° bend.

The distortion is observed in both X–Z and X–Y plane of the radiation pattern 3D/2D plots. The distortion is mainly caused due to the proximity of the upper/forearm with the antenna when placed on the wrist/elbow as the bending angle reduces, which acts as an obstruction for the radiating region of the antenna. The leg bending activity, considers five positions of the leg from 180° to 30° bending angle between the upper and lower leg. For the backward leg movement, the radiation patterns for the antenna nodes also incur distortion especially after 60° angle. The reason for distorted radiation patterns of the body-worn antennas placed on the back region of the ankle and knee is the vicinity of the upper leg and lower leg during limb bending activity. Maximum distortion is observed when minimum possible bending angle is formed between the upper and lower arm/leg which in the current study is 30°.

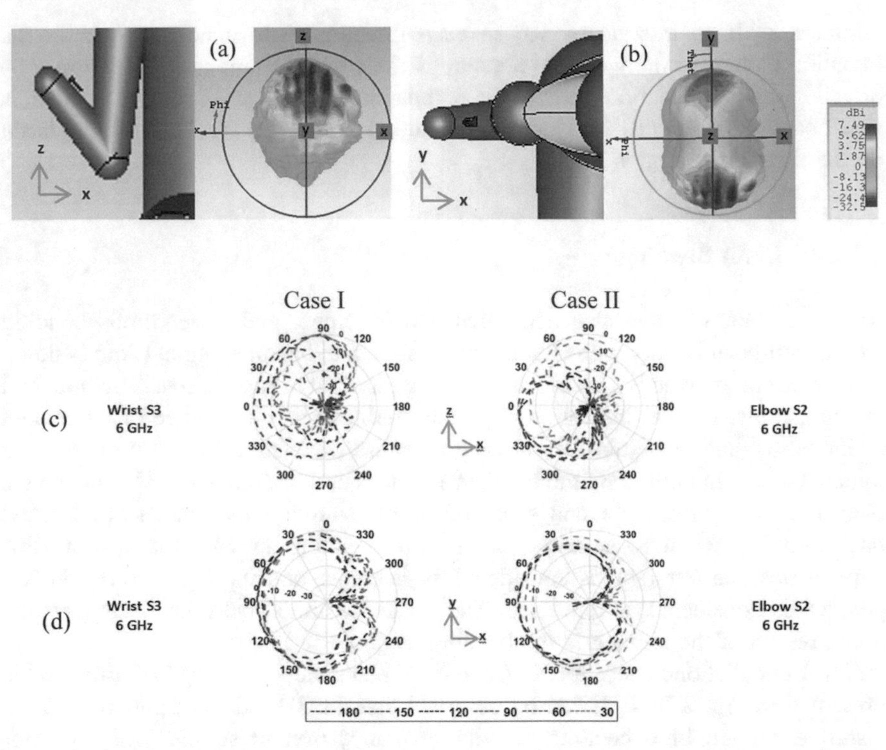

Fig. 2.7 Human arm model schematic performing right arm sideways bending movement at 30° (wrist: 3 GHz) and 3D radiation pattern. **a** X–Z plane, **b** X–Y plane. Arm bending activity: **c** X–Z and **d** X-Y plane normalized radiation pattern for 6 GHz frequency: wearable antenna placed on Case I: wrist and Case II: elbow performing limb bending movement (180° to 30° angle between the forearm and the upper arm) (Reproduced courtesy of The Electromagnetics Academy [26])

2.3.2 60 GHz Antenna Array for Wearable Smart Glasses

A 60 GHz patch antenna array with parasitic elements for wearable smart glasses is proposed [29]. The fabricated antenna has a microstrip feedline and is fed by a 1.85 mm end-launch connector. To demonstrate the antenna performance in a practical wearable environment, the radiation characteristics were analysed on a 60 GHz head phantom. The SAR and power density (PD) were calculated to evaluate the electromagnetic exposure on the head phantom. Taconic TLY (ε_r = 2.2, tanδ = 0.0009) substrate with a thickness of 0.127 mm was used in the design of the antenna (Fig. 2.8a). When the input power of each port is 350 mW (total 1.4 W), the peak SAR is 4.8 W/kg and the peak PD value is 9.48 W/m^2. If the input power is higher than 350 mW, the peak PD of the head phantom can exceed the limitation level of 10 W/m^2. The return loss is presented in Fig. 2.8b. The radiation pattern of the antenna when placed in proximity with the human head is shown in Fig. 2.8c–d [29].

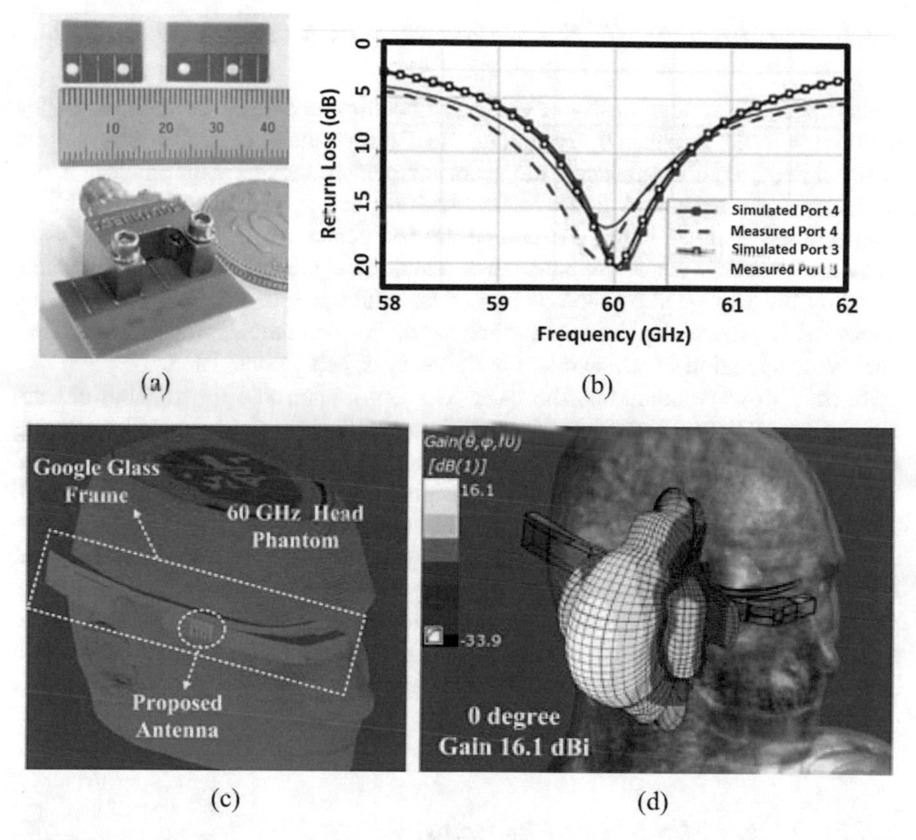

Fig. 2.8 60 GHz patch antenna array with parasitic elements for smart glasses: **a** Fabricated antenna, **b** simulated and measured return loss characteristics of the antenna. Simulated antenna performance on the 60 GHz head phantom and Google Glass frame: **c** Voxel model of the head phantom, **d** 3D radiation pattern (Reprinted with permission from IEEE [29])

2.4 Statistical On-Body Measurement Results

There have been great number of studies on propagation channel characterization for wireless body area networks at UWB and mmWave frequencies. Many issues have been investigated such as antenna performance near the human body, understanding the path loss for body channels and its variation with movements, radio propagation modes on the human body, etc. Measurements show that the channel characteristics are highly dependent on the antenna type and link geometry.

2.4.1 Electromagnetic Simulation Based Channel Modeling

Electromagnetic simulation-based channel modelling is considered as a promising option for WBAN channel modelling. Realistic channel measurements require environments, equipment, and volunteers for performing the experiments. It can also lead to the measured data with uncertainties and inaccuracies due to cabling, unintentional changes in the position of the test person or the antennas. To overcome such issues several simulations based numerical modelling tools are available based on full-wave and asymptotic numerical solutions [30–32]. The choice of the numerical approach depends on the nature of the communication links and frequency of operation of the wireless body network being considered.

In the full-wave solutions, the basic idea is to determine the channel characteristics accurately in the given scenario solving Maxwell's equations using numerical approaches. In the WBAN context, the most used full-wave numerical approaches are: Method of Moments (MoM), Finite Element Method (FEM), Finite-Difference Time-Domain (FDTD), and Finite Integration Technique (FIT) [32]. Asymptotic techniques, such as raytracing (RT) and Uniform Theory of Diffraction (UTD) are commonly used for larger environments or in high-frequency applications, in which full-wave solutions are too complex [32]. The asymptotic technique can be used purely, or as a hybrid technique, i.e., in combination with a full-wave solution to obtain the optimal solution for propagation prediction in terms of accuracy, complexity and simulation time.

2.4.2 Tissue Mimicking Phantoms

Tissue-mimicking materials (TMMs) are required to test and validate the antenna performance for on-body and implantable in-vivo communication. This is useful for a wide variety of biomedical applications such as sensor design, microwave imaging etc. Biological tissues have large range of dielectric properties which also change with frequency. In contrast to measurements directly performed on human bodies and for which the outcome may lead to fluctuating results due to inter-individual differences and body movements, the use of experimental phantoms can provide reproducible results. Experiments performed in a controlled environment give stable results which is important for studying the on-body propagation channel characteristics. Human phantoms should be stable, flexible, and realistic which can be created in standard research labs are the key requirements that needs to be taken into account.

2.4.2.1 UWB Phantom

To accurately model the antenna in presence of the phantom, the dielectric properties of the phantom should be carefully determined in the 3.1–10.6 GHz range. The compact microstrip monopole antenna (25 × 10 × 1.6 mm^3) is designed on 1.6 mm thick AR350 substrate with $\varepsilon_r = 3.5$ for which the schematic is shown in Fig. 2.9a. Two-thirds of the muscle permittivity is used as a target value for the phantom [33]. For the measurements, the antenna was placed 1 mm above two-thirds muscle equivalent parallelepiped phantom modelling a human arm (Fig. 2.9b). The phantom is made of distilled water, agar, TX-151, sodium chloride, and polyethylene powder. The relative permittivity and conductivity of the phantom are adjusted using polyethylene powder and sodium chloride. Agar is used to maintain the shape of the phantom, sodium azide is a preservative, and TX-151 improves the phantom stickiness.

For the numerical modelling, the complex dielectric permittivity ε^* of the phantom is expressed as a Debye's dispersion equation with the following optimized parameters: relaxation time $\tau = 12.5 \times 10^{-12} s$, static permittivity $\varepsilon_s = 37.1$, and optical permittivity $\varepsilon_\infty = 12.2$.

$$\varepsilon^* = \varepsilon_0(\varepsilon' - j\varepsilon'') = \varepsilon_0\left(\varepsilon_\infty + \frac{\varepsilon_s - \varepsilon_\infty}{1 + j\omega\tau}\right) \tag{2.1}$$

where ε_o, is the free space permittivity (8.85 × 10^{-12} F/m), ω depicts the angular frequency, and τ is the relaxation time. Theoretical permittivity and conductivity model is in very good agreement with target values over the considered frequency range (3–11 GHz), confirming thereby that the choice of the Debye model is appropriate (Fig. 2.9c). The phantom has been characterized using the dielectric probe kit 85070E (Agilent Tech., CA, USA). The measured complex permittivity is in a satisfactory agreement with the numerical results.

The reflection coefficients of the designed UWB antenna are presented in Fig. 2.9d for the free space and on-body (fabricated arm phantom) scenarios. These results show that the S_{11} parameters are similar and are less affected by the presence of the arm.

2.4.2.2 mmWave 60 GHz Skin-Equivalent Phantom

The electromagnetic absorptions by the human body at 60 GHz mainly takes place in the skin tissues due to a penetration depth of around 0.5 mm [34–37]. A single layer homogeneous torso phantom with electric properties of dry skin is proposed in [34]. The skin-equivalent phantom of the torso has been taken out of a high-resolution whole-body model and thus, employs a realistic body shape. The overall dimensions of the phantom are 288 × 100 × 40 mm^3. The overall number of cell volumes (voxels) in the computational domain and subsequently the computation and time requirements are reduced by using an adaptive meshing [34].

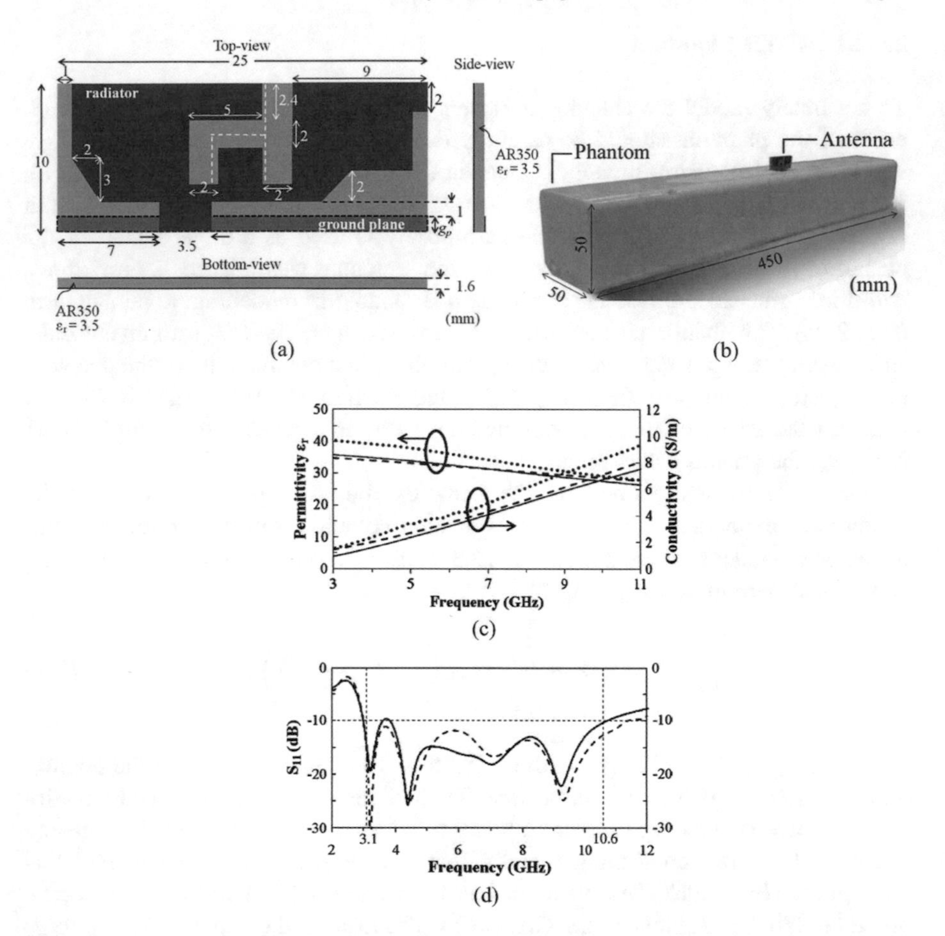

Fig. 2.9 **a** Compact UWB antenna schematic. **b** Antenna prototype mounted on the two-thirds muscle equivalent phantom. **c** Measured and computed characteristics of the phantom. —Target… Measurement___Calculation. **d** Simulated reflection coefficient of the optimized antenna for_____antenna in free space and —antenna mounted on the phantom (Reprinted with permission from IEEE [33])

The dielectric properties of the skin are well characterized up to 110 GHz based on the extrapolation of data obtained through measurements up to 20 GHz [36]. Debye model with a single relaxation time is considered to exhibit good accuracy for modelling the permittivity data in 55–65 GHz frequency range. The optimized values for the best fit in 55–65 GHz range are: $\varepsilon_s = 34.8$, $\varepsilon_\infty = 4.1$, and $\tau = 6.9$ 10^{-12}s [36] (Fig. 2.10a).

A wideband semisolid skin-equivalent phantom with dielectric properties close to human skin in the V and W-band is proposed in [37] (Fig. 2.10b). The constituents of the phantom are deionized water, agar, and gelatin powder. Gelatin powder is made from collagen and dry matter of the skin is collagen; gelatin powder is used to tune the phantom for the desired dielectric properties that correspond to

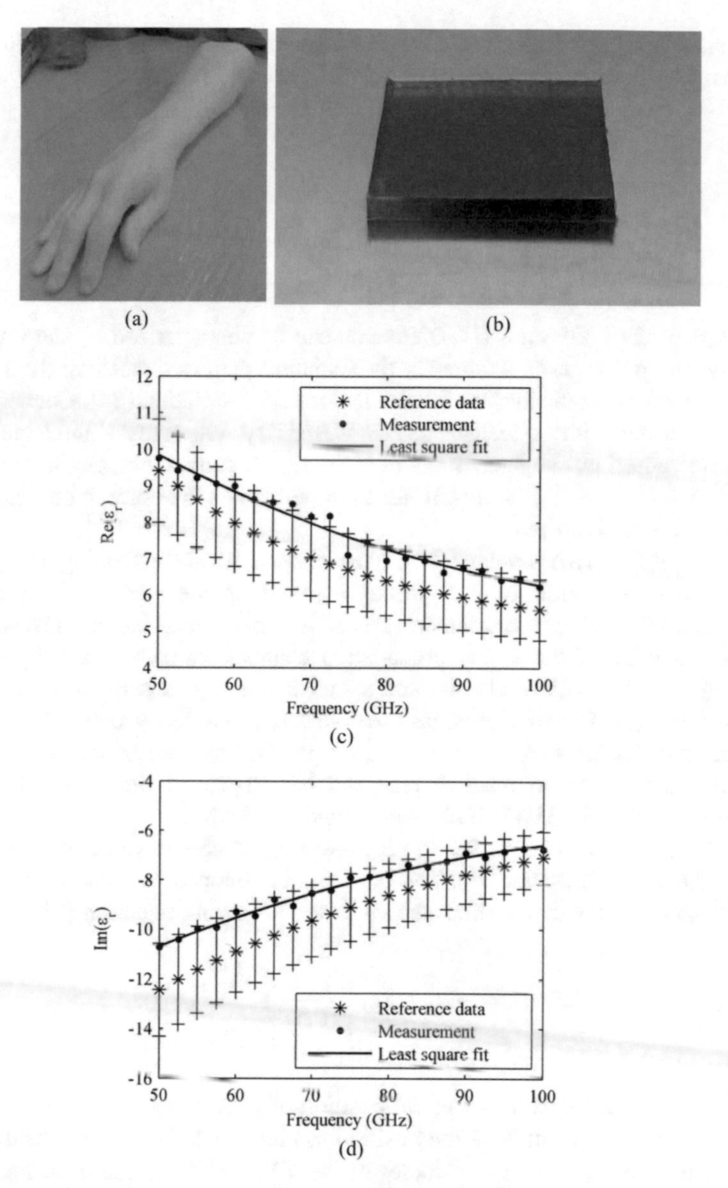

Fig. 2.10 **a**, **b** Skin equivalent phantoms for 60 GHz communication [36, 37], Complex relative permittivity of proposed skin equivalent phantom in V and W band, **c** real part, **d** imaginary part (Reprinted with permission from IEEE [37])

human skin. To fabricate the phantom, deionized water is heated to 80 °C, and afterwards agar and gelatin powder are added to the water. After melting these constituents, the mixture is poured into a mould and cooled for a few hours to room temperature for solidification. The phantom has been fabricated and validated by an open-ended waveguide method. Complex relative permittivity of the skin equivalent phantom in V and W-band for the real part and imaginary part is shown in Fig. 2.10c, d respectively.

2.4.3 On-Body Propagation Analysis for UWB Communication

The effect of the body on a UWB antenna can be characterized by the impedance detuning, the polarization mismatch, the radiation pattern distortion, the reduction of the efficiency and the system fidelity factor (SSF) variation [38]. Concerning the time-domain behavior, a realistic UWB-WBAN system study should include the dispersion caused by the lossy biological tissues. A convenient way to assess time performance is the SFF [38, 39]. It has been defined as the correlation between the input and the received pulse.

The CPW-fed TSA antenna has been chosen for investigating the on-body channel characterization in [40] through measurements performed in anechoic chamber and the indoor environment. Analysis is performed based on transient and spectral behavior of the UWB propagation channel. A path loss (PL) model is extracted from the measured data, and a statistical study is performed on the time delay parameters. The two antennas are connected to a vector network analyser to measure the transmission response (S_{21}) in the frequency range 3–10 GHz. The TSA antenna is compact in size and has improved time delay behaviour, making it suitable for UWB body area networks (BAN).

The path loss model signifies the local average received signal power (P_r) relative to the transmit power (P_t). The path loss is obtained as mean path gain over the measured frequency band, as shown in the following equation [2]:

$$PL(d(p)) = -20.\log_{10}\left\{\frac{1}{10}\frac{1}{N_f}\sum_{j=1}^{10}\sum_{n=1}^{N_f}\left|H_j^p(n)\right|\right\} \tag{2.2}$$

where $PL(d(p))$ is the path loss at the position of p, at which the distance between the Tx and Rx is a function of the position p, thus the distance is denoted by $d(p)$. N_f is the number of frequency samples of the VNA. $H_j^p(n)$ is the measured S_{21} for the position p, jth snapshot, and nth frequency sample. The path loss at any location can be described as [2]:

$$PL_{\mathrm{dB}}(d) = PL_{\mathrm{dB}}(d_0) + 10\,\gamma\,\log_{10}\left(\frac{d}{d_0}\right) + X_\sigma \tag{2.3}$$

where, PL_0 is the path loss at reference distance d_0 set to 1 m, γ is the path loss (PL) exponent and X_σ is a zero-mean Gaussian distributed random variable in dB with the standard deviation σ. The power delay profile (PDP) is the squared magnitude of the impulse response, defined as [2]:

$$P(\tau, t) = h(\tau, t)h^*(\tau, t) = \sum_{k=1}^{K} a_k^2 \delta(\tau - \tau_k) \tag{2.4}$$

The PDP gives the intensity of a signal received through a multipath channel as a function of time delay where the time delay is the difference in the travel time between multipath arrivals. The radio channel is usually characterised by the first and second central moment of the PDP respectively *i.e.,* mean excess delay τ_m and rms delay spread σ_τ (which can be used as a figure of merit for estimating data rates for multipath channels) describes the time dispersive properties of the channel. The most important parameter to characterize the time-dispersion behaviour of the wireless propagation channel is the root-mean-square delay spread, which corresponds to the second central moment of the power delay profile (PDP), that can be expressed as [2]:

$$\sigma_\tau = \sqrt{\frac{\sum_k (\tau_k - \tau_m)^2 .\ |h(\tau_k; d)|^2}{\sum_k |h(\tau_k; d)|^2}} \tag{2.5}$$

where τ_k are the multipath delays relative to the first arriving multipath component, τ_m is the mean excess delay, $h(\tau_k;d)$, takes into account the channel impulse response (CIR) amplitude dependence on the Tx-Rx separation distance d.

The PL exponent (γ) is 3.0 for the TSA when measurements are carried out in the anechoic chamber. When measurements are performed in the indoor environment, the reflection from the surrounding scatters increases the received power, causing reduction of the path loss exponent, γ is 2.6. To improve the accuracy of the path loss model, a zero mean, normal distributed statistical variable is introduced to consider the deviation of the measurements from the calculated average path loss (Fig. 2.11a, b). In the anechoic chamber the standard deviation of the normal distribution is $\sigma = 8.2$ for the TSA. In the indoor environment the values obtained is 6.7 for TSA.

The root mean square spread delay (σ_τ) is an important parameter for multipath channels because it imposes a limit to the data rate achievable [40]. Figure 2.12 shows the cumulative distribution of the delay spread (obtained applying a threshold of −20 dB) fitted to a Nakagami distribution. Higher delay spread is observed for the measurements performed in the indoor environment when compared with the results obtained from the anechoic chamber. In [41], the performance of the body-centric UWB radio channel is investigated using the TSA antenna in the 3–9 GHz range by considering several static and pseudo-dynamic on-body links. System-level modelling of potential multiband orthogonal frequency-division multiplexed UWB system has been conducted. Results show 78% in the static case and 75% and 61% for stable and unstable transmitter locations in the pseudo

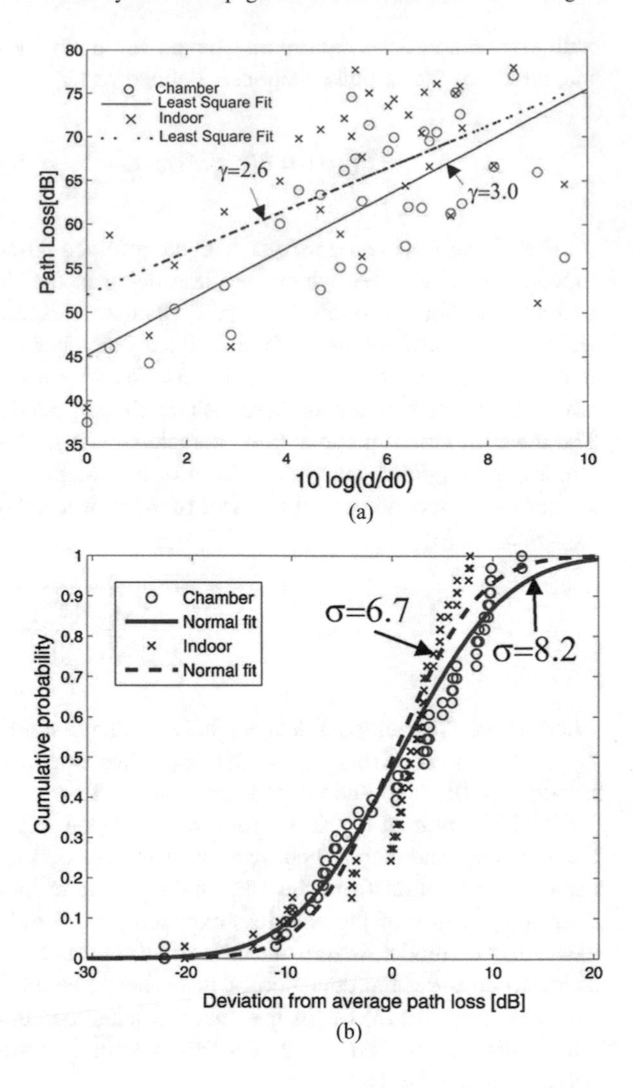

Fig. 2.11 **a** Measured and modelled path loss for on-body channel versus logarithmic Tx-Rx separation distance. **b** Deviation of the measurements from average path loss fitted to a normal distribution (Reprinted with permission from IEEE [40])

dynamic in-motion scenarios (respectively) of the specified on-body radio links, the bit error rate (BER) is equal to or less than 0.1%.

Experimental analysis for path loss, shadow fading and delay dispersion parameters of on-body UWB WBAN channel is carried out in [42] in a laboratory environment in the UWB frequency range (3.1–10.6 GHz). Both line-of-sight (LOS) and non-line-of-sight (NLOS) scenarios of WBAN channel are considered for measurement. Three human subjects of thin, average and fat built were considered during the measurements. The PL exponent ranges from 2.1 to 3.5 for LOS links and 2.6 to 1.8 for NLOS links. It is observed that the path loss exponent increases from thin to fat person in LOS scenario, and decreases in the case of NLOS scenario. In all observations the shadow fading term follows lognormal

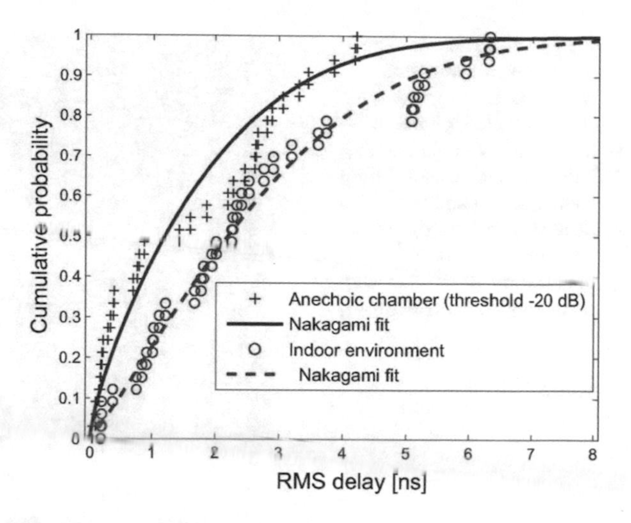

Fig. 2.12 Cumulative distribution of the rms delay spread fitted to Nakagami distribution applying a threshold of −20 dB (Reprinted with permission from IEEE [40])

distribution in log scale. The rms delay spread analysed at −20 dB PDP threshold ranges from 8.3 nsec to 10.82 nsec and 12.94 nsec and 13.92 nsec for LOS and NLOS links respectively.

Time-domain analysis is performed with antennas placed on several locations of the whole-body voxel model for impulse radio applications in [33]. Three transmission scenarios between four antennas are considered: the transmitting antenna is mounted on the left wrist (TX1), the three receiving antennas are placed on the left arm (RX1), left ear (RX2), and left leg (RX3). The distortion of the received pulses depends on the antenna position, and their fidelity equals 75%, 46%, and 81%, for RX1, RX2, and RX3, respectively. The propagation path loss for various on-body links is computed. The wrist-to-arm link, being the shortest, has the lower loss, with an average loss of 53.9 dB. The wrist-to-head link has the higher average loss of 71.7 dB due to the longer link length. In the wrist-to-calf link, an average loss of 60.5 dB is observed, and the variability is much lower compared to the other links.

2.4.4 On-Body Propagation Analysis at 60 GHz

On-body electromagnetic propagation at 60 GHz is characterized indeed by several peculiar features, e.g., strong reflection and absorption due to the high dielectric contrast between the skin permittivity and free space [43]. It is important to characterize the on-body propagation channel for optimization of wearable antenna designs in terms of surface waves, attenuation rates, path gain, absorption in the human body, impact of the polarization on the performances of the link. The on-body propagation at 60 GHz is studied using a skin-equivalent phantom to provide analytical-based fundamental models of path gain by considering vertical and horizontal elementary dipoles as reported in [44]. Propagation near a flat phantom (Fig. 2.13a) is studied theoretically for vertical and horizontal elementary dipoles. For

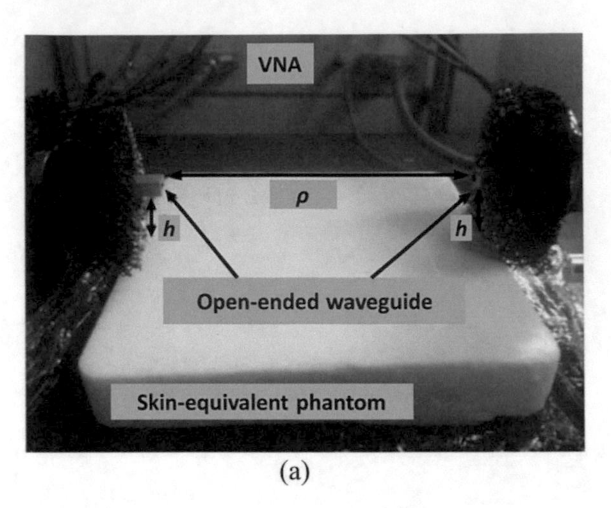

(a)

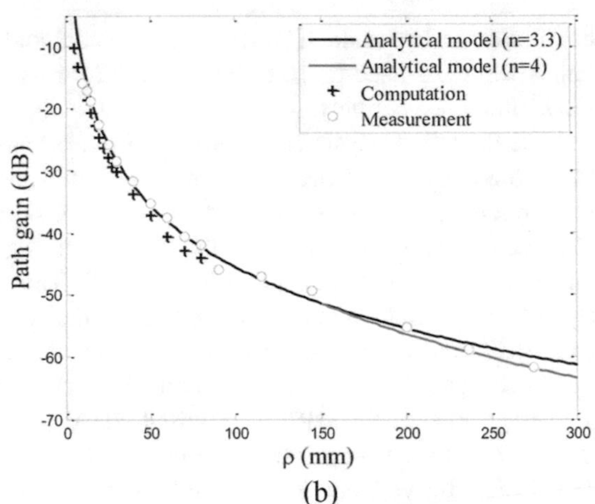

(b)

Fig. 2.13 **a** Measurement setup: Two open-ended waveguides close to a skin-equivalent phantom. **b** Vertical (V) polarization. Path gain model (solid line) computed (+) and the measured results (o) for on body phantom for V polarization (Reprinted with permission from IEEE [44])

a vertically (V) polarized wave, in an intermediate region, the path gain at 60 GHz can be modelled according to the height of the dipole above the phantom. If $h > 3$ mm, the power decay exponent $n = 4$ mm. If $h < 3$ mm, the path gain can be modelled almost similarly as done at lower microwave frequencies, with $n = 3.5$. For a horizontally (H) polarized wave, the intermediate zone with $n = 3.5$ is not present, and a power decay exponent of $n = 4$ is found for the whole range of distances.

Using linearly polarized 7-dBi gain open-ended waveguides, the on-body propagation has been studied experimentally for vertical and horizontal polarizations. The experimental results are in a good agreement with analytical models, thereby showing that the on-body propagation can be accurately represented by the analytical path gain model (Fig. 2.13b). The path gain is highly sensitive to the antenna/body separation as seen from Table 2.3. It increases with h and becomes even higher than the free-space one for H-polarization at $h = 5$ mm. It was also demonstrated that the antenna efficiency summarized in Table 2.4 of a horizontally

Table 2.3 Path gain for different antenna/body separation and different polarizations at 60 GHz [44]

H (mm)	S_{21} (dB) for ρ = 80 mm			
	VV polarization		HH polarization	
	Meas	Sim	Meas	Sim
0	−44.4 ± 0.7	−45.7	−46.6 ± 1.7	−47.1
1	−40.4 ± 0.7	−44.6	−40.3 ± 0.9	−43.1
5	−34.4 ± 0.6	−37.2	−27.8 ± 0.7	−30.1

VV both antennas are vertically polarized.
HH both antennas are horizontally polarized.

Table 2.4 Radiation efficiency for three antenna/body separations and for two antenna polarizations at 60 GHz [44]

H (mm)	V polarization		H polarization	
	Meas	Sim	Meas	Sim
0	–	0.28	–	0.81
1	0.49	0.48	0.88	0.82
5	0.65	0.59	0.86	0.85

polarized antenna changes very slightly with h, whereas it is strongly affected for a vertically polarized antenna.

Creeping wave formulation for both vertical and horizontal polarized electromagnetic waves traveling on a circular path around a cylinder are discussed in [45, 46]. A linear path gain has been derived and values of the approximated power decay exponent have been summarized for different frequencies and values of the radius of the cylinder. Measurements on the torso of a real human body at 60 GHz show that the path loss of the field radiated by an electromagnetic source at the surface of the torso can be modelled in the curved zone by assuming that the torso has a circular section of radius equal to its perimeter divided by 2π.

2.5 UWB Dynamic On-Body Communication Channels

The presence of the human subject leads to significant variation in the channel characteristics in comparison to direct propagation in free space or indoor environment. The movement of the human body also makes the radio channel more challenging to characterize. Kumpuniemi et al. [47, 48] have presented pseudo-dynamic radio channel models for ultra-wideband (UWB) wireless body area network communications. To produce the models, frequency domain on-body measurements are conducted in an anechoic chamber within the frequency band of 2–8 GHz. The study is conducted for two UWB antenna types (dipole and double loop). The antennas are mounted on the left and right wrist (L. Wrist and R. Wrist, respectively) and the left ankle (LA) of the test person. A walking sequence with five positions is modelled and the antennas are placed on both wrists and the left ankle of a person. At first, the amplitudes of the first arriving paths were solved for all links and positions. In the second phase, the resulting channels are divided into three classes: line-of-sight, (partially) obstructed line-of-sight and non-line-of-sight.

2.5.1 Classification and Statistical Analysis of the On-Body Channel During Physical Exercises

Channel classification aids in understanding the propagation phenomenon occurring between two wearable links and gives an estimate of the good and bad links depending on the location of the on-body antenna. This work presents classification of the channel using kurtosis and statistical analysis of the path loss and rms delay spread for various upper and lower limb activities [49].

Various human activity measurements are performed in an indoor environment in the 4–8 GHz range as shown in Fig. 2.14a [50]. The wearable antenna used is a compact and miniaturized tapered slot antenna (TSA) (Fig. 2.14a, inset) which has good performance in free space and on-body in the desired frequency band [50]. The size of the antenna is $12 \times 18\ \mathrm{mm}^2$ and is inspired by [40]. The position of the wearable antennas (transmitter (Tx) and receiver (Rx)) considered is the wrist for the upper limbs and ankle for the lower limb activities. For limb bending exercises shoulder/thigh region is chosen as the Tx location and wrist/ankle as the Rx. These wearable locations give maximum displacement during physical activity, through which the variation in channel parameters over distance will be more prominent. Three orientations of the wearable antennas are considered as shown in Fig. 2.14b i.e. outer (OUT), inner (IN), front (FR) of the wrist/ankle (WR/AK) giving rise to different types of on-body channel links which are denoted as: OUT_WR/OUT_AK, IN_WR/IN_AK, FR_WR/FR_AK. The antennas are connected to the 2-port vector network analyser (VNA) using low loss flexible cables to provide freedom in limb movement activities. The sweep is performed over 1601 frequency points for 4–8 GHz range. Data is recorded at different positions for each activity (five repetitions) and the angles are estimated through digital protractor with shoulder/thigh or elbow/knee joint as reference. Figs. 2.15 and 2.16 depict various schematics of the upper and lower limb activities respectively for various types of activities carried out, both limb movement, single limb movement and limb bending.

2.5.2 On-Body Links Channel Classification

Kurtosis is one such channel parameter which aids in evaluation of the channel in terms of LOS/NLOS, gives an idea of certain channel properties, hence, provides guidelines for the design and development of channel/time of arrival estimation algorithms [27, 52]. Kurtosis κ is mathematically defined as follows:

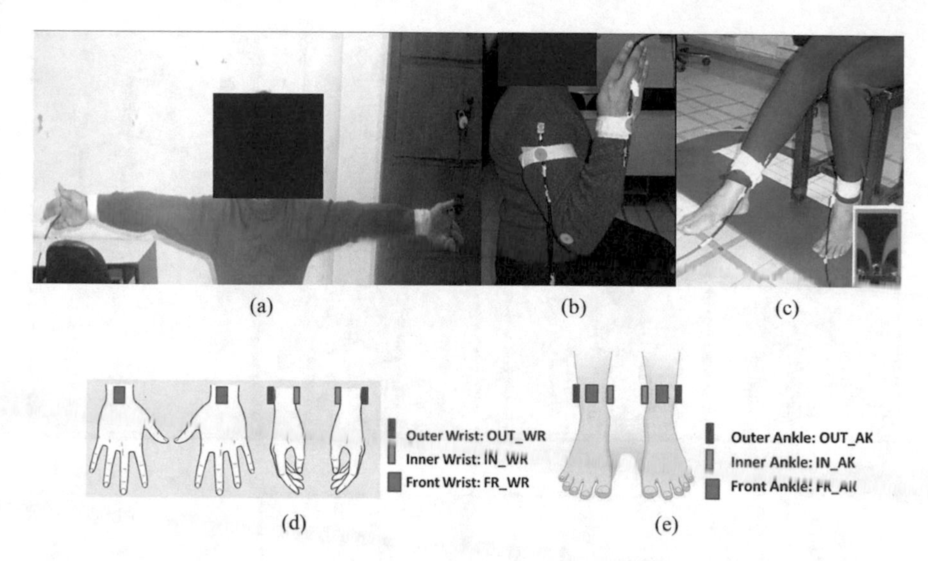

Fig. 2.14 Activity monitoring in an indoor environment: **a** Upper limbs, both arms movement; **b** lower limb, single leg movement; **c** upper limb bending movement, inset TSA antenna (Reprinted with permission from IEEE [50, 51]). Orientations of the wearable antenna for the **d** wrist and **e** ankle region of the human subject (Reprinted with permission from IEEE [51])

$$\kappa(x) = \frac{1}{\sigma^4}\frac{\sum_i (x_i - \bar{x})^4}{N} \tag{2.6}$$

where σ is the standard deviation of the variable x and $\bar{x}$ is the mean value of x. N is the number of samples of x. A channel impulse response (CIR) with high κ, classifies as LOS links and for NLOS links, due to high multipath and low magnitude, κ will be lower. As seen in Figs. 2.17 and 2.18, for both arms/legs movement activity, generally κ in the range of 50–200 are in LOS situation due to high peakedness of the CIR. For NLOS/partial NLOS links, κ is very low ranging from 5 to 20 for total NLOS and 20–40 for partial NLOS. Table 2.5 summarizes the channel classification results in terms of percentage of NLOS and LOS links formed while performing various types of upper and lower limb activities.

2.5.3 *Upper Limbs Activity*

2.5.3.1 Both Arms Movement

The wearable antennas are placed on the left (L.) wrist and right (R.) wrist for the activities considered as shown in Fig. 2.15a, A1–A4. As observed from Table 2.5, for the scenario when both the Tx and Rx wearable antennas are placed

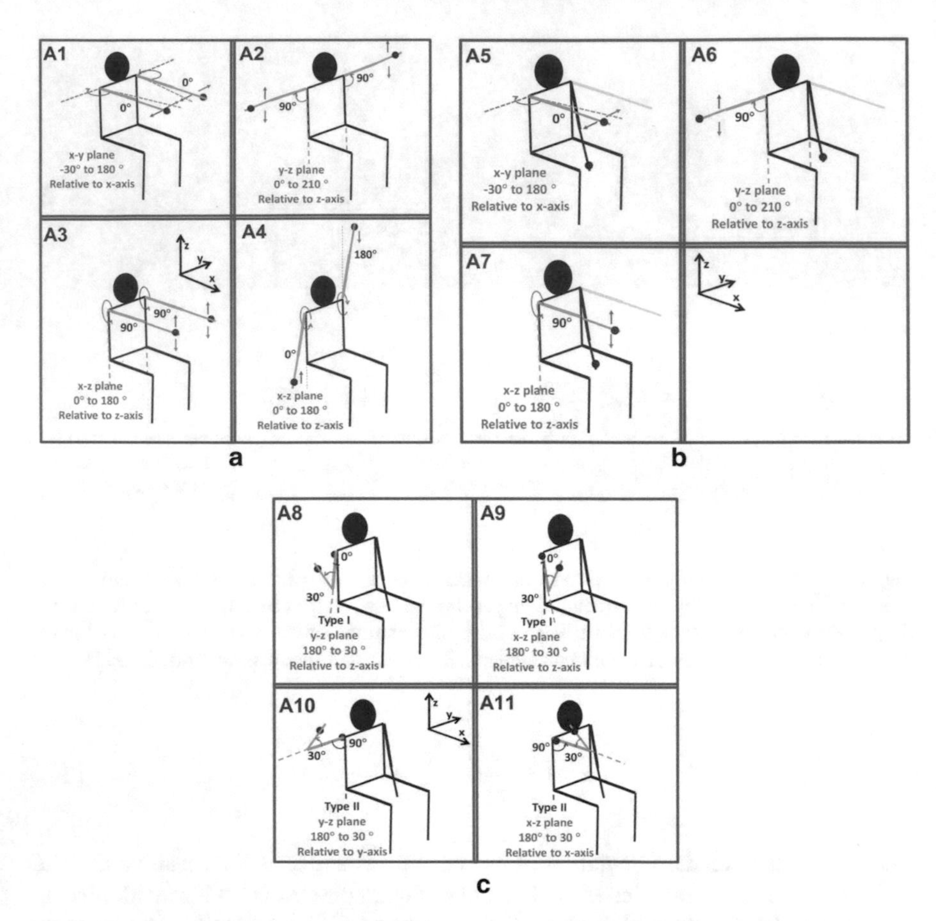

Fig. 2.15 Upper limb activity: **a** both limb movement **b** single limb movement **c** limb bending (Reprinted with permission from IEEE [51])

in OUT_WR orientation, 25% LOS links are observed and 75% NLOS links are observed. The IN_WR orientation has maximum number of LOS scenarios while performing upper limb activities, due to the higher occurrence of direct line-of-sight path. For the FR_WR orientation, there are more LOS links than NLOS by 5%. Figure 2.17 depicts the classification of the channel type for both arms movement activity based on kurtosis κ values.

2.5.3.2 Single Arm Movement

For single arm movement, the wearable antennas are placed at the same location as that of both arm movements activity. But only, right arm is performing the activity and the left arm is at rest which can be observed in Fig. 2.15b, A5-A7. Total NLOS situations are observed for single arm movement for OUT_WR and IN_WR

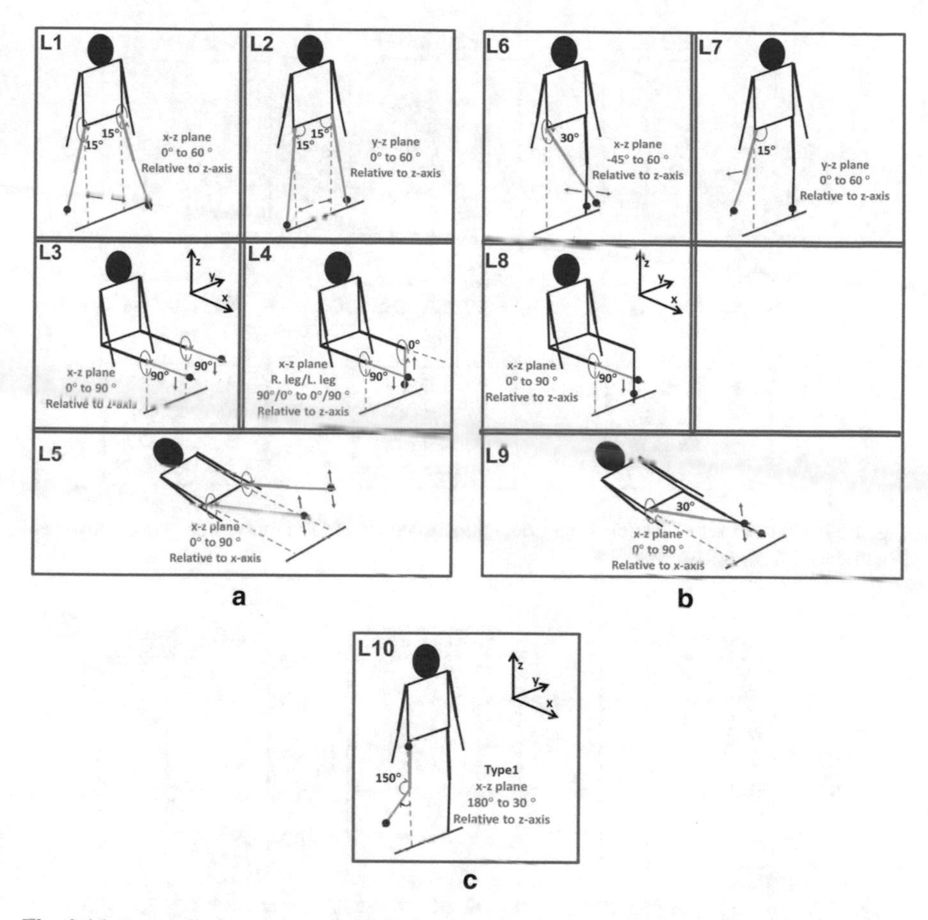

Fig. 2.16 Lower limb activity: **a** both limb movement **b** single limb movement **c** limb bending (Reprinted with permission from IEEE [51])

antenna orientation due to obstruction caused by the torso/thigh region between the nodes placed on the L./R.wrist and the orientation. For the case of FR_WR orientation of the wrist, there are 20% chances of LOS scenarios for the activities performed and rest 80% links formed are NLOS/partial NLOS scenarios.

2.5.3.3 Arm Bending Movement

During arm bending activity, one of the antennas is placed on the outer region of the shoulder, the other antenna is placed on the wrist in three different orientations: OUT_WR/IN_WR/FR_WR. Four different limb bending activities have been considered as shown in Fig. 2.15c, A8-A11. Generally, for 60–80% LOS links are observed and 20–40% NLOS links are observed for different limb bending activities.

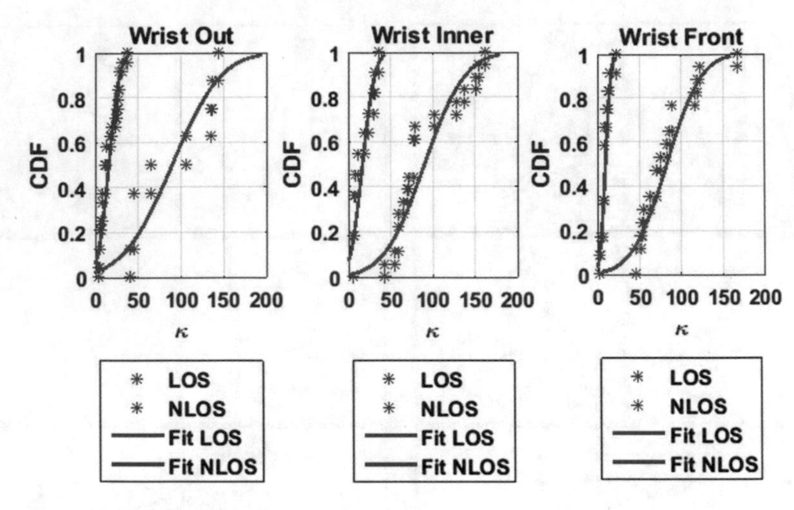

Fig. 2.17 Channel classification based on kurtosis: upper limb activity, both arms movement (Reproduced from [49])

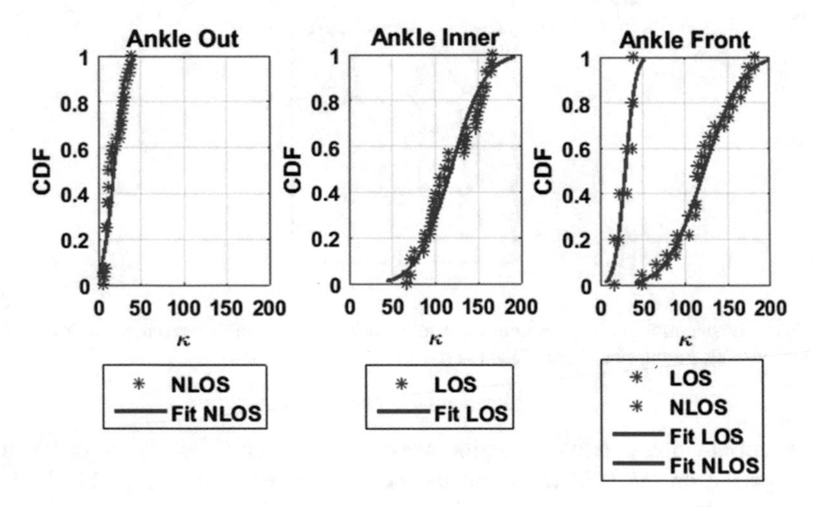

Fig. 2.18 Channel classification based on kurtosis: lower limb activity, both legs movement (Reproduced from [49])

2.5.4 Lower Limbs Activity

2.5.4.1 Both Legs Movement

The wearable antennas are placed on the ankle region and various activities are performed in standing, sitting, and sleeping posture which can be seen in Fig. 2.16a, L1-L5. For OUT_AK orientation total NLOS links are observed due to obstruction caused by the legs, for IN_AK total LOS links are observed due to the direct path propagation between the Tx and Rx, FR_AK orientation leads to a 85% LOS links and 15% partial NLOS links which occur due to the transition from LOS→NLOS while performing a particular limb movement activity (e.g. L1). Figure 2.18 depicts the classification of the channel type for both legs movement activity based on kurtosis κ values.

2.5.4.2 Single Leg Movement

In this scenario, the left leg is in static position with the wearable antenna placed on the ankle region and the right leg is performing various movements as shown in Fig. 2.16b, L6-L9. Similar classification is observed as that of both leg movement activities with OUT_AK having total NLOS links, IN_AK with total LOS links and FR_AK generally having LOS links and few NLOS links which is presented in Table 2.5.

2.5.4.3 Leg Bending Movement

The wearable antenna is placed on the right leg, one on the outer region of the thigh and the other antenna on the ankle region in three different orientations: OUT_AK/ IN_AK/FR_AK (Fig. 2.16c). As observed from Table 2.5, the OUT_AK has total LOS links, IN_AK has 40% LOS links due to the respective orientations of the wearable antenna placed on the ankle. FR_AK has no LOS links due to the obstruction caused by the shank/thigh, hence the channel parameter pattern is dependent on the antenna placement.

Table 2.5 Percentage channel classification: Kurtosis based (Reproduced from [49])

	OUT		IN		FR	
	LOS	NLOS	LOS	NLOS	LOS	NLOS
Upper limbs activity						
Both arms (A1–A4)	25	75	60	40	55	45
Single arm (A5–A7)	0	100	0	100	20	80
Arm bend (A8–A11)	100	0	80	20	60	40
Lower limbs activity						
Both legs (L1–L5)	0	100	100	0	85	15
Single leg (L6–L9)	0	100	100	0	80	20
Leg bend (L10)	100	0	40	60	0	100

2.5.5 Path Loss and Rms Delay Spread Statistical Analysis

Monitoring of physical activities is challenging and must consider various parameters such as body effects, location of the wearable device/antenna on the body, the type of physical activity performed. The analysis is application specific as actions range from routine day-to-day activities, rehabilitation, physiotherapy, sports activities to strenuous training. The aim of this work is to target monitoring of rehabilitation, physiotherapy, and general fitness exercises. Channel parameters, namely, PL and σ_τ have been proposed as good indicators to monitor the physical activity of the human subject.

Two channel parameters PL and σ_τ are statistically analysed for all the activities and orientations considered. The mean and standard deviation (STDEV) values of PL and σ_τ have been presented in Tables 2.6 and 2.7, respectively for different activity types and orientations of the on-body antenna. Tables 2.8 and 2.9 depict the mean and standard deviation values for the leg movement activities. From the analysis it can be observed that the magnitude of PL and σ_τ is generally lower in comparison to upper limb activities. This is because the on-body antennas on the upper limbs are significantly influenced by the torso region which is the major cause of NLOS path between the Tx-Rx. For the lower limb, the legs are only the main source of obstruction and in this case, the ankle region has a small diameter in comparison to the thigh region which leads to less signal attenuation level and low multipath occurrence. Also due to the distance factor between the Rx and Tx nodes, lower magnitude of PL and σ_τ are obtained for the antenna placed on the ankle joint in comparison to the wrist. The general spread of arms is much larger than the legs while performing physical activities in scenarios where wrist-to-wrist or ankle-to-ankle links are being studied. The PL and σ_τ generally range from 30 to 50 dB and 0.3–3 nsec for the ankle joint for the LOS links, whereas for the wrist, the values can range upto 40–60 dB and 0.5–6 nsec. In both cases total NLOS links have high magnitude with an average of PL and σ_τ in the range 60–65 dB and 10–30 nsec, respectively.

Table 2.6 Path loss magnitude: arm activity (Reproduced from [49])

	OUT		IN		FR	
	Mean	STDEV	Mean	STDEV	Mean	STDEV
Both arms movement: (A1–A4)						
LOS	48.7	9.5	46.8	8.4	49.7	8.0
NLOS	63.1	1.1	63.8	1.2	62.8	1.3
Single arm movement: (A5–A7)						
LOS	0.0	0.0	0.0	0.0	57	2.2
NLOS	63.8	1.2	62.8	1.1	62.4	1.6
Bending arm movement: (A8–A11)						
LOS	52.5	4.4	45.8	7.2	53	5.9
NLOS	61.0	1.1	60.5	1.5	61.1	1.1

Table 2.7 Rms delay spread: arm activity (Reproduced from [49])

	OUT		IN		FR	
	Mean	STDEV	Mean	STDEV	Mean	STDEV
Both arms movement: (A1–A4)						
LOS	2.5	1.9	1.5	1.0	2.2	1.4
NLOS	23.0	6.2	24.0	8.8	21.5	7.3
Single arm movement: (A5–A7)						
LOS	0.0	0.0	0.0	0.0	4.7	1.1
NLOS	30.7	3.1	28.8	4.3	26.0	10.6
Bending arm movement: (A8–A11)						
LOS	1.7	1.4	0.8	0.4	2.0	1.6
NLOS	10.3	3.1	9.9	4.1	12.2	3.7

Table 2.8 Path loss magnitude: leg activity (Reproduced from [49])

	OUT		IN		FR	
	Mean	STDEV	Mean	STDEV	Mean	STDEV
Both legs movement: (L1–L5)						
LOS	0.0	0.0	36.9	8.1	40.3	8.2
NLOS	61.9	1.4	0.0	0.0	60.6	1.3
Single leg movement: (L6–L9)						
LOS	0.0	0.0	43.3	7.6	51.2	6.4
NLOS	61.7	1.6	0.0	0.0	62.5	1.2
Bending leg movement: (L10)						
LOS	57.5	1.2	58.2	1.2	0.0	0.0
NLOS	0.0	0.0	60.6	0.5	62.5	1.3

Table 2.9 Rms delay spread: leg activity (Reproduced from [49])

	OUT		IN		FR	
	Mean	STDEV	Mean	STDEV	Mean	STDEV
Both legs movement: (L1–L5)						
LOS	0.0	0.0	0.8	1.0	0.9	1.3
NLOS	23.8	7.5	0.0	0.0	17.2	7.8
Single leg movement: (L6–L9)						
LOS	0.0	0.0	0.9	0.4	1.9	1.5
NLOS	22.5	8.1	0.0	0.0	15.4	8.3
Bending leg movement: (L10)						
LOS	2.9	0.7	4.8	1.2	0.0	0.0
NLOS	0.0	0.0	9.9	4.9	20.6	6.5

2.5.6 On-Body Channel Links Analysis During Daily Physical Activities

A detailed study is carried out regarding the variation in the on-body channels when the human subject movements vary while performing walking activity [53]. A female human subject of height 1.6 m, weight of 60 kg with a body mass index (BMI) of 23 has been chosen for performing various physical activities. The on-body antennas were positioned 2–5 mm away from the body at different locations. The pictorial representation of the sitting and walking activity is depicted in Fig. 2.19a, b respectively. Measurements were performed in an indoor office environment in the 4–8 GHz of the UWB band. Miniaturised tapered slot UWB antennas (inset Fig. 2.19b) inspired from [40] are used as wearable antennas (transmitter/receiver (Tx/Rx)).

Standing→Sitting→Standing

The sensor location chosen is the right shoulder (R. SH) and right lower thigh region (R. TH). The schematic of the sitting pattern is shown in Fig. 2.19c. The channel parameters PL and σ_τ variation is presented in Fig. 2.20a, b. Maximum distance between the shoulder and lower thigh region leads to highest value of PL and σ_τ when in standing position and decreases as the subject goes from standing to sitting position. For sitting position P4, lowest PL and σ_τ values are observed due to least distance between the body-worn sensors while the subject is bending forward. The distance increases slightly as the subject leans backward while sitting as seen for P5. For sitting to standing activity similar values of PL and σ_τ are observed. Hence, completing a standing→sitting→standing activity set.

Standing→Walking

The sensor locations chosen are the left ankle (L. AK) and the right ankle (R. AK). The schematic of the walking pattern is shown in Fig. 2.19d. The channel parameters PL and σ_τ variation is presented in Fig. 2.20c, d. Minimum distance between the two legs is observed for standing position P1, leading to lowest value of PL and σ_τ. Large distance between the two legs leads to higher value of PL and σ_τ for the walking position P2. The values of PL and σ_τ reduces for P3 and P4 in comparison to P2 due to decrease in distance between the ankles during walking. The channel parameters increase due to an increase in distance between the legs during walking with a maximum distance for P6. P7 shows decrease in distance between the legs which tends to further increase from P7 to P9. P7 to P9 shows increase in the channel parameter values PL and σ_τ with P9 returning to initial walking position P1.

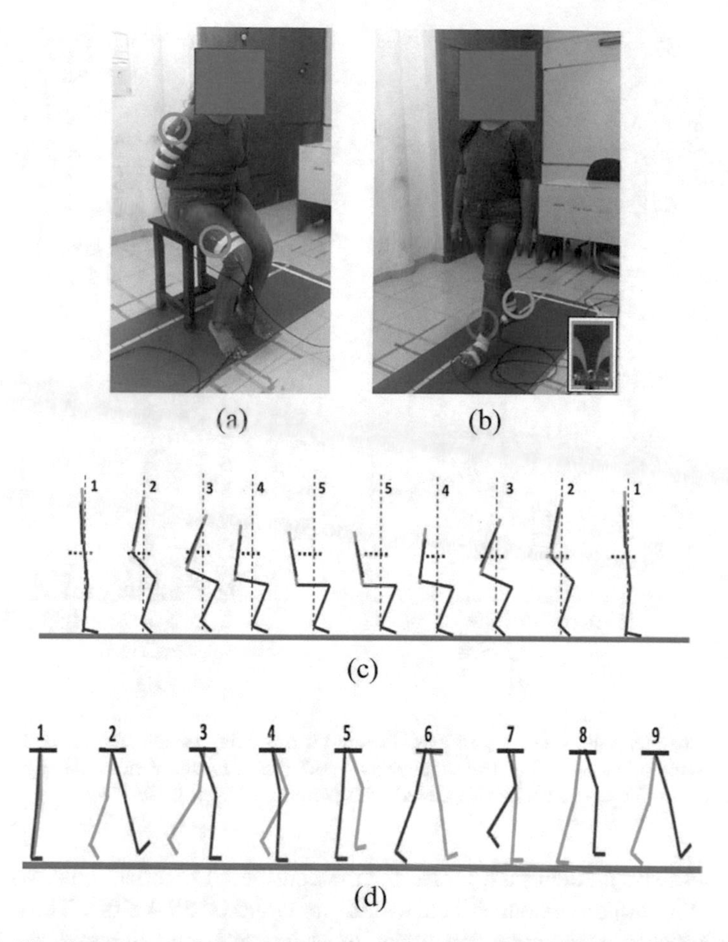

Fig. 2.19 Pictorial representation of the daily physical activity performed: **a** sitting (Tx/Rx: Right Shoulder/Thigh facing outwards). **b** Walking (Tx/Rx: Right/Left ankle facing inwards). (Inset) Compact and low-cost miniaturized UWB tapered slot antenna used as wearable antennas. Schematic representation of the daily physical activities performed during **c** Sitting→Standing→Sitting and **d** Standing→Walking (Reprinted with permission from IEEE [53])

2.6 Dynamic 60 GHz On-Body Propagation Channels

Variation of the on-body millimetre-wave propagation channels with body movements has been studied experimentally in [54]. Two quarter-wavelength monopole antennas with circular ground plane of a 25 mm diameter were connected to direct access ports of the network analyser with two 2 m long coaxial cables. Five antenna positions and eight links between them were investigated. During the measurement, the test subject performed several random movements emulating various everyday activities, such as walking, eating, writing, operating computers, exercising. It is observed that the attenuation in the channels is much stronger than at lower

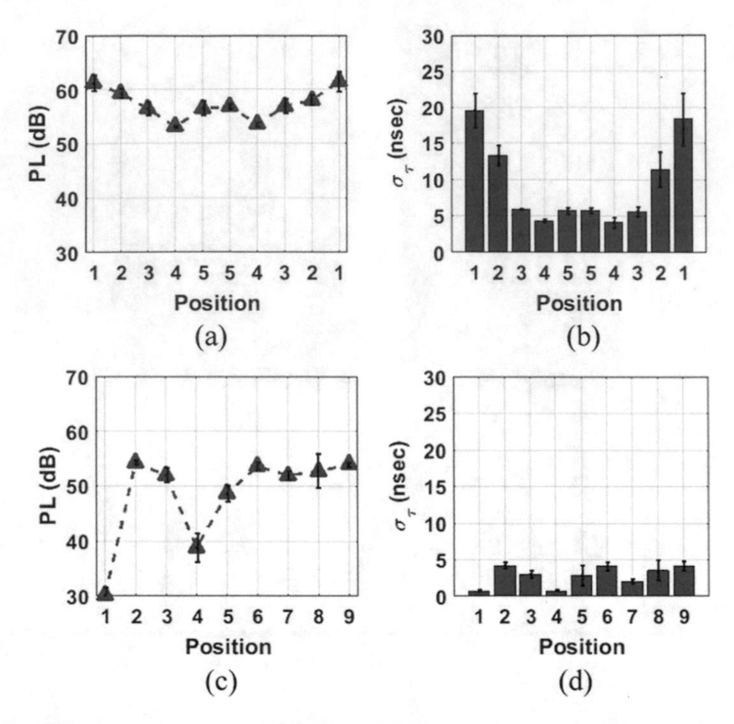

Fig. 2.20 Channel parameters **a** path loss (PL) and **b** rms delay spread (σ_τ) of the Standing → Sitting → Standing activity. Channel parameters **c** path loss (PL) and **d** rms delay spread (σ_τ) of the Standing → Walking activity (Reprinted with permission from IEEE [53])

frequencies. The median path losses for the considered practical links with omnidirectional monopole antennas lied within the range of 55–88 dB. Thus, the separations between communication nodes must be small and line-of-sight between them is essential. Large signal fading is mostly caused by shadowing, antenna pointing error, polarisation mismatch and distance changes. These variations are slow with typical level crossing rates (LCR)s < 1 s^{-1} and average fade duration average fade duration (AFD)s of > 0.1 s. Multipath fading is much less prominent than at lower frequencies and only becomes significant when the signal level is low. The short-term variations of the signal are approximated by a Cauchy–Lorentz distribution near the median value.

On-body propagation losses at 60 GHz have been computed by studying the toe-hop movement during walking movement [55]. Three kinds of toe position with Tx antenna (front, center and back condition) are considered and the links are analysed with respect to various on-body links positioned A–E during walking. The Tx antenna is placed on the right toe. The locations of the Rx antenna is placed on the front right pocket of pants (A), side right pocket of pants (B), left chest region (C), top of right shoulder (D) and right side of head (E). The measurement results show the magnitude of human body attenuation between Tx and Rx path. The measured attenuation of position A, B and E is less than 27 dB in all walking situation. The result of position C is less than permissible value except for the back-toe condition. The position D is not the communication area of the toe node, hence leads to high attenuation which makes it the worst communication link. Channel capacity is simulated through Monte Carlo simulations ensures a channel capacity of 10 Mbit/s with a probability of more than 99.6% in a walking situation [56].

2.7 Conclusion

On-body links have been analysed for two UWB and 60-GHz mmWave communication. Various aspects have been presented in the chapter from on-body antenna design to channel modelling of the on-body links. The design features vary for both the communication technologies studied, leading to different methodologies and techniques to obtain the desired antenna parameters. Results and analysis show that the on-body scenarios are different from free space characterization having significant influence of the antenna characteristics in terms of impedance matching, radiation patterns, gain and efficiency. Apart from the antenna performance, the presence of human subject causes variation in the channel parameters, channel modelling during static and dynamic scenarios, hence, effecting the performance of the wearable devices and the communication system. The non-line-of-sight links caused due to blockage by the human body links influence the channel to a greater extend in comparison to line-of-sight links.

References

1. Yan S, Soh PJ, Vandenbosch GAE (2018) Wearable ultrawideband technology—a review of ultrawideband antennas, propagation channels, and applications in wireless body area networks. IEEE Access 6:42177–42185
2. Hall PS, Hao Y (2012) Antennas and propagation for body-centric wireless communications. Artech House
3. Bharadwaj R, Swaisaenyakorn S, Parini CG, Batchelor JC, Alomainy A (2017) Impulse radio ultra-wideband communications for localization and tracking of human body and limbs movement for healthcare applications. IEEE Trans Ant Propag 65(12):7298–7309
4. Zhadobov M (2018) Millimeter-wave technologies for body-centric applications. In: 43rd International conference on infrared, millimeter, and terahertz waves (IRMMW-THz), Nagoya, p 1
5. Wang JC, Leach M, Wang Z, Lim EG, Ma KL, Huang Y (2015) State-of-the-art of 60 GHz antennas in wireless body area network. In: 2015 International SoC design conference (ISOCC), Gyungju
6. Cotton SL, D'Errico R, Oestges C (2014) A review of radio channel models for body centric communications. Radio Sci 49(6):371–388
7. Bharadwaj R, Koul SK (2019) UWB channel analysis using hybrid antenna configuration for BAN localization applications. In: 2nd Indian conference on antennas and propagation (InCAP 2019), 19–22 Dec 2019, Ahmedabad, India

8. Bharadwaj R, Koul SK (2021) Study of the influence of human subject on the indoor channel using compact UWB directive/omni-directional antennas for wireless sensor network applications. Wirel Ad-Hoc Netw 118:102521
9. Yang D, Hu J, Liu S (2018) A low profile UWB antenna for WBAN applications. IEEE Access 6:25214–25219
10. Sambandam P et al (2020) Compact monopole antenna backed with fork-slotted EBG for wearable applications. IEEE Ant Wirel Propag Lett 19(2):228–232
11. Yeboah-Akowuah KP, Chen Y (2017) A Q-slot monopole for UWB body-centric wireless communications. IEEE Trans Ant Propag 65(10):5069–5075
12. Doddipalli S, Kothari A (2019) Compact UWB antenna with integrated triple notch bands for WBAN applications. IEEE Access 7:183–190
13. Nikolaou S, Abbasi MAB (2017) Design and development of a compact UWB monopole antenna with easily-controllable return loss. IEEE Trans Ant Propag 65(4):2063–2067
14. Wu XY, Akhoondzadeh-Asl L, Wang ZP, Hall PS (2010) Novel Yagi-Uda antennas for on-body communication at 60 GHz. In: 2010 Loughborough antennas and propagation conference, Loughborough, pp153–156
15. Wu XY, Nechayev Y, Hall PS (2011) Antenna design and channel measurements for on-body communications at 60 GHz. In: 2011 XXXth URSI general assembly and scientific symposium, Istanbul, pp 1–4
16. Puskely, Pokorny M, Lacik J, Raida Z (2015) Wearable disc-like antenna for body-centric communications at 61 GHz. IEEE Ant Wirel Propag Lett 14:1490–1493
17. Razafimahatratra S et al (2015) On-body propagation characterization with an H-plane substrate integrated waveguide (SIW) horn antenna at 60 GHz. In: European microwave conference (EuMC), Paris, pp 211–214
18. Zarifi D, Farahbakhsh A, Zaman AU (2017) A gap waveguide-fed wideband patch antenna array for 60-GHz applications. IEEE Trans Ant Propag 65(9):4875–4879
19. Lee H, Li ES, Jin H, Li C, Chin K (2018) 60 GHz wideband LTCC microstrip patch antenna array with parasitic surrounding stacked patches. IET Microw Ant Propag 13(1):35–41
20. Jin H, Che W, Chin K, Shen G, Yang W, Xue Q (2017) 60-GHz LTCC differential-fed patch antenna array with high gain by using soft-surface structures. IEEE Trans Ant Propag 65 (1):206–216
21. Li Y, Luk K (2015) 60-GHz substrate integrated waveguide fed cavity-backed aperture-coupled microstrip patch antenna arrays. IEEE Trans Ant Propag 63(3):1075–1085
22. Lee S, Choi J (2018) A 60-GHz Yagi-Uda circular array antenna with omni-directional pattern for millimeter-wave WBAN applications. In: IEEE International symposium on antennas and propagation and USNC/URSI national radio science meeting, Boston, MA, pp 1699–1700
23. Koohestani M, Moreira AA, Skrivervik AK (2014) Feeding structure influence on performance of two UWB antennas near a human arm. In: The 8th European conference on antennas and propagation (EuCAP 2014), The Hague
24. Koohestani M, Pires N, Skrivervik AK, Moreira AA (2012)Influence of dielectric loading on the fidelity factor of an Ultra-Wideband monopole antenna. In: 2012 Loughborough antennas and propagation conference (LAPC)
25. Leduc C, Zhadobov M (2017) Impact of antenna topology and feeding technique on coupling with human body: application to 60-GHz antenna arrays. IEEE Trans Ant Propag 65 (12):6779–6787
26. Bharadwaj R, Parini CG, Koul SK, Alomainy A (2019) Effect of limb movements on compact UWB wearable antenna radiation performance for healthcare monitoring. Prog Electromag Res C 91:15–26
27. Bharadwaj R, Parini C, Alomainy A (2015) Experimental investigation of 3-D human body localization using wearable ultra-wideband antennas. IEEE Trans Ant Propag 63(11):5035–5044

28. Rehman, MU, Gao Y, Chen X, Parini CG, Ying Z (2007) Effects of human body interference on the performance of a GPS antenna. In: The second European conference on antennas and propagation, EuCAP 2007, Edinburgh, 1–4
29. Hong Y, Choi J (2018) 60 GHz patch antenna array with parasitic elements for smart glasses. IEEE Ant Wirel Propag Lett 17(7):1252–1256
30. Bharadwaj R, Koul SK (2016) Numerical analysis of ultra-wideband propagation for body-centric communication. In: 2016 Asia-Pacific microwave conference (APMC), 5–9 Dec 2016, New Delhi, India, pp 1–4
31. Wang Q, Wang J (2009) Performance of on-body chest-to-waist UWB communication link. IEEE Microw Wirel Compon Lett 19(2):119–121
32. Särestöniemi M, Hämäläinen M, Iinatti J (2017) An overview of the electromagnetic simulation-based channel modeling techniques for wireless body area network applications. IEEE Access 5:10622–10632
33. Chahat N, Zhadobov M, Sauleau R, Ito K (2011) A compact UWB antenna for on-body applications. IEEE Trans Ant Propag 59(4):1123–1131
34. Ur-Rehman M, Malik NA, Yang X, Abbasi QH, Zhang Z, Zhao N (2017) A low profile antenna for millimeter-wave body-centric applications. IEEE Trans Ant Propag 65(12):6329–6337
35. Garrett J, Fear E (2014) Stable and flexible materials to mimic the dielectric properties of human soft tissues. IEEE Ant Wirel Propag Lett 13:599–602
36. Chahat N, Zhadobov M, Sauleau R (2012) Broadband tissue-equivalent phantom for ban applications at millimeter waves. IEEE Trans Microw Theory Tech 60(7):2259–2266
37. Lacik J, Hebelka V, Velim J, Raida Z, Puskely J (2016) Wideband skin-equivalent phantom for V- and W-band. IEEE Ant Wirel Propag Lett 15:211–213
38. Koohestani M, Moreira AA, Skrivervik AK (2014) Fidelity concepts used in UWB systems. In: 2014 IEEE Antennas and propagation society international symposium (APSURSI), pp 824–825
39. Koohestani M, Moreira AA, Skrivervik AK (2015) System fidelity factor evaluation of wearable ultra-wideband antennas for on-body communications. IET Microw Ant Propag 9 (10):1054–1058
40. Sani et al (2010) Experimental characterization of UWB on-body radio channel in indoor environment considering different antennas. IEEE Trans Ant Propag 58(1):238–241
41. Abbasi QH, Sani A, Alomainy A, Hao Y (2010) On-body radio channel characterization and system-level modeling for multiband OFDM ultra-wideband body-centric wireless network. IEEE Trans Microw Theory Tech 58(12):3485–3492
42. Goswami D, Sarma KC, Mahanta A (2015) Experimental determination of path loss and delay dispersion parameters for on-body UWB WBAN channel. In: IEEE International conference on signal processing, informatics, communication and energy systems (SPICES), Kozhikode, pp 1–4
43. Valerio G, Chahat N, Zhadobov M, Sauleau R (2013) Theoretical and experimental characterization of on-body propagation at 60 GHz. In: 7th European conference on antennas and propagation (EuCAP), pp 583–585
44. Chahat N, Valerio G, Zhadobov M, Sauleau R (2013) On-body propagation at 60 GHz. IEEE Trans Ant Propag 61(4):1876–1888
45. Petrillo L, Mavridis T, Sarrazin J, Lautru D, Benlarbi-Delaï A, Doncker PD (2014) Analytical creeping wave model and measurements for 60 GHz body area networks. IEEE Trans Ant Propag 62(8):4352–4356
46. Petrillo L, Mavridis T, Sarrazin J, Benlarbi-Delaï A, Doncker PD (2015) Statistical on-body measurement results at 60 GHz. IEEE Trans Ant Propag 63(1):400–403
47. Kumpuniemi T, Hämäläinen M, Yekeh Yazdandoost K, Linatti J (2017) Human body shadowing effect on dynamic UWB on-body radio channels. IEEE Ant Wirel Propag Lett 16:1871–1874

48. Kumpuniemi T, Hämäläinen M, Tuovinen T, Yazdandoost KY, Linatti J (2014) Radio channel modelling for pseudo-dynamic WBAN on-body UWB links. In: 2014 8th International symposium on medical information and communication technology (ISMICT), Firenze, pp 1–5
49. Bharadwaj R, Koul SK (2021) Channel analysis of on-body channel links during physical activities, Internal Report
50. Bharadwaj R, Koul SK (2019) Monitoring of limb movement activities during physical exercises using UWB channel parameters. In: IEEE MTT-S international microwave biomedical conference (IMBioC), Nanjing, China, pp 1–3
51. Bharadwaj R, Koul SK (2021) Assessment of limb movement activities using wearable ultra-wideband technology. IEEE Trans Ant Propag 69(4):2316–2325
52. Marano S, Gifford WM, Wymeersch H, Win MZ (2010) NLOS identification and mitigation for localization based on UWB experimental data. IEEE J Sel Areas Commun 28(7):1026–1035
53. Bharadwaj R, Koul SK (2020) Wearable ultra wideband technology for daily activity recognition. In: 2020 IEEE MTT-S international microwave biomedical conference (IMBioC), Toulouse, France, pp 1–3
54. Nechayev YI, Wu X, Constantinou CC, Hall PS (2013) Millimetre-wave path-loss variability between two body-mounted monopole antennas. IET Microw Ant Propag 7(1):1–7
55. Akimoto K, Motoyoshi M, Kameda S, Suematsu N (2018) Measurement of on-body propagation loss for directional millimeter-wave WBAN. In: 11th Global symposium on millimeter waves (GSMM), Boulder, CO, USA, pp 1–3
56. Akimoto K, Motoyoshi M, Kameda S, Suematsu N (2019) Evaluation of intra-Wban channel capacity for toe-hop millimeter-wave WBAN. In: 12th Global symposium on millimeter waves (GSMM), Sendai, Japan, pp 1–3

Chapter 3
Indoor Off-Body and Body-to-Body Communication: UWB and mmW Technologies

3.1 Introduction

Body-centric wireless communication (BCWC) is an attractive research and development topic is contributing to improvement in various domains such as healthcare, personal activity monitoring, communication between users aimed for overall well-being of the society [1, 2]. The miniaturization of devices, advanced digital signal processing and use of robust algorithms has led to enhancement of the system performance and create user-friendly wearable wireless devices [3, 4].

Ultra-wideband (UWB) technology has attracted much attention for providing innovative solutions and technological advancements. Its key features include low-power, compact light weight antennas/devices, high data rate technology that provides immunity to multipath interference and has robustness to jamming because of its low-probability of detection [2, 5, 6]. With properties such as low interference to and from other existing wireless systems, low sensitivity to fading, easier wall-and-floor penetration, increased security etc., UWB technology proves to be a natural choice for body-centric communication [2, 7].

Recent advances in miniaturization of electronic devices and the bandwidth availability in the millimeter band have allowed the development of 60 GHz communications [8]. Millimeter-wave (mmWave) technologies has generated great deal of interest for body centric and wearable systems providing high bandwidth capabilities in the 59–66 GHz range [8–10]. Operating wearable systems within this part of the mmWave spectrum will be attractive due to the small size of, the lower interference and greater frequency reuse.

Various body-centric links can be formed in a realistic wireless network scenario such as on-body, off-body, or body-to-body [2]. In off-body channels, although one end of the communications link is not worn, shadowing by the human body will still be an important factor, especially in low multipath environments. Body-to-body channels will be particularly vulnerable to shadowing as they can suffer from dual node shadowing where both persons' bodies obstruct the main

S. K. Koul and R. Bharadwaj, *Wearable Antennas and Body Centric Communication*, Lecture Notes in Electrical Engineering 787,
https://doi.org/10.1007/978-981-16-3973-9_3

signal path causing the link to be lost altogether even at very short separation distances of a few meters [11, 12]. Shadowing caused by the human body can have significant effect on the communications channel. The wireless link may be subject to continuous random shadowing of the dominant signal (including free-space LOS or strong specular) components. Channel modeling and statistical analysis plays an important role for the design and development of compact antenna, algorithms and protocols, and transceiver circuitry to be used in body-centric communications [11–13]. Hence, it is important to study UWB and mmWave interactions with the human body for off-body and body-to-body channels which can be commonly found in an indoor environment.

This chapter focusses on the communication between the human bodies and characterization of the channel across the UWB and mmWave frequency ranges. Various channel link types such as off-body, body-to-body have been explored and results related to channel modeling, large-scale and small-scale fading are discussed.

3.2 The Indoor Propagation Environment

The indoor radio channel depends on various factors which include building structures, layout of rooms, objects and the type of construction materials used due to which there is reduction in signal strength and attenuation of the signal [14–16]. The radio channel can be difficult to model because the channel varies significantly with the environment. In environments such as inside of buildings with a high metallic content, additional multipath clusters can also be created which contribute to and in some cases dominate signal reception in body-centric channels [15]. Due to the varied nature of the propagation channel and presence of various objects, large-scale and small-scale propagation losses occur which are best described with the help of statistical models [16–18]. A complete channel model should comprise of the path loss model and the multipath model. The path loss is defined as the reduction in power density of an electromagnetic wave as it propagates through free space. The multipath model describes how the signal energy is dispersed over the multipath components.

3.2.1 Path Loss Model

One of the most important aspects of statistical characterization is the derivation of a model describing thc fluctuations of the received signal with respect to the distance. Models of this kind are called large-scale propagation models and the output of these models is usually the estimation of the path loss at a certain distance. The path loss model signifies the local average received signal power (P_r) relative to the transmit power (P_t). The path loss [2, 14,15], representing the attenuation suffered by the signal as it travels through the wireless channel is given by the difference of the transmitted and received power and is expressed as:

$$PL(dB) = 10\log\frac{P_t}{P_r} \tag{3.1}$$

Based on various theoretical and measurement studies for indoor environments, it has been shown that the average received signal decrease logarithmically with distance. The average path loss presented in dB for a distance d between the transmitter and receiver is expressed as:

$$PL_{dB}(d) = PL_{dB}(d_0) + 10\gamma\log\left(\frac{d}{d_0}\right) \tag{3.2}$$

where γ is the path loss exponent that indicates the rate at which path loss increases with distance d, and d_0 is a reference distance set in the measurement, d_0 is normally set to 1 m for indoor channels; for on-body channel characterisation, it is usually set to 0.1 m.

3.2.2 Multipath Model

In a typical complex indoor environment, a signal, as it travels through the wireless channel, undergoes many kinds of propagation effects such as reflection, diffraction, and scattering, apart from LOS communication due to the presence of buildings, walls, doors, furniture and other such obstructions [2, 19, 20]. The most important parameter, predicted by the propagation models based on the above three phenomena, is the received power. The physics of the above phenomena's may also be used to describe small scale fading and multipath propagation. Multipath results when the transmitted signal arrives at the receiver by more than one path. The multipath signal components combine at the receiver to form a distorted version of the transmitted waveform. Therefore, there would be multipath interference, causing multipath fading. Adding the effect of movement of either Tx or Rx or the surrounding clutter to it, the received overall signal amplitude or phase changes over a small amount of time.

The multipath components can combine constructively or destructively depending on phase variations of the component signals. The destructive combination of the multipath components can result in a severely attenuated received signal. Multipath fading degrades the performance of the wireless communication systems because, in the propagation environment, the signal arriving at a receiver experiences the effects of various propagation-dependent mechanisms. Therefore, accurate channel characterization is required to provide a reliable simulation model. The amplitude of the fading can follow different distributions, such as Rician, Rayleigh, Nakagami, Log-normal, Gamma, Normal and Weibull [2–4].

3.2.3 UWB Multipath Channel

In a cluttered indoor environment, the radio channel propagation is complex due to the presence of various objects made of different materials, which leads to signal refractions, reflections, and interferences. Owing to the large bandwidth of a UWB signal, multipath components are normally resolvable but can be difficult to identify in highly cluttered environments. The Saleh-Valenzuela (S-V) channel model is widely used in research for UWB systems as it takes into consideration the clustering phenomenon of the multipath components. In the S-V model, multipath components arrive at the receiver in clusters. Cluster arrivals are Poisson distributed and so are the subsequent arrivals in each cluster. The impulse response of the model can be presented as [2, 21]:

$$h(t) = \sum_{l=0}^{L} \sum_{k=0}^{K} \beta_{k,l} \delta\left(t - T_l - \tau_{k,l}\right) \tag{3.3}$$

where L: cluster of multipath channels, K: ray number within each cluster, $\beta_{k,l}$: the gain of the kth path in the lth cluster, T_l: the excess delay of the lth cluster, $\tau_{k,l}$: the delay of the kth path within the lth cluster.

3.2.4 Human Body Influence on Body-Centric Propagation Channels

Due to the dynamic nature of body-centric communications, signal transmission is often subject to random shadowing occurrence. This is caused by the transmitted electromagnetic waves impinging upon non-homogeneous obstacles, including the human body, which obscure the direct LOS signal path [2, 20, 21]. Apart from the direct LOS link, the body centric communications channels have occurrence of diffracted and creeping waves. Multipath clusters are formed due to signal components being reflected and scattered by the complicated geometrical and dielectric properties of the body as the signal propagates between the Tx and Rx. Off-body and body-to-body communications, communicate within short range between the Tx and Rx, hence, the type of environment in which propagation is taking place will also contribute to the channel characteristics. The signal can be affected by the indoor environment, metallic objects, different furniture along with the human body/bodies, overall leading to a complex propagation phenomenon. Various propagation aspects contribute to forming the channel multipath clusters such as reflection from objects in the vicinity of the Tx-Rx link, diffracted waves propagating around the body and signal scattered from the body and local surroundings. Like off-body scenarios, body-to-body links also suffer from shadowing issues as both ends of the links have body worn antennas leading to dual-body shadowing events depending on the orientation of the human subjects [2, 10].

The propagation channel is mainly characterised by the large-scale and small-scale fading parameters namely, path loss and rms delay spread. The path loss is obtained as mean path gain over the measured frequency band, as shown in the following equation [2, 12]:

$$PL(d(p)) = -20.\log_{10}\left\{\frac{1}{10}\frac{1}{N_f}\sum_{j=1}^{10}\sum_{n=1}^{N_f}\left|H_j^p(n)\right|\right\} \tag{3.4}$$

where *PL*(*d*(*p*)) is the path loss at the position of *p*, at which the distance between the Tx and Rx is a function of the position *p*, thus the distance is denoted by *d*(*p*). N_f is the number of frequency samples of the VNA. $H_j^p(n)$ is the measured S_{21} for the position *p*, *j*th snapshot, and *n*th frequency sample. High PL is observed for obstructed links leading to attenuation of the signal whereas lower magnitude of PL is observed for direct path links.

The power delay profile (PDP) is the squared magnitude of the impulse response [15]:

$$P(\tau, t) = h(\tau, t)h^*(\tau, t) = \sum_{k=1}^{K} a_k^2\delta(\tau - \tau_k) \tag{3.5}$$

The power delay profile (PDP) gives the intensity of a signal received through a multipath channel as a function of time delay where the time delay is the difference in the travel time between multipath arrivals. The rms Delay spread is defined as [15]:

$$\sigma_\tau = \sqrt{\frac{\sum_k (\tau_k - \tau_m)^2 .\ |h(\tau_k; d)|^2}{\sum_k |h(\tau_k; d)|^2}} \tag{3.6}$$

where τ_k are the multipath delays relative to the first arriving multipath component and *d* is the separating distance between the Tx and Rx. This parameter helps to distinguish between LOS and NLOS links as the rms delay spread values for NLOS scenarios are much larger.

3.3 UWB Channel Modelling and Characterization

3.3.1 Off-Body Link

Modeling of the ultra-wideband (UWB) indoor off-body communication channel in low and high dense multipath environments, has been reported in [22]. A statistical model for UWB off-body indoor communication channels in the 3.5–6.5 GHz is proposed. The signal power gain is modeled by a *log*-linear dual-breakpoint model, which depends on the body orientation as well as the distance between the transmitter and the on-body sensor. The general trend implies that in large distances, the received signal strength is less sensitive to changes in the body orientation angle, especially if propagation occurs in dense multipath environments. Large-scale channel parameters such as radial path loss, the azimuth decay coefficient and the lit and shadow-region critical angles are investigated. Results show that the front and the back azimuth decay coefficient can be considered independent of distance, whereas the azimuth decay coefficient has lower values at large distances. The body orientation has a more significant impact in the received power, at shorter distances rather than at longer distances which is more prominent in dense multipath environments. The small-scale fading of the total received signal power was found to follow a normal distribution. The rms delay spread computed is exponentially distributed for the corridor and the anechoic chamber environment, whereas it maintains normal fit for the office environment.

UWB off-body channel measurement campaigns in indoor environments are presented in [23]. Measurements were performed by using isolated UWB antenna and body-mounted antennas on different positions. In [23] a human phantom has been considered, aiming at the characterization of the delay and the angular properties of the indoor off-body channel (3–10 GHz). The phantom body orientation is varied from 0° to 180° and the Rx antenna is facing the human phantom, leading to LOS and NLOS links which in turn affects the delay dispersion values and number of clusters in the channel. Multipath and cluster parameters are estimated by means of space alternating generalized expectation maximization (SAGE) and a clustering algorithm. These estimates allow an analysis of the variations of the channel properties due to the presence of the body. The delay spread is characterized as a function of the body orientation and the antenna position on the body and the angles-of-arrival based on a Laplacian distribution is reported.

Off-body large scale channel measurements have been reported in [24] for 3–8 GHz by placing the antenna on different body locations with face-to-face (F-F) orientation leading to LOS scenarios. The PL at 1 m (PL_0) is in the range of 46–57 dB for the torso region (upper torso left (TO_L) and abdomen right (AB_R)). The rms delay spread (σ_τ), obtained for 1–3 m Tx-Rx separation distance is 1.5→6.5 nsec (TO_L) /1.9→3.9 nsec (AB_R) for −30 dB threshold of the power delay profile (PDP). In [25] off-body, PL trend is studied (5–9 GHz) using commercial UWB tags as Tx is placed on different locations of the human body and compact UWB antennas acting as Rx facing the Tx tag. γ ranges from 1 to 1.5 for 1–6 m with a deviation from average PL by 2.7–3.1 dB. Body centric antenna positioning effects for off-body UWB communication (3–6 GHz) in a conference room environment has been presented in [26]. Approaching and retreating journey was performed by the human subject for a 4 m distance, leading to LOS/NLOS links. Results showed the antenna placement on the chest/torso had maximum channel variation in comparison to wrist/shoulder location when LOS and NLOS scenarios are compared.

3.3.2 *Body-to-Body Link*

The transmission performance as well as a channel model for the UWB (3–11 GHz) inter-body communications is presented in [27]. The RF transmission between two human bodies was characterized and a channel model developed for UWB communications. Inter-body large scale channel analysis is briefly carried out in [27] for 3–11 GHz frequency range with one subject static in facing front orientation and the other at different angular positions (0°–180° at 15° interval) for 0.8–2 m with d_0 = 0.8 m. It is observed that when the antennas are placed on the center of the chest region, LOS/NLOS scenario shows lower/higher PL and maximum/minimum γ. The path loss is greatly dependent on the position of the transmitter and receiver antennas on the body, the relative orientation of the human bodies and the antenna radiation patterns. As compared to the free space path loss, the path loss when the antennas were mounted on the body can be reduced for certain body orientations. Path loss (PL) parameters (PL at reference distance, PL_0 = 30 dB and PL exponent, γ = 1.87 for dipole antenna) are reported in [28] for Face-Face (F-F) orientation of the Tx-Rx pair in the 2–8 GHz frequency range for 1–2 m with d_0 = 0.5 m.

3.3.3 *Angular Body-Centric Channel Characterization at UWB Frequencies*

Spatial body-centric UWB channel simulations and measurements have been conducted using compact and cost-effective Tapered Slot antennas in the 3–8 GHz frequency range [29, 30] (Fig. 3.1). As seen from Fig. 3.2, the directivity increases for all the cases as the front to back ratio of the radiation pattern is significantly increased due to high reflections from the body. The distance between the Tx-Rx antennas is 2 m for initial 0° position and are in face-to-face orientation. The antennas are placed vertically with respect to the ground at a height of 1.1 m. The antenna is placed on the torso region of the human male subjects having height 1.7 m and average built for both off-body and body-to-body scenario and is separated from the body at a gap of 2–5 mm [30]. For no-body influence channel link, the Tx and Rx antennas are directly fitted to the tripod stand for measurements. A vector network analyser (VNA) was used to measure the S_{21} parameters between the Tx and Rx. Five-meter-long low loss cables were used to connect the base stations and wearable antenna to the VNA. The sweep is performed over 1601 equally spaced frequency points for the frequency range of 3 to 8 GHz with center frequency of 5.5 GHz. For each scenario (body-to-body, off-body, no-body influence), the Tx antenna is static and the Rx antenna is placed at 16 different locations along the circumference of the circle at intervals of 22.5°. In order to compute the channel impulse response (CIR) inverse fast Fourier transform (IFFT) is applied to the S_{21} parameters.

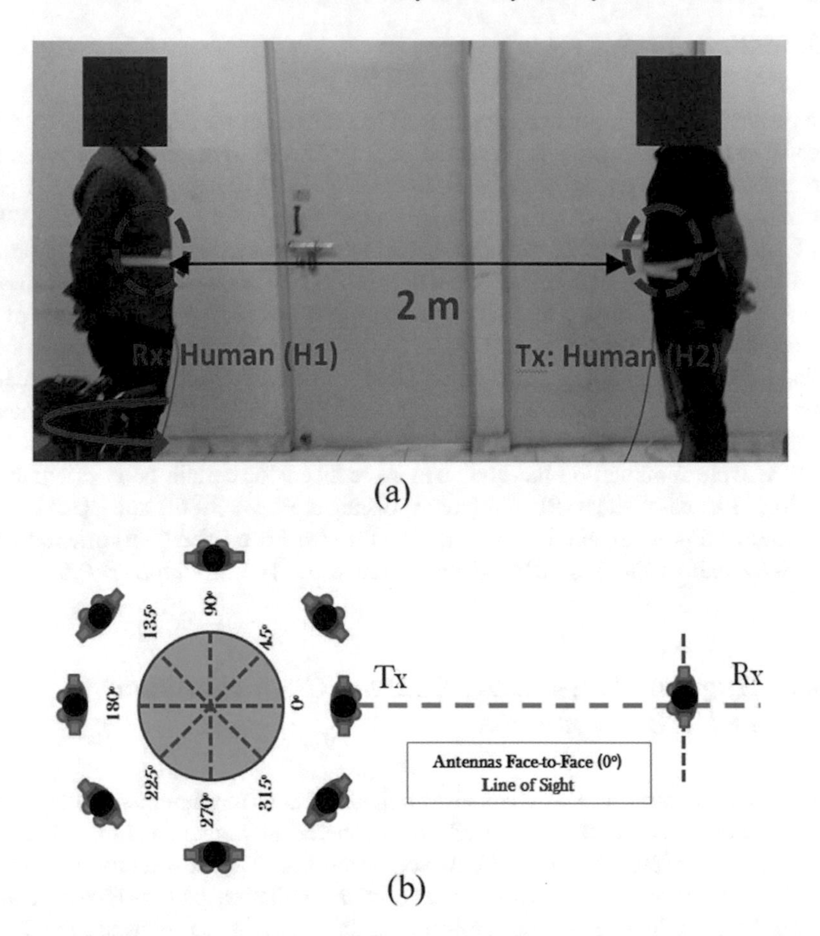

Fig. 3.1 **a** Measurement set up for body-to-body channel. Human subjects H1 and H2 with body worn antenna (red circle). The human body (H1) is made to rotate full 360° in steps along with the Tx antenna. **b** Schematic of the locations of the human subjects (Reprinted with permission from IEEE [30])

Figure 3.3a, b presents two scenarios of the measured CIR in which the human subject is directly facing the Rx (0°) and in the other scenario the human subject is facing backwards (180°). Figure 3.3a presents the normalized CIR for the total LOS scenario in which direct path (DP) propagation is taking placc between the Tx-Rx nodes and no human subject is obstructing the path. It can be observed that the DP is clearly depicted, and the channel has least multipath components. Figure 3.3b presents the CIR for the total NLOS scenario caused due to the presence of the human subject between the Tx-Rx nodes leading to attenuation and higher multipath/spread of the signal. It can be clearly seen that the presence of the human subject has a strong influence on the UWB propagation channel.

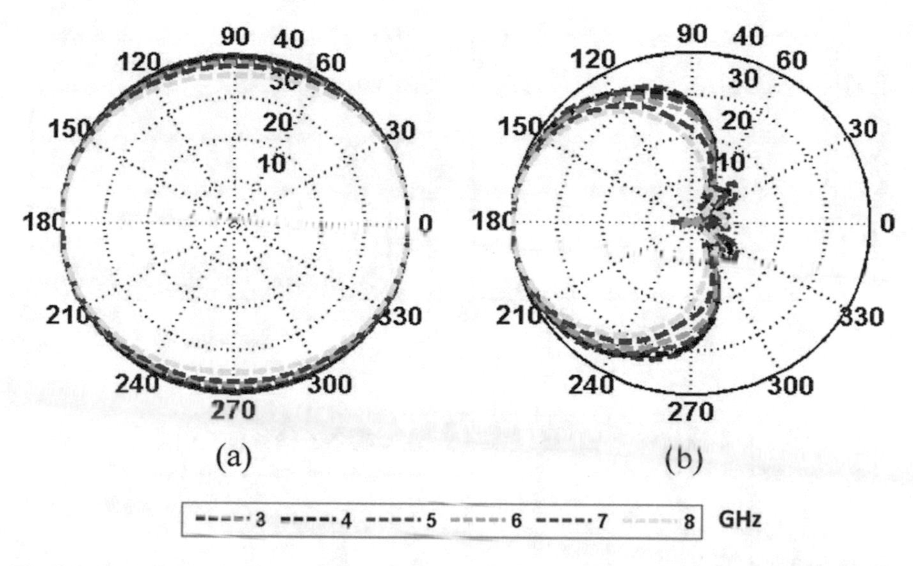

Fig. 3.2 Variation in simulated radiation pattern in the azimuth plane **a** no-body influence, **b** body present for the frequency range 3 to 8 GHz (Reprinted with permission from IEEE [30])

Figure 3.3c presents the rms delay spread trend as the Rx angular orientation changes from 0° to 360°. It can be observed for body-to-body scenario and off-body scenario, the values initially increase from 0° to 225°/180° and further decreases till 360°. This variation is due to the change in the Tx-Rx link which is going from LOS situation to partial LOS/NLOS and further to total NLOS situation. From 180° onwards the Tx-Rx link gradually returns from NLOS to → partial NLOS (PNLOS) to → LOS scenario. Tables 3.1 and 3.2 lists the average magnitude of the delay parameters for three different threshold levels (−15 dB, −20 dB and −25 dB).

For LOS and partial LOS scenarios, σ_τ values for body-to-body link are much less than off-body case, as in the body-to-body communication link, both the wearable antennas have directive radiation patterns. For the case where there is no-body influence, only LOS scenarios are present, the σ_τ values are similar for all the positions of the Rx antenna. For NLOS situation which occur only in body-to-body and off-body link, the rms delay values are much higher with mean values of 12 nsec respectively for −20 dB threshold. The values for NLOS scenarios are comparable for both the communication links.

3.3.4 Body-to-Body and Off-Body Links: Experimental Investigation

Experimental investigations and analysis are carried out for various body-centric channels which form the basis of understanding various communication links [31].

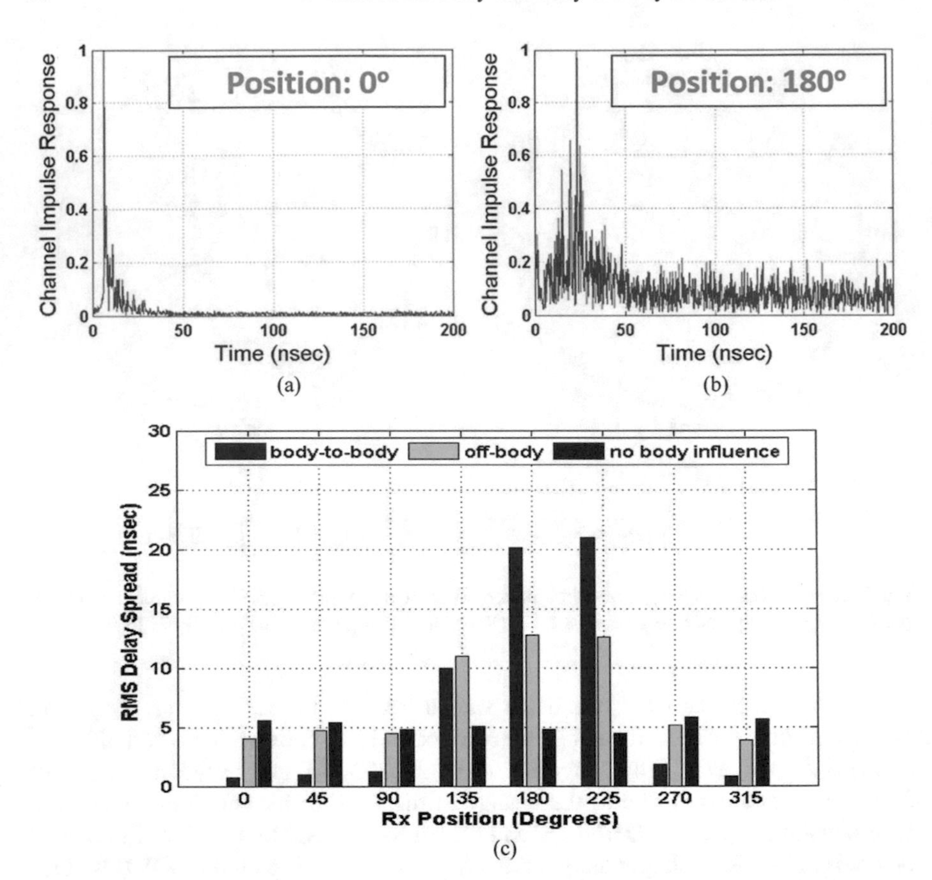

Fig. 3.3 Channel impulse response for **a** LOS (Human Subject: 0° orientation) and **b** NLOS links (Human Subject: 90° orientation). **c** Rms delay spread for various angular positions of the Rx antenna for the three scenarios: body-to-body, off-body and no-body influence (Reprinted with permission from IEEE [30])

Table 3.1 Average rms delay spread (Reprinted with permission from IEEE [30])

	B-B	B-B	Off-B	Off-B	No-B
PDP Threshold (dB)	LOS	NLOS	LOS	NLOS	LOS
−15	0.85	8.48	3.86	8.23	4.61
−20	1.40	12.56	4.21	12.12	5.76
−25	2.08	16.54	5.03	16.34	6.73

B-B body-to body, *Off-B* off-body, *No-B* no-body

Measurements are performed in an indoor office environment and the frequency range chosen (3–8 GHz) adequately covers the low and high frequencies of the UWB spectrum. Compact and cost-effective tapered slot antenna (TSA) in the 3–8 GHz frequency range is used in this study. The TSA antenna acts as the transmitter and receiver for the body-centric links and no-body scenario in the

Table 3.2 Rms delay spread standard deviation (Reprinted with permission from IEEE [30])

	B-B	B-B	Off-B	Off-B	No-B
PDP Threshold (dB)	LOS	NLOS	LOS	NLOS	LOS
−15	0.15	3.18	0.36	1.74	0.29
−20	0.50	3.26	0.47	4.02	0.34
−25	1.54	3.84	1.21	4.69	0.37

B-B body-to-body, *Off-B* off-body, *No-B* no-body

indoor environment (Fig. 3.4). The Tx antenna is static, and the Rx antenna is moved from 0.4 to 5 m at intervals of 0.2 m with 20 measurement points in total. The antennas are placed vertically with respect to the floor at a height of 1.1 and 1.9 m below the ceiling. The human subjects have average built and height (1.65 m) with a BMI in the range of (22.1–24.5). The antenna is placed near the center of the torso (waist region) of the human subjects with a gap of (2–5) mm from the body. The torso region of the human subject is chosen to place the compact wearable antennas as it has large volume and thickness making the propagation of the signals more complex and challenging. The antennas are connected to the two-port VNA using low loss cables for measuring the S_{21} magnitude and phase to obtain the channel impulse response for each measured communication link.

Different orientations of the Tx and Rx compact UWB antennas are considered for all the three types of communication links that are being compared. The antennas are placed in the following orientations: *Tx-Rx: F-(F/B/*S_R*/*S_L*), B-(F/B/*S_R*/*S_L*),* S_R*-(F/B/*S_R*/*S_L*),* S_L*-(F/B/*S_R*/*S_L*)*; where F/B/S_R/S_L represent Face, Back, Side Right, Side Left respectively. The three types of communication links considered are C1: body-to-body, C2: off-body and C3: no-body scenario. Two of the link combinations *Tx-Rx: F-(F/B/*S_R*/*S_L*)* and *B-(F/B/*S_R*/*S_L*)* have been detailed below as shown in Fig. 3.5a, b respectively.

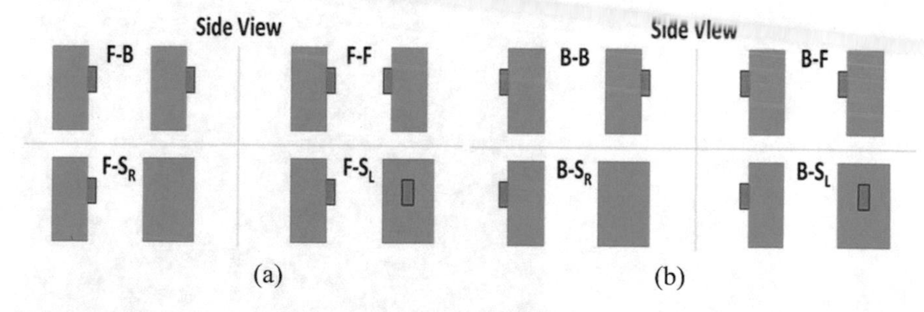

Fig. 3.4 Schematic (side view) of different positions of the human subject for body-to-body links. **a** *Tx:(F)-Rx:*(F/B/S_R/S_L) **b** *Tx:(B)-Rx:*(F/B/S_R/S_L) where F/B/S_R/S_L represent face, back, side right, side left respectively (Reprinted with permission from IEEE [31])

Fig. 3.5 Measurement set up for various Tx-Rx links: **a** body-to-body, **b** off-body **c** no-body (Reprinted with permission from IEEE [31])

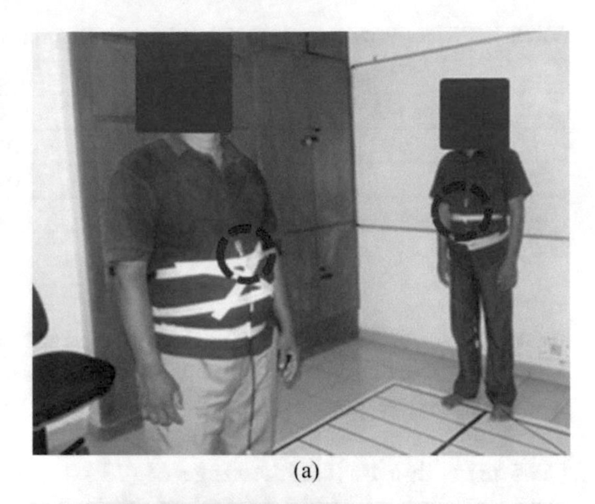

(a)

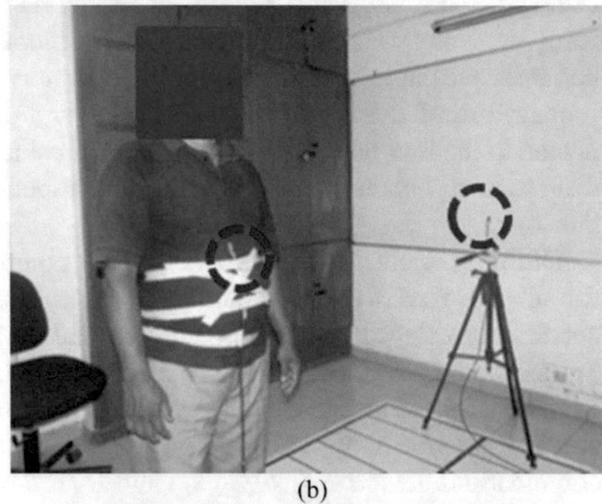

(b)

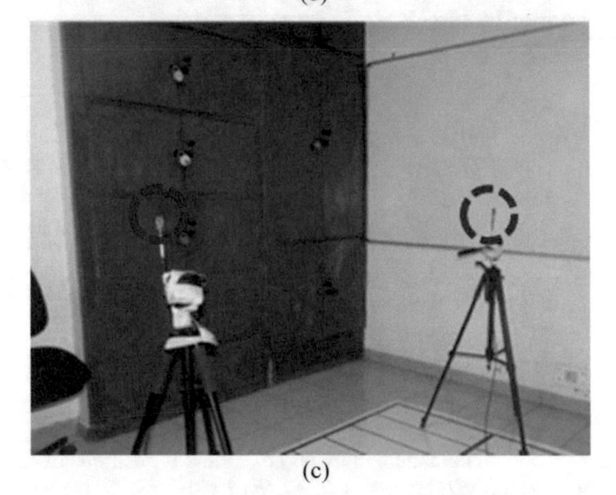

(c)

3.3.4.1 Path Loss Magnitude

Body-to-Body

Path loss magnitude is much higher for scenarios when the human subject with the Rx antenna is facing back (*F-B* link) leading to more attenuation of the signal (Fig. 3.6a) which can be seen for (*C1: F-F/B/S_R/S_L*) links. Total NLOS situations are observed for *C1: F-B* and *C2: F-B* links leading to deterioration of the signal strength. As observed for body-to-body case (*C1: B-F/B/S_R/S_L*) all the four channel links are in NLOS situation showing higher magnitude of PL and lower variation of the PL magnitude with distance (Fig. 3.6b). Among all the NLOS situations *C1: B-F* link has least PL magnitude which is due to the orientation of the antenna and the fact that only one of the wearable antennas (Tx) which is facing back, is mainly causing an obstruction for the propagation path whereas the Rx antenna is facing forward. *C1: B-B* link has maximum PL due to the fact that both the human subjects are facing back leading to maximum signal attenuation.

Off-Body

Generally, 1–2 dB difference of magnitude is observed for NLOS body-to-body scenario in comparison to NLOS off-body links, with body-to-body scenario having slightly more PL magnitude as observed in *Tx: (F)-Rx: (F/B/S_R/S_L)* links. For *F-F* link, the magnitude is lower for body-to-body scenario in comparison to off-body scenario due the directive nature of the wearable antennas when placed in proximity to the body. In the current work for off-body, F-F orientation, PL_0 obtained is 46 dB. Considering *Tx: (B)-Rx: (F/B/S_R/S_L)* links, *C2: B-B* PL magnitude is much lower than *C1: B-B* though both are representing total NLOS links as observed in Fig. 3.6b. This is because *C2: B-B* link has only the Rx with the body-worn antenna in comparison to B-to-B where both Rx and Tx have body-worn antennas. Links, *C2: B-S_R* and *C2: B-S_L* have lower PL in comparison to total NLOS scenario of the off-body case (*C2: B-B*) as the channel type is LOS/partial LOS type due to the position of the human subject (H1) who is facing sideways (left and right). *C2: B-F* has lower PL magnitude than rest of links considered for off-body scenario due to LOS scenarios.

The slope of the NLOS links for PL magnitude versus distance is much smaller in comparison to partial NLOS/LOS scenario and total LOS scenario. The range of magnitude variation for 0.4–5 m is around 4–5 dB for NLOS links whereas for LOS scenarios the range over which PL varies over the measured distance is around 10 dB. Hence, slope for LOS situations is much higher in comparison to total NLOS situations. For body-to-body links, LOS/NLOS scenarios show lower/higher PL magnitude and higher/lower γ, due to the blockage caused by the human subjects in NLOS scenarios. Off-body also follows similar trend, as the presence of the human subject leads to variation in the channel behavior.

No-Body

For all the scenarios of case *C3* (no-body influence) where the human subject is not present, the path loss parameters reported are PL_0:48–50, dB and γ: 1.2–1.3 for F-F

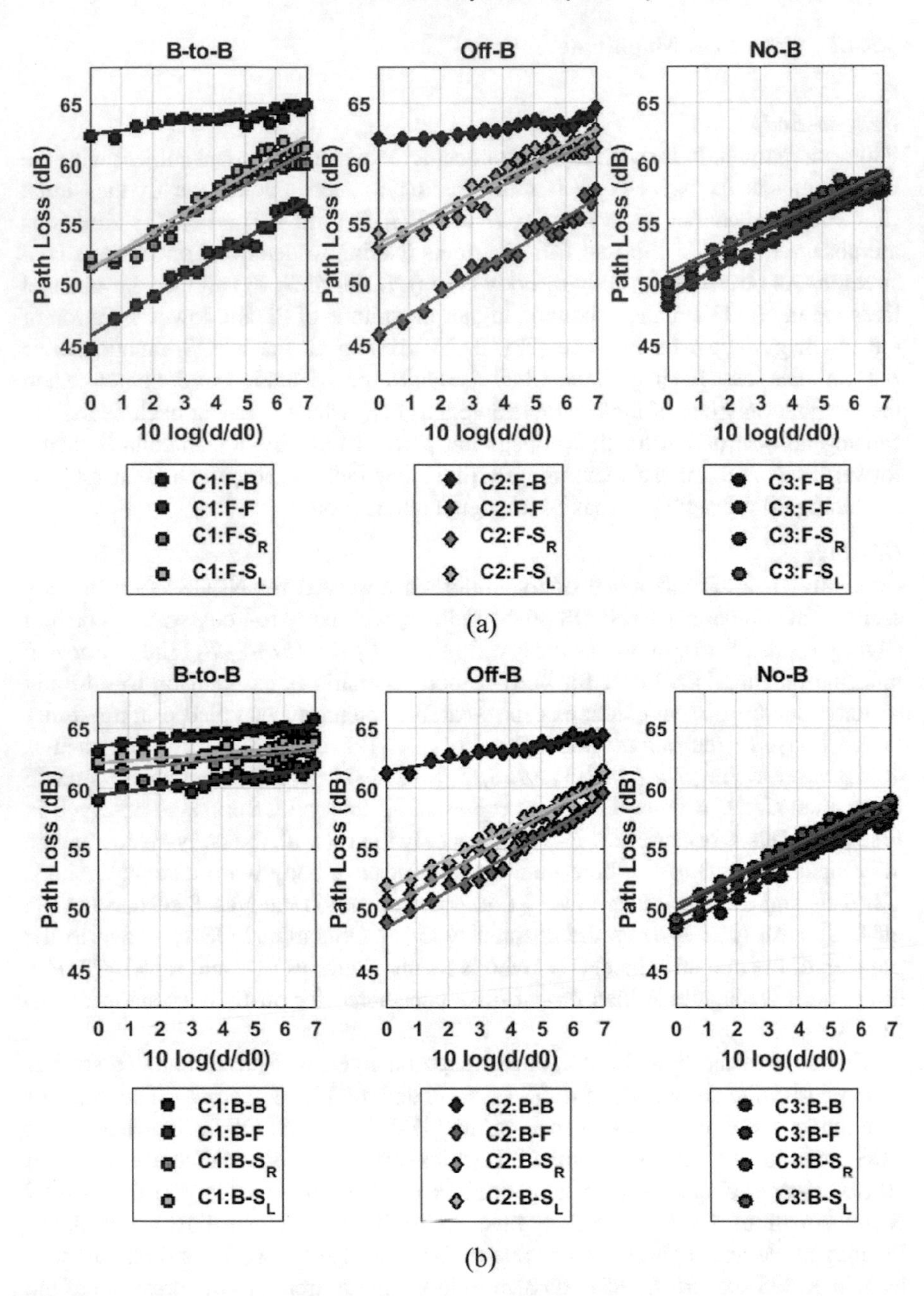

Fig. 3.6 Path loss versus distance for various communication links: C1: body-to-body, C2: off-body and C3:no-body. Positions of the Tx/Rx considered are: **a** *Tx:(F)-Rx:(F/B/S_R/S_L)*, **b** *Tx:(B)-Rx:(F/B/S_R/S_L)* (Reprinted with permission from IEEE [31])

orientation (Fig. 3.6). The values of PL are generally much lower in comparison to the body-to-body and off-body scenarios due to total LOS links formed. The only exception is the *C1: F-F* and *C2: F-F* scenario where the PL magnitude is much lower than the no-body influence case due to the orientation of the antennas and also the directive nature of the radiation pattern when placed on the body.

3.3.4.2 Rms Delay Spread

Rms delay spread gives an indication of multipath in the channel and information regarding time dispersion characteristics. σ_τ is computed at -20 dB threshold of the measured power delay profile.

Body-to-Body

As observed in Figs. 3.7a and 3.8a for *Tx: (F)-Rx:(F/B/*S_R*/*S_L*)* links, high rms delay spread values (20–25 nsec) for NLOS situation *C1: F-B* are observed as the human subject is facing backwards leading to high distortion and spread of the signal. Least σ_τ (0.5 to 5 nsec) is observed for *C1: F-F* as both the human subjects are facing each other with directive radiation patterns leading to less delay spread. For sideways position of the human subject, mid-range delay spread (2–15 nsec) values are observed. High σ_τ is observed for all the four situations as the Tx antenna is always facing backwards for *Tx: (B)-Rx: (F/B/*S_R*/*S_L*)* which is leading to obstruction of the signal path for all the scenarios which is clearly depicted in Figs. 3.7b and 3.8b. But maximum σ_τ (17–27 nsec) are observed for *C1: B-B* link as both the human subjects are facing back to each other leading to higher amount of signal spread. Lower values of σ_τ (8–16 nsec) are observed for *C1: B-F* scenario and mid-range values (12–22 nsec) for *C1: B-*S_R and *C1: B-*S_L which is highly dependent on the position of the human subject and placement of the antenna.

Off-Body

Similar trend of σ_τ values are observed for off-body scenario with slight reduction (by 2 to 3 nsec) in delay spread values for NLOS situation observed for *Tx: F-Rx: (F/B/*S_R*/*S_L*)* links. For LOS and partial LOS/NLOS situations higher values of σ_τ (by 2–4 nsec) are observed in comparison to body-to-body as in this scenario, only one human subject (H2) with the wearable antenna (Rx) is present leading to directive radiation pattern. The Tx antenna is having more omni-directional radiation pattern due to which propagation of the signal occurs in various directions before reaching the body-worn Rx antenna. As depicted in Figs. 3.7b and 3.8b, for *Tx: B-Rx: (F/B/*S_R*/*S_L*)* off-body links, F-F orientation, σ_τ at −20 dB PDP threshold is 1.6→3 nsec for 1→3 m of Tx-Rx separation distance which are like that reported in [25] For the trend of σ_τ observed for off-body scenario is very different from that of the body-to-body scenario. For off-body scenario, only the Rx antenna (body worn antenna) has directive radiation pattern leading to variation in results in comparison to body-to-body links. NLOS situation is only observed for *C2: B-F* with slight reduction in σ_τ (2–3 nsec) in comparison to body-to-body scenario. For

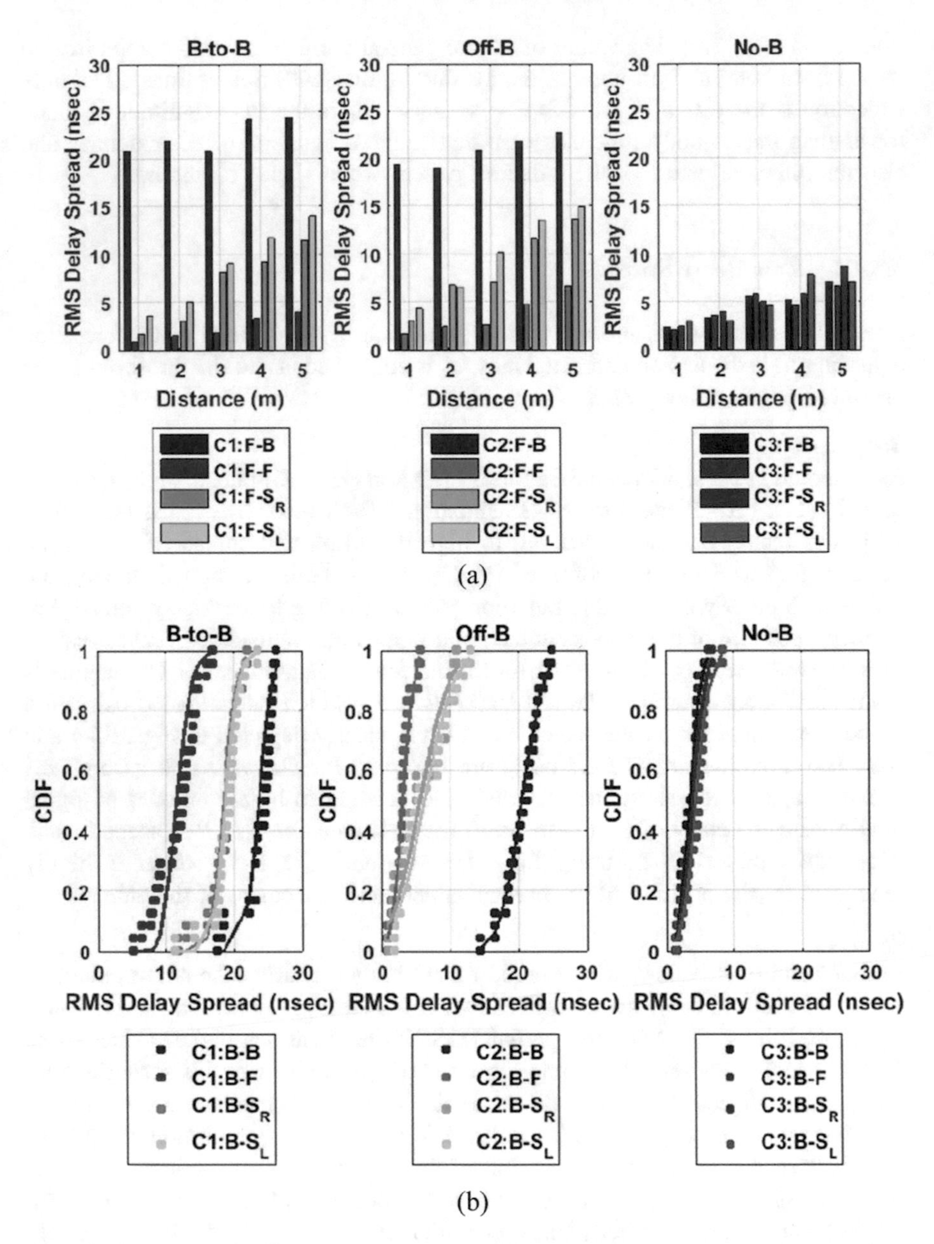

Fig. 3.7 Cumulative distribution function for rms delay spread for various communication links. C1: body-to-body, C2: off-body and C3: no-body. Positions of the Tx/Rx considered are: **a** *Tx: (F)-Rx:* $(F/B/S_R/S_L)$, **b** *Tx: (B)-Rx:* $(F/B/S_R/S_L)$ (Reprinted with permission from IEEE [31])

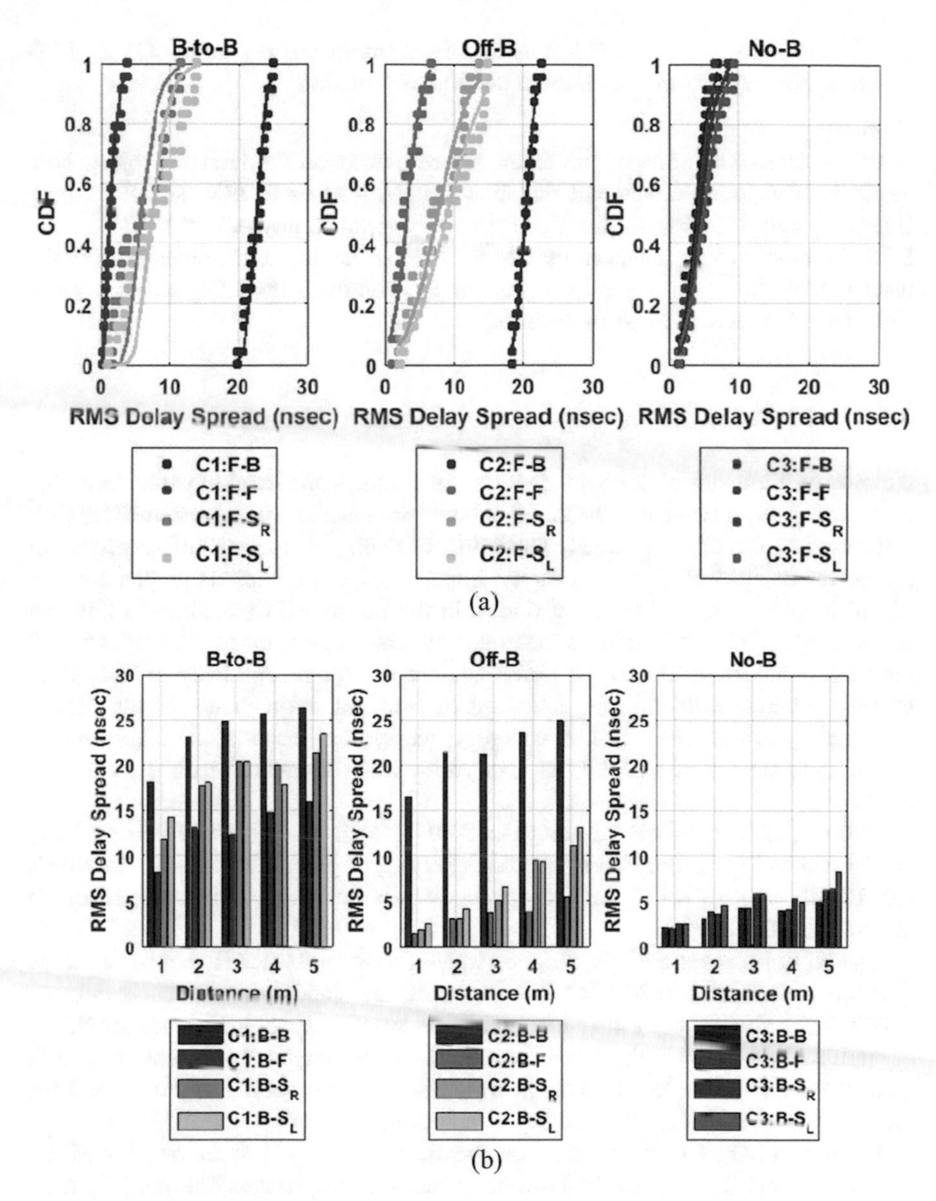

Fig. 3.8 Rms delay spread vs. distance for various communication links. C1: body-to-body, C2: off-body and C3: no-body. Positions of the Tx/Rx considered are: **a** *Tx: (F)-Rx:* $(F/B/S_R/S_L)$, **b** *Tx: (B)-Rx:* $(F/B/S_R/S_L)$ (Reprinted with permission from IEEE [31])

C2: $B\text{-}S_R$/*C2:* $B\text{-}S_L$, LOS and partial LOS situations are observed leading to higher values of σ_τ in comparison to total LOS scenario (*C2: B-F*).

For NLOS scenarios variation of σ_τ with distance is not as linear as that of LOS and partial LOS situations. Also the change in delay spread magnitude from 0.4 to

5 m is much lower for NLOS situations in comparison to partial LOS or LOS situation due the presence of human body as obstruction.

No-body
In this scenario, as none of the antennas are placed on the human subject, both antennas have omni-directional radiation pattern leading to LOS situations only (Figs. 3.7 and 3.8). But σ_τ are higher in comparison to off-body or body-to-body LOS situations which is especially true for *F-F* orientation of the antennas. For the results of no-body influence in an indoor environment, the LOS scenario gives (F-F) (σ_τ: 2–8 nsec for 1–5 m distance).

3.3.4.3 Multi Path Components

Number of multipath components (MPCs) follow the same trend of variation as that of the rms delay spread, due the fact that both parameters give indication of the time dispersive nature of the channel. The MPCs are computed at −20 dB threshold of the power delay profile of the channel impulse response. Various graphs are presented in Fig. 3.9a, b showing variation in the no. of MPCs leading to different channel type. Table 3.3 lists the mean and standard deviation of the MPCs for all the links considered. The MPCs differ significantly for several body-to-body links when compared with off-body links. Apart from variation in no. of the MPCs, considering similar channel link category, the type of channel classification i.e. LOS/NLOS or partial LOS/NLOS also varies when compared with the off-body links.

The no. of MPCs is in the range of 100–160 for body-to-body and off-body NLOS channel type. For NLOS scenarios (as categorized in Table 3.3), the MPCs are slightly more (10–15 MPCs) for body-to-body scenario in comparison to off-body NLOS links.

The NLOS links show less variation in MPCs over distance. It can be observed that for a mean of 146 MPCs (B-B scenario), the standard deviation is only 15 MPCs which shows very little variation of MPCs with respect to distance for NLOS scenarios. This is since the wearable antenna is placed on the human subject is present for all the captured readings which is the main source of MPCs and the threshold level of the PDP taken into account for computing no. of MPCs.

For partial LOS/NLOS scenarios (as mentioned in Table 3.3) the range of MPCs are (5–80) and (10–70) for body-to-body and off-body links. The no. of MPCs increase significantly with an increase in distance. As observed in Table 3.3 the standard deviation is more for body-to-body links in comparison to off-body links. For every 1 m increase in distance, the MPCs generally increase linearly by (1.2–2) for body-to-body links and by (1.2–1.5) for off-body links.

For total LOS situation, (as categorized in Table 3.3) the average of MPCs is low in the range of 5–30 for 1–5 m distance. The MPCs generally increase by 1.1 to 1.4

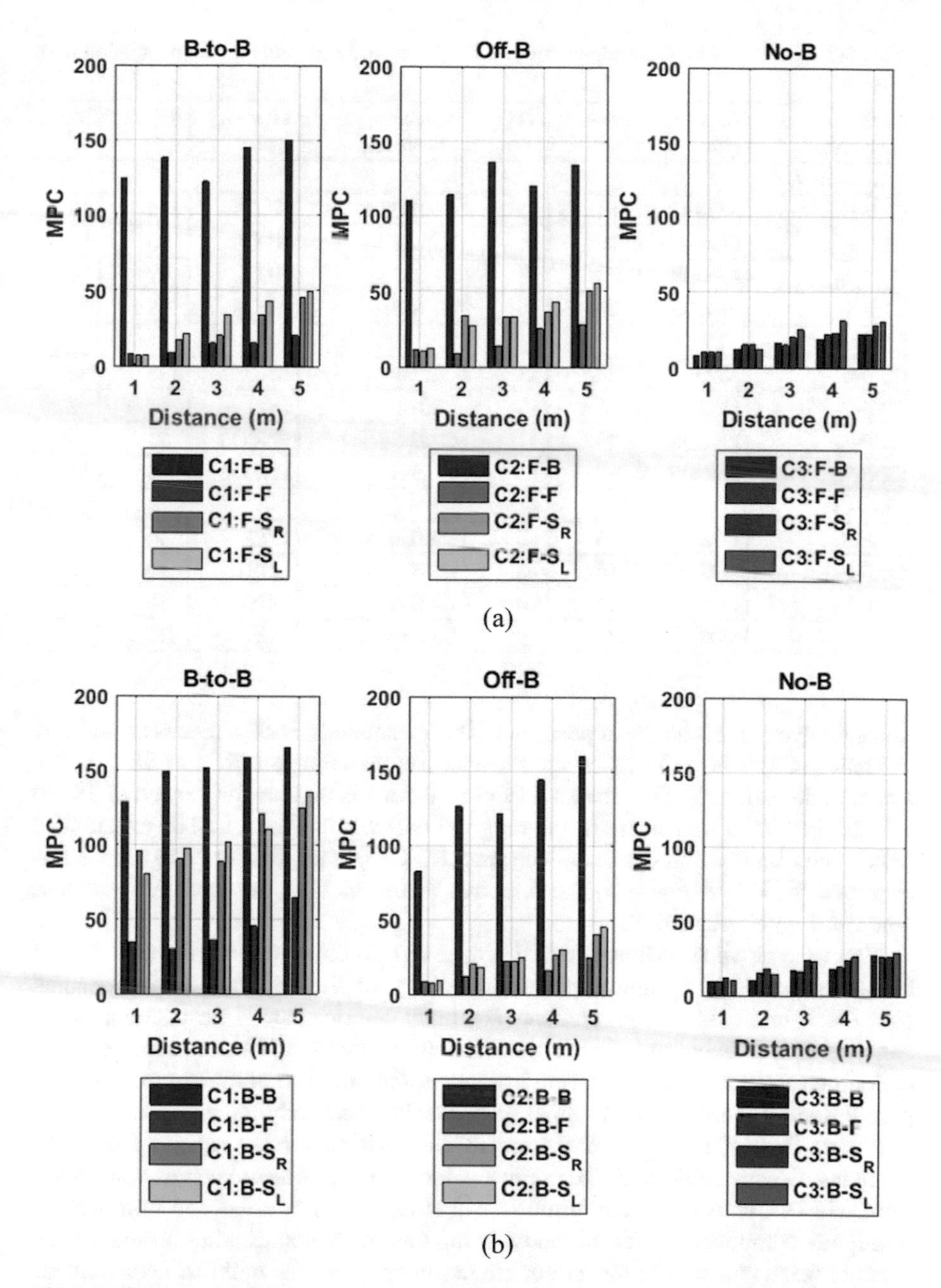

Fig. 3.9 Number of multipath components versus distance for various communication links: C1: body-to-body, C2: off-body, and C3: no-body. Positions of the Tx/Rx considered are **a** *Tx: (F)–Rx: (F/B/S_R/S_L)*, **b** *Tx: (B)–Rx: (F/B/S_R/S_L)* (Reprinted with permission from IEEE [31])

Table 3.3 Number of MPCs: body-to-body, off-body, no-body (Reprinted with permission from IEEE [31])

Antenna orientation	Link type	Mean	STD	Antenna orientation	Link type	Mean	STD
Body-to-body							
C1:F-B	NLOS	140	16	C1:B-B	NLOS	146	15
C1:F-F	LOS	10	6	C1:B-F	NLOS	40	14
C1:F-S_R	PNLOS	23	13	C1:B-S_R	NLOS	87	26
C1:F-S_L	PNLOS	29	16	C1:B-S_L	NLOS	98	29
Off-body							
C2:F-B	NLOS	123	13	C2:B-B	NLOS	126	30
C2:F-F	LOS	16	8	C2:B-F	LOS	15	6
C2:F-S_R	PNLOS	29	11	C2:B-S_R	PNLOS	21	10
C2:F-S_L	PNLOS	30	13	C2:B-S_L	PNLOS	24	11
No-body							
C3:F-B	LOS	16	6	C3:B-B	LOS	14	6
C3:F-F	LOS	16	6	C3:B-F	LOS	16	6
C3:F-S_R	LOS	18	6	C3:B-S_R	LOS	18	6
C3:F-S_L	LOS	22	9	C3:B-S_L	LOS	19	7

times for every 1 m change in position of the Rx antenna. For the results of no-body influence (C3) with LOS links, the channel has an average MPCs of 18–20. The standard deviation (STDV) for total LOS situations is high as for a mean of 18–20 MPCs, the STDV values are in the range of 6–9 which shows higher variation of MPCs with increase in distance. The least MPCs (mean: 10) for LOS scenario is observed for *C1: F-F* due to the directive nature of both the wearable antennas when in F-to-F orientation.

From the detailed analysis carried out, it can be observed that the presence of human subject/s is a significant factor leading to the variation in the channel parameters in an indoor environment. The body-centric channel parameter analysis carried out is suitable in indoor environments such as home/office as per classification presented in [12]. In such environments, the presence of human subjects will play a dominant role in the channel type. For body-to-body links, F-F orientation will have least effect of the environment due to the directive nature of both the antennas. Other contributing factors in the indoor environment such as wall, floor, roof, window, wooden/plastic furniture will have secondary effect and form weaker multipath components. For no-body case, the main contributing factor of the channel behaviour will be the indoor environment, as all the multipath components arising in the CIR will be due to the surroundings.

3.3.5 *Spatial Variation of Path Loss for Off-Body Links: Application Specific*

This work presents an in-depth analysis on the variation of the UWB path loss (PL) statistics in the x–y plane and z-axis for the off-body links in relation to the placement of the base stations (BSs) keeping the focus on the indoor body centric localization and sensing applications [32]. The variation in PL statistics is dependent on various factors such as the orientation/placement of the wearable nodes, location of the human subject and the position of the BSs which has been evaluated in this work. For the mobile station (MS)-BS pair in which both line-of-sight (LOS)/non-(N) LOS links are present in the x–y plane due to change in orientation of the human subject over the localization region, are analysed separately and show different PL statistics in comparison to total LOS/NLOS region. The change in PL magnitude with respect to distance in the x–y plane or z-axis is more prominent for LOS in comparison to NLOS scenarios. PL values in the range of 45–60/65–75 dB is observed for total LOS/NLOS links. The percentage of occurrence of LOS/NLOS links in the localization region is maximum for elbow/torso region and BS1/BS4, respectively. Hence, the influence of the spatial distribution of the base stations and the wearable nodes is an important factor in WBANs channel modelling and classification.

Channel characterization and modelling is an important aspect for building algorithms and aiding in smart wearable device development [2, 3]. Among various channel parameters, path loss (PL) analysis is a commonly studied to characterize the links present in WBANs. Work carried out in open literature related to UWB path loss (PL) analysis is mainly confined to a human subject tracing a path in an indoor environment over few meters, which gives the trend of the PL with respect to distance over a linear range [23–27]. Existing work such as reported in [24] is confined to off-body PL analysis for line-of-sight (LOS) links in an indoor environment; [26] presents analysis for retreating and forward journey of the human subject, hence analysing both LOS/non-(N)LOS links over distance in an indoor setup; [23, 24, 30, 31] presents the PL trend for various human subject orientations and variation with distance. PL analysis over distance in hospital environment has been presented in [33] for aiding body-centric communication systems under hospital environments. UWB commercial tags have been used in [25] to understand the PL trend in an indoor environment.

In real life scenarios, the situation is quite different, and the base station/wearable antenna orientation will vary over a coverage area leading to different types of links in the localization region such as LOS, NLOS and partial LOS/NLOS, which needs to be considered [31]. In order to accurately determine and track a person's position and activity in an indoor environment channel information (large-scale and small-scale fading parameters) and time of arrival positioning techniques are applied [19]. The motivation for the work undertaken is to examine the channels links for the different wearable antenna locations with respect to the BS placement to interpret PL parameters variation, for various indoor tracking applications. The investigations will prove beneficial in aiding localisation algorithms and channel information which is embedded into the wearable devices for localisation, tracking and sensing applications.

The experimental campaign was carried out in an indoor environment and a 1.68 m tall male with average built was considered as the human subject. The environment consists of various furniture such as chair, tables, equipment's, wooden doors, glass windows, and the basic structure made of cement and bricks. The base stations were placed in cuboid-shape configuration and eight different wearable antenna locations were chosen for analysis as shown in Fig. 3.10a. The wearable locations chosen on the upper limbs were shoulder joint, elbow joint and the wrist joint. For the torso region, the center of the chest and waist was chosen as locations to place the wearable antennas. The human subject stands at 49 different positions with spacing of 15 cm in an area of 1.5×1.5 m^2 for each wearable antenna location which is depicted in Fig. 3.10b [19]. The spacing between the wearable antennas and human subject was 2–5 mm as shown in Fig. 3.10c.

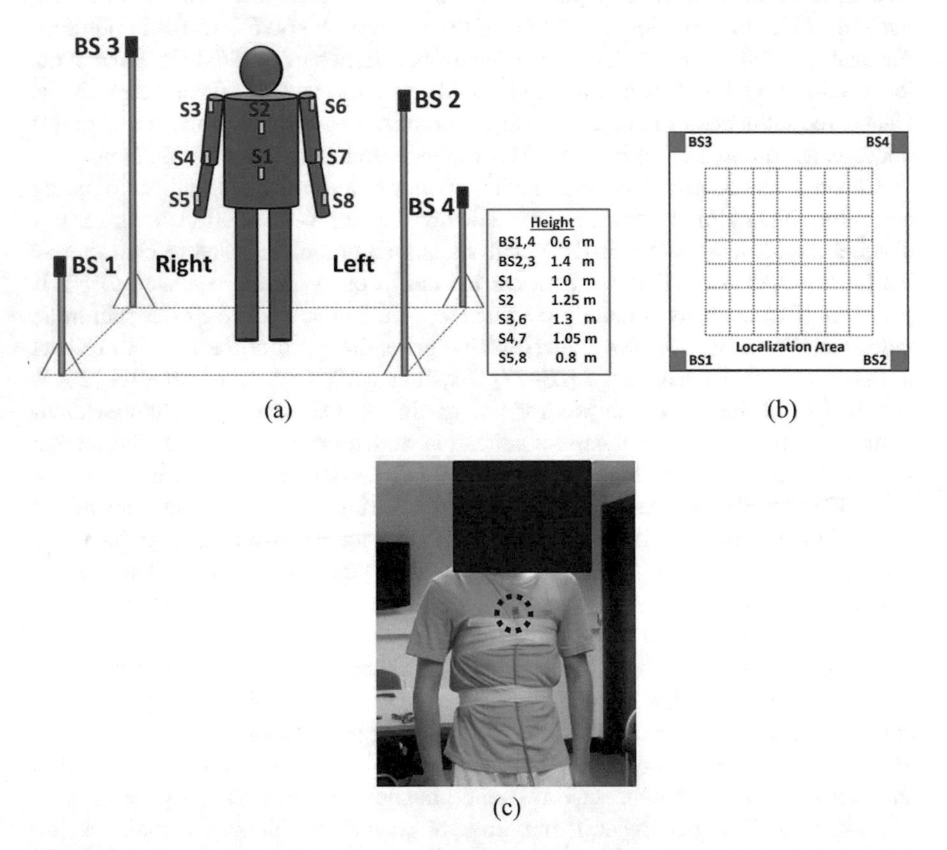

Fig. 3.10 **a** Localisation schematic of cuboid shape BS configuration consisting of four BSs (BS1–BS4) with the human subject standing in the centre with the eight different wearable antenna locations (S1-S8) and the corresponding heights. **b** BSs and the 7 by 7 localisation grid with spacing of 15 cm (Top view). **c** Human subject with the wearable antenna placed on the torso region (Reproduced from [32])

3.3.5.1 Statistical Analysis of Path Loss Magnitude

Detailed analysis of the statistical parameters (mean PL magnitude and standard deviation (STDV) for each body-worn antenna location (S1-S8) and base station (BS1–BS4) has been presented in Table 3.4. Figure 3.11a–c present the cumulative distribution function (CDFs) for PL variation corresponding to LOS (Range 1) and NLOS (Range 2) region of the localization area with respect to the BSs for selected cases. Colour graph showing the variation for BS1 and BS3 for chest/R. shoulder/L. wrist location is presented in Fig. 3.12a–c.

Torso: The antennas are placed on the chest and waist region of the human torso. Both the body-worn locations exhibit similar channel type links and PL magnitude with respect to the BSs of the cuboid-shape configuration. BS1 and BS2 have direct path propagation with respect to the wearable nodes, hence have low PL values with an average of 51–55 dB and STDV in the range of 2–4 dB. For BS3 and BS4, total obstructed paths are observed for both the wearable antenna locations, leading to high PL magnitudes in the range of 63–69 dB. The BS4 exhibits highest PL magnitude because of the vertical height difference between the wearable locations on the torso and BS4 position at the extreme left corner of the cuboid-shape BS configuration. Least STDV values are observed for BS4 which is in the range of 0.6 to 1 dB. The CDF for different BSs-wearable antenna links is shown in Fig. 3.11a

Table 3.4 Path loss parameters (dB) (Reproduced from [32])

		BS1		BS2		BS3		BS4	
		R1	R2	R1	R2	R1	R2	R1	R2
Waist: S1	$\overline{PL}$	52.8	–	54.1	67.2	58.0	68.2	–	68.0
	σ_s	3.1	–	4.1	2.2	3.0	1.5	–	1.0
Chest: S2	$\overline{PL}$	54.7	–	51.5	–	–	66.9	–	69.0
	σ_s	2.1	–	3.4	–	–	1.6	–	0.6
R. Shoulder: S3	$\overline{PL}$	56.7	–	56.6	63.3	55.0	63.9	–	68.2
	σ_s	2.5	–	2.5	2.0	2.1	2.9	–	0.9
R. Elbow: S4	$\overline{PL}$	54.1	–	54.2	64.0	55.0	68.4	–	69.0
	σ_s	2.8	–	3.2	2.2	2.5	1.8	–	1.7
R. Wrist: S5	$\overline{PL}$	53.0	–	55.6	64.0	58.0	66.3	–	68.6
	σ_s	2.9	–	2.4	2.1	2.6	1.7	–	1.2
L. Shoulder: S6	$\overline{PL}$	59.3	65.7	52.8	–	57.8	68.6	57.2	67.8
	σ_s	1.5	1.7	2.9	–	1.3	1.8	0.9	1.3
L. Elbow: S7	$\overline{PL}$	57.7	63.8	53.7	–	–	67.7	56.3	65.8
	σ_s	2.4	1.6	3.1	–	–	1.2	2.5	1.1
L. Wrist: S8	$\overline{PL}$	58.7	64.4	56.2	62.0	–	70.5	58.6	64.3
	σ_s	0.6	1.8	2.1	0.4	–	0.9	1.3	1.3

Range 1: R1 and Range 2: R2

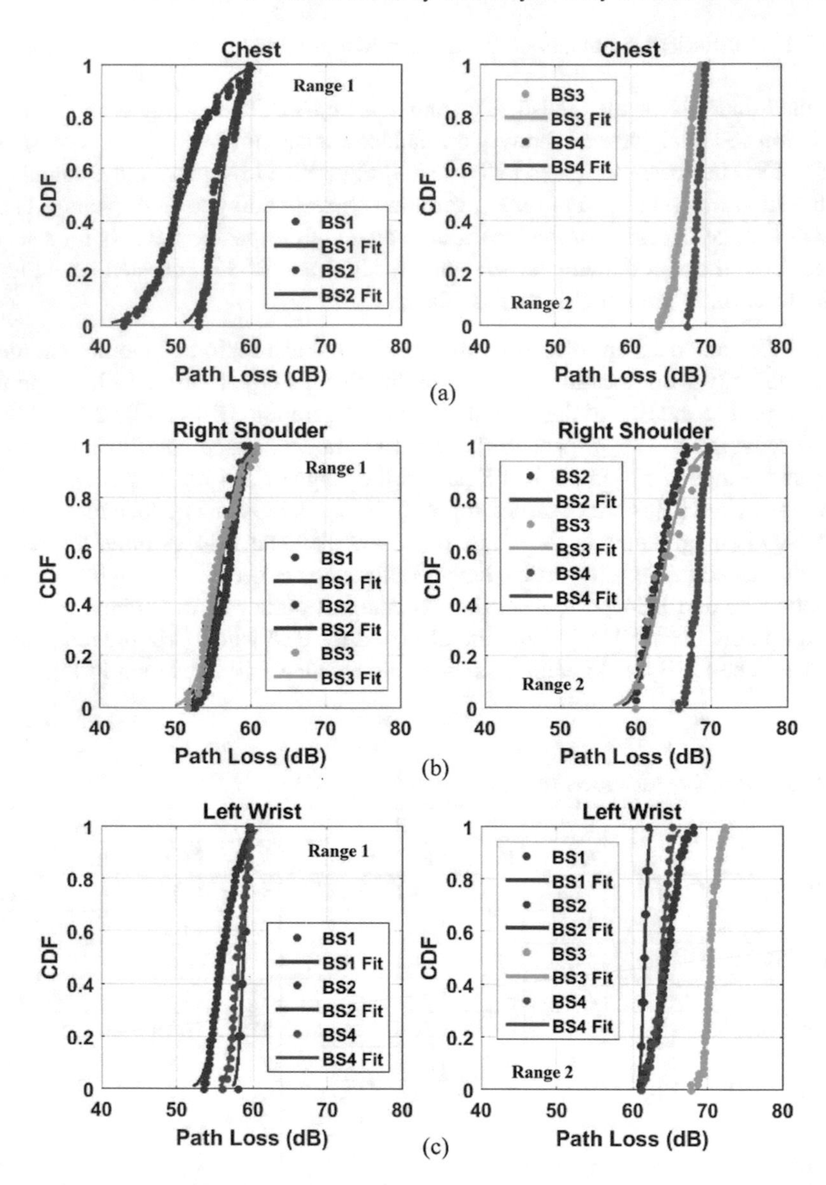

Fig. 3.11 Cumulative distribution function for path loss variation with respect to different base station locations: **a** Chest (S2), **b** Right Shoulder (S3), **c** Left Whrist (S8) (Reproduced from [32])

and the color graph presenting the variation of the PL magnitude for different ranges R1 and R2 is analysed for BS1 and BS3 (Fig. 3.12a).

Right Limb: For BS1, the R. wrist observes minimum PL values for the whole localisation area with an average of 53 dB. The average PL magnitude increases from wearable antenna location S5→S3 (51→57 dB), due to an increase in vertical

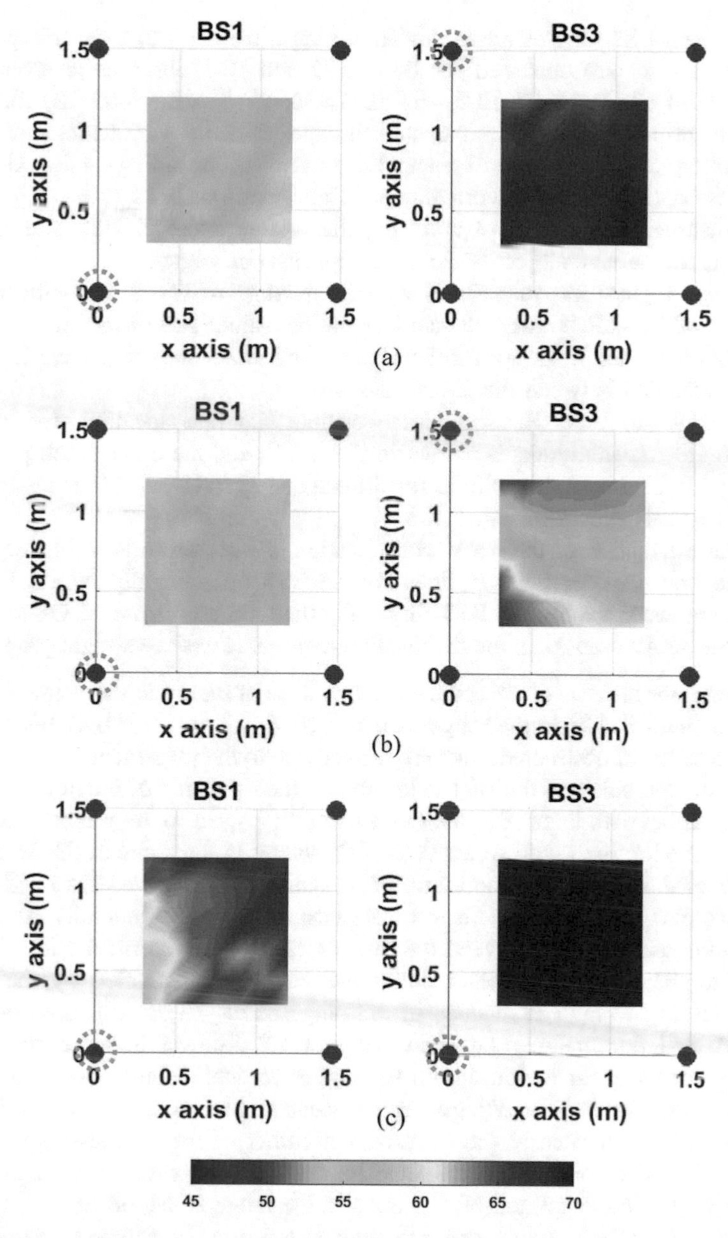

Fig. 3.12 Colour graphs showing variation of the path loss magnitude in the x–y plane with respect to different base station locations BS1 and BS3 (green circle): **a** Chest (S2), **b** Right Shoulder (S3), **c** Left Wrist (S8) located at different heights in the z-axis (Reproduced from [32])

distance from BS1. The standard deviation ranges from 2.5 to 3 dB and generally direct path links are observed for BS1. BS2 and BS3 observes an average PL magnitude of (BS2: 54–57 dB/63–64 dB)/(BS3: 55–58 dB/64–69 dB) for R1/R2 PL ranges respectively. The human subject is positioned at various locations in the grid, leading to different type of channel links formed depending on the orientation of the base station and body worn antenna. Hence, various links such as direct path, partial obstructed and total obstructed paths with respect to BS2 and BS3 are formed as the human subject walks in the localisation region.

BS4 has highest PL magnitude with an average of 66–69 dB with standard deviation of 0.9–1.7. Lower standard deviation values are observed for BS4 as generally obstructed links are formed due to the human subject present leading to major attenuation between the Tx and Rx links.

The CDF for the R. shoulder wearable antenna location for different BSs-wearable antenna links is shown in Fig. 3.11b and the color graph presenting the variation of the PL magnitude for different ranges R1 and R2 is analysed for BS1 and BS3 is shown in Fig. 3.12b.

As the orientation of the body worn antenna is more towards facing BS1, LOS links are observed for S3-BS1 link. For S3-BS3 link, initially NLOS links are observed between the S3 and BS3. Gradually the links shift from NLOS to LOS as the human subject moves in the localization region due to line-of-sight propagation.

Left Limb*:* For the case of BS1, the average values of PL are in the range of 57–58/63–66 dB with STDV in the range of 0.6–2.4/1.6–1.8 dB for R1/R2 respectively. Higher chances of obstructed links are formed due to the placement of the wearable links on the left side and the BS1 is located on the right side of the human subject. For BS2, direct path links (R1) are observed with respect to the wearable antennas placed on the left limb with an average of PL values in the range of 52–56 dB with a STDV of 2.5–3 dB. The minimum PL magnitude is observed for L. shoulder–BS2 link due to the minimum vertical distance and the generally direct path propagation taking place between the two nodes. The links formed with BS3 with respect to the wearable antenna nodes are generally obstructed by the human subject leading to high PL values in the range of 68–70 dB with low values of standard deviation ranging between 0.9 and 1.8 dB. Maximum obstruction is observed for the wrist region due to the higher vertical distance between the two links and due to the torso/thigh region causing obstruction between the two antennas. BS4 is located at the extreme left corner of the cuboid-shape configuration and has a combination of LOS/NLOS scenario with LOS links having average PL of 56–58 dB and NLOS links in the range of 64–68 dB.

The CDF for the L. wrist wearable antenna location for different BSs-wearable antenna links is shown in Fig. 3.11c and the color graph presenting the variation of the PL magnitude for different ranges R1 and R2 is analysed for BS1 and BS3 is shown in Fig. 3.12c.

Off-body path loss analysis is carried out with respect to different base stations positioned in cuboid-shape base station configuration mainly aiming for localisation, tracking, and sensing applications. Variable PL magnitude is observed for each BS location, with BS1–BS2 facing forward and BS3-BS4 positioned behind the human subject. BS1 generally has occurrence of low PL values in the range of 45–60 dB and BS4 has higher occurrence of high PL magnitude in the range of 60–75 dB. Due to change in the orientation of the human subject while moving in the localization area in the x–y plane, the BS-wearable antenna links varies from LOS to NLOS scenarios, which is reflected in the PL statistics. The base station and wearable antennas which have variable type of links such as LOS and NLOS over the localisation area, have much higher standard deviation values (3–4 dB) due to the presence of both lower and higher range of PL magnitude. The PL exponent for LOS scenarios falls in the range of 1–1.8 and that of NLOS links in the range 0.3–0.8 showing higher variation in PL over distance for LOS links. The variation of the height in the z-axis of the wearable antenna and the BSs location has significant effect on the PL magnitude with the LOS links having more prominent variation in comparison to total NLOS links. The results presented will be beneficial in aiding localisation algorithms and further develop the propagation channel understanding for various wireless body area network applications.

3.4 mmWave: 60 GHz

3.4.1 Off-Body Communication

Wideband channel measurements for off-body channel at 60 GHz are reported in [34]. A Saleh–Valenzuela impulse response is proposed as numerical model. Different scenarios and links have been considered for off body communications such as Belt, Right Wrist, and Right side of the Head. The base station was 1.2 m from the ground, and the measurements have been conducted for three distances d between the transmitting antenna and the user body: 1, 2, and 3 m in an indoor environment. For each distance d, the channel has been sampled for four body orientations: “front,” “right side,” “back,” and “left side” with respect to the transmitter position. Measurements of wideband off-body channels have been conducted using a Rohde and Schwarz ZVA vector network analyzer (VNA) up to 75 GHz with quasi-omnidirectional antennas FLANN Microwave ref. MD249-AA. The subject under study is a male of 1.85 m height, 75 kg mass, and 93 cm body perimeter.

The first cluster has been modeled in the LOS and NLOS cases. It has been shown that “Belt” node has always the lower power, while “Wrist” and “Head” nodes have similar powers. The power of the first cluster has been modeled by a lognormal distribution in the LOS case and in the NLOS case for the “Belt node,” while a Weibull distribution has been found more appropriate to describe NLOS in “Wrist” and “Head” nodes. The shape of the first cluster presented a sharper decay for the LOS channels compared to NLOS.

A fast computation and accurate analytical model for off-body propagation is derived in [35]. The work discusses the off-body model propagation from an external source to a receiver located on the body. The model is developed for normal incident plane wave by describing the human body with a circular cylinder. The total received electric field around the human body can be written as a creeping wave in the shadow region and as a Geometrical Optics (GO) result for the lit region. It is also shown that at 60 GHz, the shadow boundary width is negligible. The model shows perfect agreement with the experimental results conducted on a perfectly conducting cylinder. Measurements of the creeping wave path gain have been also conducted on a real body to assess the validity of the cylinder assumption. The results have shown a path gain of about 5 dB/cm for transverse magnetic (TM) case and 3 dB/cm for transverse electric (TE) case. The standard deviation between the measurements and the cylindrical model is about 3.5 dB for both TM and TE cases.

The first-order characteristics of dynamic off-body communications channels at 60 GHz within indoor and outdoor environments have been investigated in terms of path loss, large-scale and small-scale fading. LOS and NLOS channel conditions have been investigated in which the Tx-Rx were facing each other, and the human subject was facing at 180° to the Rx [36].

The experiments were carried out in the 59–66 GHz band at an operating frequency of 60 GHz. The measurement system consisted of a Hittite HMC6000LP711E millimeter wave transmitter (TX) module and Hittite HMC6001LP711E millimeter wave receiver (RX) module, both containing on-chip, low profile antennas with +7.5 dBi gain. The base station antenna was positioned at a height of 1.25 m and the wearable antenna positioned on the torso region of the human subject (height 1.83 m, mass 78 kg) were at 1.45 m. The measured azimuthal radiation patterns for the Tx antenna in free space and on the human subject is shown in Fig. 3.13.

A. ***Path loss***

The parameter estimates for PL_0 and n over all the considered environments are given in Table 3.5 along with the body shadowing factor (BSF) which is defined as the difference between PL_0 for the LOS and NLOS scenarios. Figure 3.14 shows the path loss model fits for the LOS and NLOS in the hallway environment respectively. PL_0 for the NLOS case was greater than that for the LOS due to the shadowing effects caused by the test subject's body. The body shadowing effects were more predominant in the anechoic chamber and car park environment. For both the LOS and NLOS scenarios, the estimated path loss exponents for the anechoic chamber and car park environments were greater than those for the hallway and open office environments due to the additional multipath present in the latter environments.

B. ***Large-scale fading***

Received signal power with the superimposed path loss fit and large scale fading for the (a) LOS and (b) NLOS in the hallway environment were obtained using maximum likelihood estimation (MLE). Figure 3.14 shows the large-scale fading overlaid on the received signal power for the LOS and NLOS in the hallway

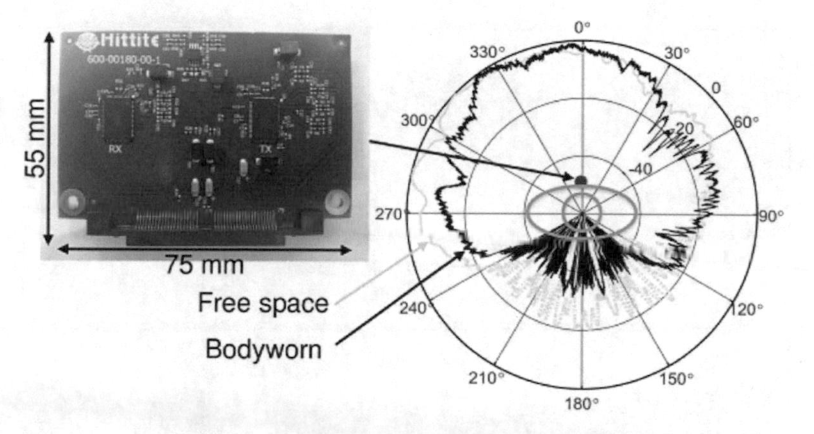

Fig. 3.13 Measured free-space and body-worn azimuthal radiation patterns when the Tx was positioned at the center of the chest region (After [36], IEEE)

Table 3.5 Difference between lowest recorded signal power and noise threshold (D_{SN}) path loss parameters (After [36], IEEE)

Environments	Path loss				
	D_{SN} (dB)	LOS		NLOS	
		n	PL_o(dB)	n	PL_o(dB)
Hallway	12.7	1.13	42.3	1.18	58.5
Office	20.0	1.18	45.1	1.07	57.4
Anechoic	2.2	1.41	43.0	1.86	75.5
Car park	2.6	1.53	48.7	1.98	88.8

environment, respectively. Figure 3.15 depicts the CDFs of the gamma distribution fitted to the empirical data for the LOS and NLOS walking scenarios in each environment. Table 3.5 provides the parameter estimates for the gamma distribution over all four environments.

C. ***Small-scale fading***

Similar to the large-scale fading, the parameter estimates for the Rice and Nakagami-m fading models were obtained using MLE. As expected, the K factor for the LOS case was greater than unity for all the considered environments indicating that a strong dominant signal component existed. The K factors for the anechoic chamber and car park environments were larger than those for the hallway and open office environments due to varying levels of scattered signals found in each environment. As shown in Table 3.6, the m parameter estimates were greater than unity for all the environments which indicates the fluctuations of the signal envelope observed in NLOS scenarios were less severe than those experienced in Rayleigh fading channels ($m = 1$). Figure 3.16 shows the fits of the Rice and Nakagami-m fading models for the LOS and NLOS scenarios, respectively.

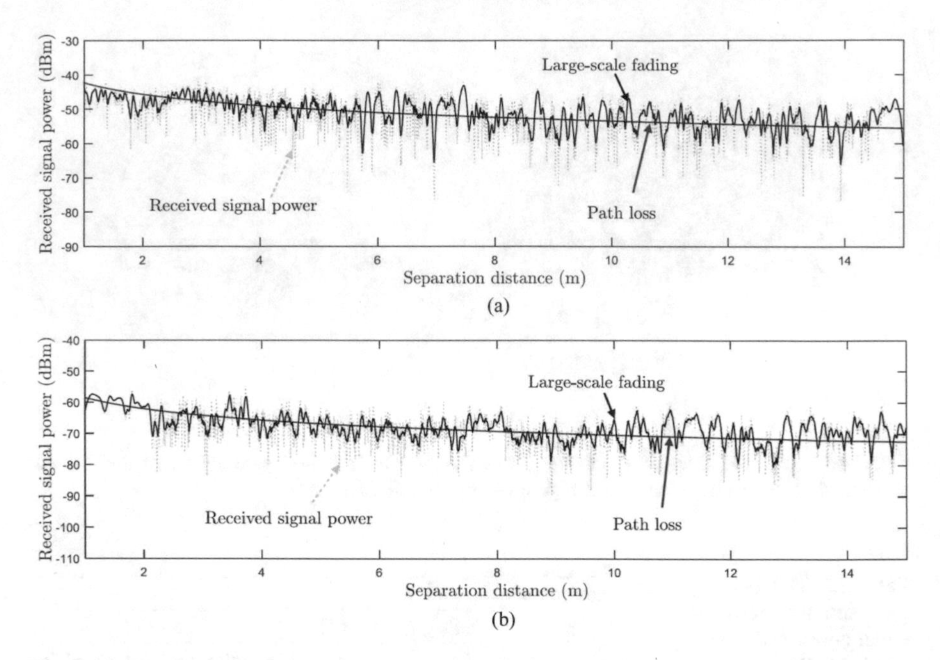

Fig. 3.14 Received signal power superimposed with path loss fit and large scale fading **a** LOS and **b** NLOS in the hallway environment (After [36], IEEE)

The shape of the CDF for the LOS case was much more constricted than that observed for the NLOS condition. This puts forward that LOS channels in sparse environments will not suffer from significant fluctuations due to small-scale fading.

3.4.2 Body-to-Body Communication at 60 GHz

Simulations related to body centric channels using animation software and a full 3D ray tracing algorithm to characterize the channel for mmWave applications is reported in [37, 38]. Using soldier-to-soldier communications as an example, several important channel metrics such as root mean square angle and delay spread are investigated. The algorithm can calculate all reflections, penetrations, and diffractions; animation sequence and CAD environment model, assignment of dielectric properties to material layers, antenna orientations, speed, transmit power other related channel simulation properties were taken into consideration. Channel has been analyzed for dynamic human body movements in complex environments. Simulations of mmWave body-to-body signal propagation were performed, and the results have shown that rms angle spread was greatest within indoor environments, while the most significant delay dispersion was found outdoors and in large building structures. Angle of arrival (AOA) statistics such as the root mean square (rms) angle spread are presented alongside important wideband channel parameters

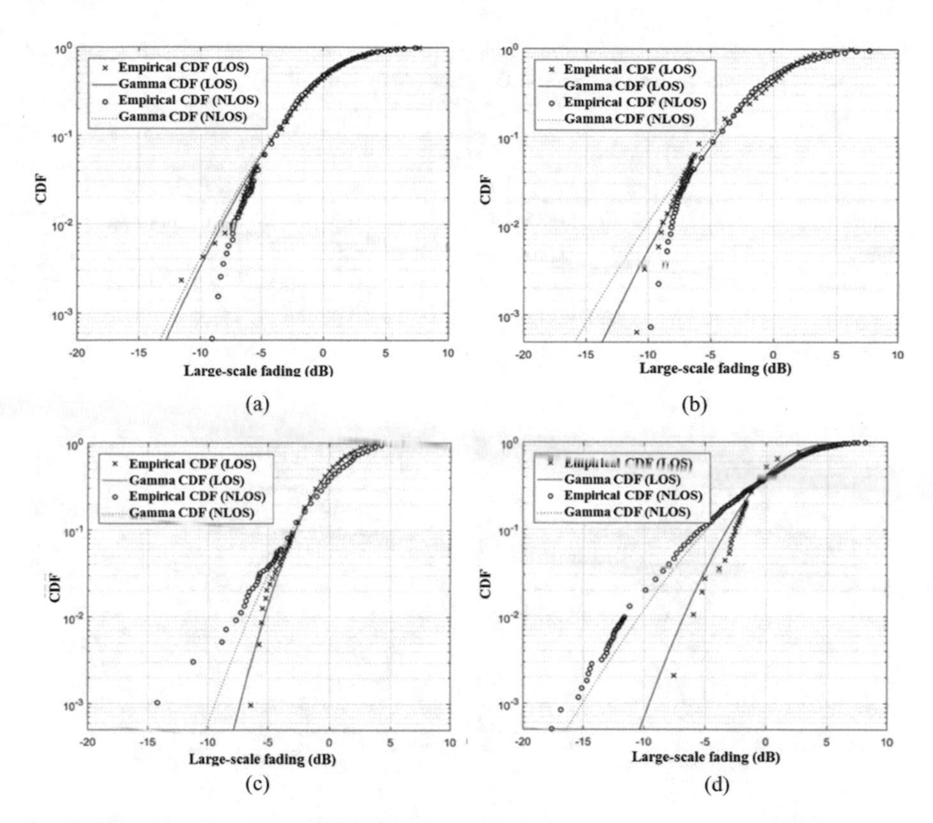

Fig. 3.15 Empirical CDFs of the large-scale fading with corresponding gamma CDF fit: **a** Hallway, **b** Open office, **c** Anechoic chamber, and **d** Car park (After [36], IEEE)

such as power delay profiles and rms delay spread. The latter two can be used to investigate the potential impact of inter-symbol interference (ISI) and its consequences for wideband receiver design.

The inter-user interference between wireless body area networks worn by two moving persons in an indoor environment at 60 GHz and compared with 2.4 GHz is investigated experimentally in [39]. Both omni-directional antennas (monopoles) and directional antennas (horns) were used in the measurements (60 GHz) with realized gains of 4.9 dBi and 18.7 dBi respectively in the presence of skin phantom (Fig. 3.17a, b). The radiation patterns are presented in Fig. 3.17c, d. The interference power level variation and carrier-to-interference ratio were measured and characterized. Multiple antenna placements on the body were also investigated.

The measurements were performed in a rich multipath propagation indoor lab environment and two subjects performed random movements/rotations within an area of 1×2.5 m^2. The random body movements included walking, waving arms, rotating trunks, squatting, bending trunks, running, etc. The received signal power during the activities was measured with a Rohde and Schwarz ZVA 67 Vector

Table 3.6 Body Shadowing factors along with the parameter estimates for the Gamma, Rice and Nakagami-m distributions for LOS and NLOS cases (After [36], IEEE)

Environments	Large-scale fading					Small-scale fading				
	BSF (dB)	LOS		NLOS						
		α	β	α	B	s	σ	K (dB)	m	Ω
Hallway	16.2	7.49	0.15	7.06	0.16	0.94	0.30	6.7	2.67	1.07
Office	12.3	7.01	0.15	5.77	0.20	0.93	0.34	5.7	2.35	1.10
Anechoic	32.5	17.39	0.06	10.27	0.11	0.98	0.16	12.5	1.94	1.12
Car park	40.1	10.30	0.11	5.11	0.23	0.97	0.17	11.9	2.74	1.10

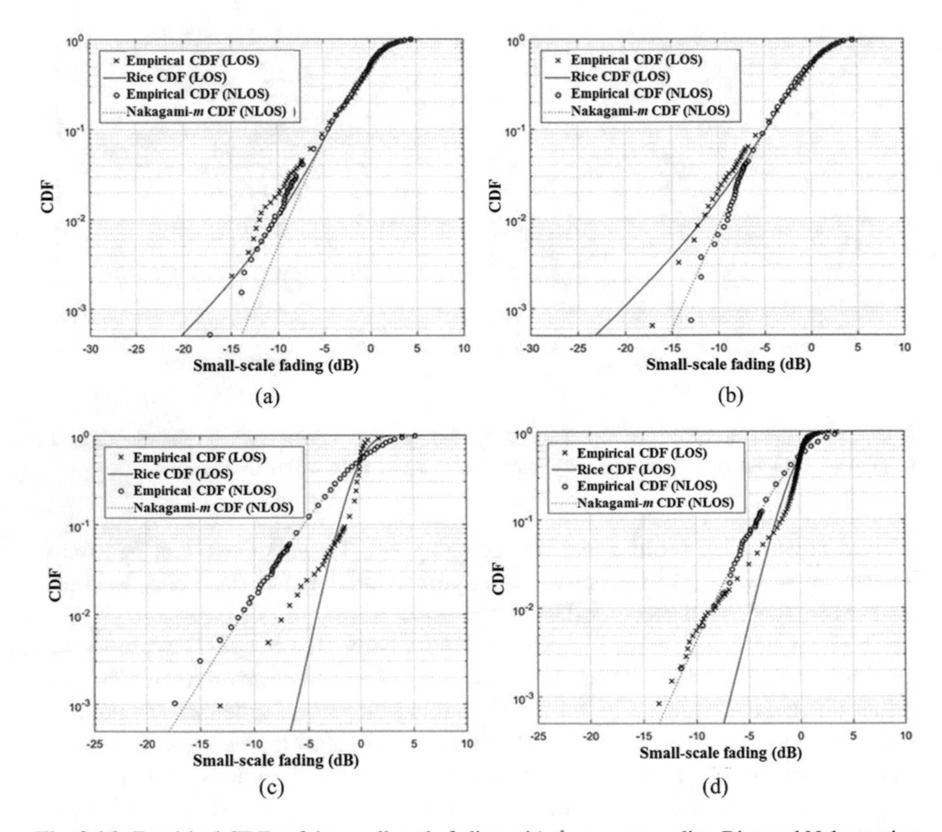

Fig. 3.16 Empirical CDFs of the small-scale fading with the corresponding Rice and Nakagami-m CDF fits: **a** Hallway, **b** Open office, **c** Anechoic chamber, and **d** Car park (After [36], IEEE)

Network Analyzer (VNA). For the 60 GHz measurement setup, the VNA was set to 60 GHz continuous wave (CW) with 60,001 sweep points.

As shown in Fig. 3.18, the channel on Subject 1 is defined as the wanted-signal channel. The other two channels between the two subjects are defined as the

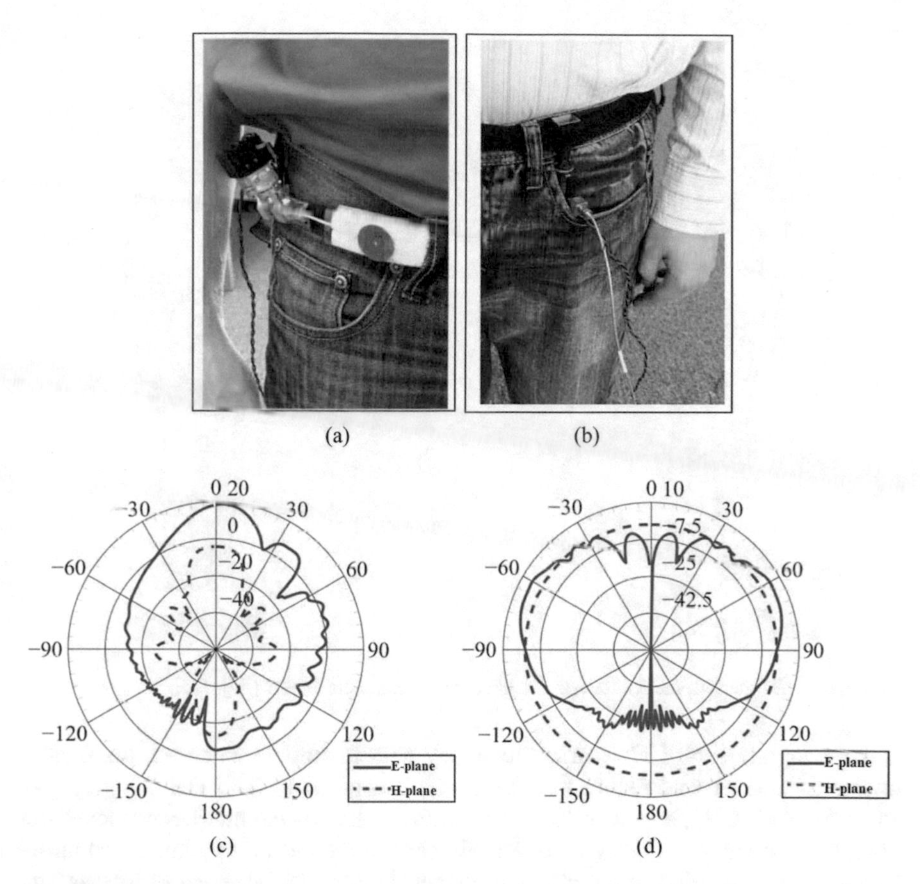

Fig. 3.17 Placement of **a** monopole and **b** horn antennas on the human subject. Simulated radiation pattern for **c** monopole and **d** horn antenna placed on the skin phantom at 60 GHz (After [39], IEEE)

interference channels. Two body channels were measured: the head-abdomen channel and the head–chest channel. Various notions related to antenna type and placement is given below:

$$D^{n}_{P_D T_D}, S^{m}_{P_S T_S}; P_D, P_S \in \{Head,\ Abdomen,\ Chest\}; T_D, T_S \in \{H, M\}, m, n \in \{1, 2\}$$

where *D, S* stand for the destination and source of the link, respectively; P_D and P_S represent the antenna position for the source and destination of the link, respectively; T_D, T_S represent the antenna type. Abbreviations are used to represent antenna type: *H* for horns and *M* for monopoles. m and n label the mth and nth subjects, respectively. If $m = n$, it means that this channel is the wanted signal; else it is an interference signal.

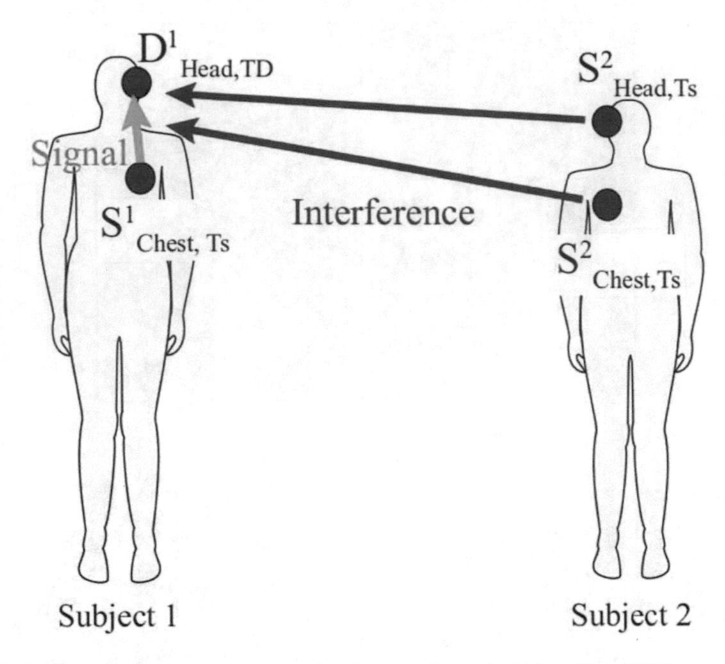

Fig. 3.18 CIR measurement scenario for head–chest channels (After [39], IEEE)

By adopting a 60 GHz carrier frequency with monopole antennas, the median interference level was reduced up to 20 dB compared to 2.45 GHz. Using horn antennas at 60 GHz, a further 20 dB reduction of the median interference level was observed for the same WBAN-to-WBAN separation and orientation distributions leading to as reduction in interference. Results indicate higher level of interference when placed on the wrist.

3.4.3 *Near-Body Shadowing at 60 GHz*

A numerical model based on the indoor channel model standardized by IEEE802.11ad coupled with simple diffraction models has been proposed in to study near body shadowing at 60 GHz [40, 41]. The diffraction model assumes that the human body can be modeled by a circular cylinder. This spatial region is extended from 5 to 30 cm away of thc user body and is relevant for various situations during use of devices such as smartphones, laptops, or tablets [41]. This near-body zone has been split up into two regions: (1) front and (2) back, with respect to the Tx since the physical phenomena are different. It has been shown that this latter is spatially distributed as a Two Wave Diffuse Power (TWDP) distribution for both the front and back regions, which models the fading by considering two specular components with diffuse scattered fields. The model has been studied for the case of the conference room of the indoor channel IEEE 802.11ad.

The mean attenuation of the channel has been spatially studied. Between these two regions, the specular components are reduced by about 10–15 dB in the back region and the diffuse power is also reduced by about 3 dB.

The cylindrical model has been widely used with electric properties of the body model considered to be of human skin [42] and having least impact of clothing [43]. The base station and the human subject are located in an indoor conference room environment. The communication device is held by the user in close proximity. The front region is defined as the set of points where the base station and the receiving antenna are in line-of-sight (LOS), while the locus given by the non-line-of-sight (NLOS) configurations is the back region. The lit and shadow region have been defined by considering the behavior of the electric field (Geometrical Optics (GO) and creeping wave). The diffraction model in the front (or lit) region is calculated using GO [44].

Theoretical Model

In the front and back region the electric field is calculated using Geometric Optics [44]. The main impact on mean frequency channel response μ_{II} is due to the line-of-sight ray and the reflection off the cylinder. The other rays arriving on the body have lower power due to the higher traveled distance. This situation is well suited to be modeled by a two-wave diffuse power (TWDP) distribution [45] which models the received amplitude V.

$$V = V_1 e^{j\varphi_1} + V_2 e^{j\varphi_2} + \sum_i \tilde{V} e^{j\varphi_i} \tag{3.7}$$

where $V_{1,2}$ denotes the magnitude of the specular components 1 and 2, and $\varphi_{1,2}$ are the random associated phases. The second part of (3.7) describes the non-specular components.

Experimental Approach

Experiments were performed in an indoor environment with dimensions ($7 \times 4 \times 2.5$ m^3) with the user located at 2 m from the Tx and the Rx was located randomly from 5 to 30 cm away from the user body. The antennas were placed at a height of 1.2 m and the Rx pair was located 5 to 30 cm away from the human body. An omnidirectional antenna (FLANN microwave ref. MD249-AA) was used at the Rx side. The experimental mean attenuation over the bandwidth and TWDP distribution in the front region shows similar trend for the experimental and simulated results.

3.5 Conclusion

The large-scale and small-scale fading has been analyzed in this chapter for various body-centric channel links for UWB and 60 GHz communication. It is widely known that the successful development of hardware and wireless networking protocols is highly dependent on thorough knowledge of transmission channel characteristics relative to deployment. Research involving UWB and mmWave 60 GHz short-range communications has been presented considering a range of indoor environments, static and dynamic scenarios for channel modeling and characterization. Statistical analysis and modeling of the channel parameters, such as path loss exponent and rms delay spread and best fit models for various body centric links and frequency range are presented. The amplitude, delay spread parameters related to the body-centric channel can vary to a great extend with change in distance between the transmitter and receiver, changes in posture and orientation of the antenna and the human subject. The channel parameters have found to have high influence of the antenna type and configuration, the local surroundings, and the location of the wearable antenna on the human subject.

References

1. Cavallari R, Martelli F, Rosini R, Buratti C, Verdone R (2014) A survey on wireless body area networks: technologies and design challenges. IEEE Commun Surv Tutorials 16 (3):1635–1657
2. Hall PS, Hao Y (2012) Antennas and propagation for body-centric wireless communications. Artech House
3. Cotton SL, D'Errico R, Oestges C (2014) A review of radio channel models for body centric communications. Radio Sci 49(6):371–388
4. Yan S, Soh PJ, Vandenbosch GAE (2018) Wearable ultrawideband technology—a review of ultrawideband antennas, propagation channels, and applications in wireless body area networks. IEEE Access 6:42177–42185
5. Bharadwaj R, Swaisaenyakorn S, Parini CG, Batchelor JC, Alomainy A (2017) Impulse radio ultra-wideband communications for localization and tracking of human body and limbs movement for healthcare applications. IEEE Trans Ant Propag 65(12):7298–7309
6. Bharadwaj R, Swaisaenyakorn S, Parini CG, Batchelor JC, Koul SK, Alomainy A (2018) Base-station random placement effect on the accuracy of ultrawideband body-centric localization applications. IEEE Ant Wirel Propag Lett 17(7):1319–1323
7. Hamalainen M, Taparugssanagorn A, Tesi R, Linatti J (2009) Wireless medical communications using UWB. In: IEEE international conference on ultra-wideband, Vancouver, BC, pp 485–489
8. Zhadobov M (2018) Millimeter-wave technologies for body-centric applications. In: 43rd International conference on infrared, millimeter, and terahertz waves (IRMMW-THz), Nagoya, p 1
9. Wang JC, Leach M, Wang Z, Lim EG, Ma KL, Huang Y (2015) State-of-the art of 60 GHz antennas in wireless body area network. In: 2015 International SoC design conference (ISOCC), Gyungju, pp 171–172
10. Venugopal K, Valenti MC, Heath RW (2015) Analysis of millimeter wave networked wearables in crowded environments. In: 49th Asilomar conference on signals, systems and computers, Pacific Grove, CA, pp 872–876
11. Smith DB, Miniutti D, Lamahewa TA, Hanlen LW (2013) Propagation models for body-area networks: a survey and new outlook. IEEE Ant Propag Mag 55(5):97–117
12. Molisch AF et al (2006) A comprehensive standardized model for ultrawideband propagation channels. IEEE Trans Ant Propag 54(11):3151–3166

13. Maltsev et al (2010) Channel models for 60 GHz WLAN systems, IEEE 802.11 TGad document: IEEE 802.11-09/0334r8
14. Ghassemzadeh SS, Greenstein LJ, Kavcic A, Sveinsson T, Tarokh V (2003) UWB indoor path loss model for residential and commercial buildings. In: IEEE 58th vehicular technology conference VTC 2003-Fall (IEEE Cat No03CH37484), Orlando, FL, pp 3115–3119
15. Rappaport TSW (1996) Wireless communications principles and practice. Prentice Hall Inc, New Jersey
16. Donlan BM, McKinstry DR, Buehre M (2006) The UWB indoor channel: large and small scale modeling. IEEE Trans Wireless Commun 5(10):2863–2873
17. Bharadwaj R, Koul SK (2020) Analytical and experimental investigation of ultra wideband channel characteristics in the presence of door/window glass. Wirel Pers Commun 110:763–780
18. Bharadwaj R, Koul SK (2017) Study and analysis of ultra wideband through glass propagation channel characteristics. In: IEEE 13th international conference on wireless and mobile computing, networking and communications (WiMob), Rome, pp 1–5
19. Bharadwaj R, Parini C, Alomainy A (2015) Experimental investigation of 3-D human body localization using wearable ultra-wideband antennas. IEEE Trans Ant Propag 63(11):5035–5044
20. Alomainy A, Sani A, Rahman A, Santas JG, Hao Y (2009) Transient characteristics of wearable antennas and radio propagation channels for ultrawideband body-centric wireless communications. IEEE Trans Ant Propag 57(4):875–884
21. Bharadwaj R, Koul SK (2021) Study of the influence of human subject on the indoor channel using compact UWB directive/omni-directional antennas for wireless sensor network applications. Wirel Ad-Hoc Netw 118
22. Goulianos AA, Brown TWC, Evans BG, Stavrou S (2009) Wideband power modeling and time dispersion analysis for uwb indoor off-body communications. IEEE Trans Ant Propag 57 (7):2162–2171
23. Pasquero OP, D'Errico R (2016) A spatial model of the UWB off-body channel in indoor environments. IEEE Trans Ant Propag 64(9):3981–3989
24. Garcia-Serna RG, Garcia-Pardo C, Molina-Garcia-Pardo JM (2015) Effect of the receiver attachment position on ultrawideband off-body channels. IEEE Ant Wirel Propag Lett 14:1101–1104
25. Khan MM, Abbasi QH, Alomainy A, Hao Y (2013) Experimental characterisation of ultra-wideband off-body radio channels considering antenna effects. IET Microw Ant Propag 7(5):370–380
26. Catherwood PA, Scanlon WG (2014) Body-centric antenna positioning effects for off-body UWB communications in a contemporary learning environment. In: The 8th European conference on antennas and propagation (EuCAP 2014), The Hague, pp 1571–1574
27. Kumpuniemi T, Hämäläinen M, Yazdandoost KY and Iinatti J (2015) Measurements for body-to-body UWB WBAN radio channels, 2015 9th European Conference on Antennas Propagation (EuCAP), Lisbon, 2015, pp.1–5.
28. See TSP, Hee JY, Ong CT, Ong LC, Chen ZN (2009) Inter-body channel model for UWB communications. In: 3rd European conference on antennas and propagation, Berlin, pp 3519–3522
29. Bharadwaj R, Koul SK (2016) Numerical analysis of ultra-wideband propagation for body-centric communication. In: 2016 Asia Pacific microwave conference, APMC 2016, New Delhi, India, 5–9 Dec 2016
30. Bharadwaj R, Koul SK (2017) Study and analysis of channel characteristics of ultra-wideband communication links using wearable antennas. In: 2017 Asia Pacific microwave conference, APMC 2017, Kuala Lumpur, Malaysia, 13–16 Nov 2017
31. Bharadwaj R, Koul SK (2019) Experimental analysis of ultra wideband body-to-body communication channel characterization in an indoor environment. IEEE Trans Ant Propag 67(3):1779–1789

32. Bharadwaj R, Parini C, Koul SK, Alomainy A (2021) Influence of spatial distribution of base-stations on off-body path loss statistics for wireless body area network applications. Springer Wireless Networks
33. Cui P, Yu Y, Lu W, Liu Y, Zhu H (2017) Measurement and modeling of wireless off-body propagation characteristics under hospital environment at 6–85 GHz. IEEE Access 5:10915–10923
34. Petrillo L, Mavridis T, Sarrazin J, Benlarbi-Delaï A, Doncker PD (2017) Wideband off-body measurements and channel modeling at 60 GHz. IEEE Ant Wirel Propag Lett 16:1088–1091
35. Mavridis T et al (2014) Theoretical and experimental investigation of a 60-GHz off-body propagation model. IEEE Trans Ant Propag 62(1):393–402
36. Yoo SK, Cotton SL, Chun YJ, Scanlon WG, Conway GA (2017) Channel characteristics of dynamic off-body communications at 60 GHz under line-of-sight (LOS) and non-LOS conditions. IEEE Ant Wirel Propag Lett 16:1553–1556
37. Cotton L, Scanlon WG, Madahar BK (2009) Millimeter-wave soldier-to-soldier communications for covert battlefield operations. IEEE Commun Mag 47(10):72–81
38. Cotton SL, Scanlon WG, Madahar BK (2010) Simulation of millimetre-wave channels for short-range body to body communications. In: Proceedings of the fourth European conference on antennas and propagation, Barcelona, pp 1–5
39. Wu X, Nechayev YI, Constantinou CC, Hall PS (2015) Interuser interference in adjacent wireless body area networks. IEEE Trans Ant Propag 63(10):4496–4504
40. Mavridis T, Petrillo L, Sarrazin J, Benlarbi-Delaï A and Doncker PD (2014) Human influence on 60 GHz communication in close-to-user scenario, XXXIth URSI General Assembly and Scientific Symposium (URSI GASS) pp. 1–4
41. Mavridis T, Petrillo L, Sarrazin J, Benlarbi-Delaï A, Doncker PD (2015) Near-body shadowing analysis at 60 GHz. IEEE Trans Ant Propag 63(10):4505–4511
42. Chahat N, Valerio G, Zhadobov M, Sauleau R (2013) On-body propagation at 60 GHz. IEEE Trans Ant Propag 61(4):1876–1888
43. Guraliuc AR, Zhadobov M, Valerio G, Chahat N, Sauleau R (2014) Effect of textile on the propagation along the body at 60 GHz. IEEE Trans Ant Propag 62(3):1489–1494
44. Kravtsov YA, Zhu NY (2010) Theory of diffraction: heuristic approaches alpha science
45. Chew WC (1995) Waves and fields in inhomogeneous media. IEEE Press, New York

Chapter 4
Flexible and Textile Antennas for Body-Centric Applications

4.1 Introduction

Flexible and wearable electronics have gained great popularity recently and are drawing more and more attention from both academia and industry. Next generation of wearable electronics demands the wireless device/antennas directly worn on soft and curved human body which leads to motivation of exploring the wide range of flexible substrates and their performance on the body. Their lightweight, energy efficiency, low manufacturing cost, reduced fabrication complexity, and the availability of inexpensive flexible films/substrates (i.e., papers, polyester, polyethylene, and plastics) make flexible electronics an appealing alternative for the current electronics technology, which is based on rigid and brittle substrates [1]. Various flexible substrates can also be integrated with the clothing or directly placed on the body [1, 2]. Such substrates include textile-based antennas where substrates are made of felt, silk, jean cloth, cotton or any suitable textile with conductive materials such as adhesive copper tape, fabrics, e-threads, and wires [3–5]. Flexible antennas can be used in various applications such as mobile communications, wireless medical monitoring/diagnosing, military applications, fields of personal communication, medicine, sports, industry and entertainment (Fig. 4.1) [1].

Ultra-wideband (UWB) technology has matured as low cost, low power, high data rate, simpler hardware configuration, easy to use commercially viable technology for body centric applications [6–8]. The present limitation is of difficulty in integrating/fixing rigid antenna on the body or garments which can be overcome through flexible/textile antennas [8, 9]. Impulse radio UWB technology aims at providing high data rates and high-resolution providing sub–cm ranging accuracy due to nano second width of the UWB pulses by using low power and compact antennas making it very suitable for various short range wireless applications [8–11].

With the increased need for high speed (>1 Gbps) wireless communication systems, the millimeter wave (mmWave) band around 60 GHz has received a lot of attention for the development of short-range, high data rate technology for wireless

S. K. Koul and R. Bharadwaj, *Wearable Antennas and Body Centric Communication*, Lecture Notes in Electrical Engineering 787,
https://doi.org/10.1007/978-981-16-3973-9_4

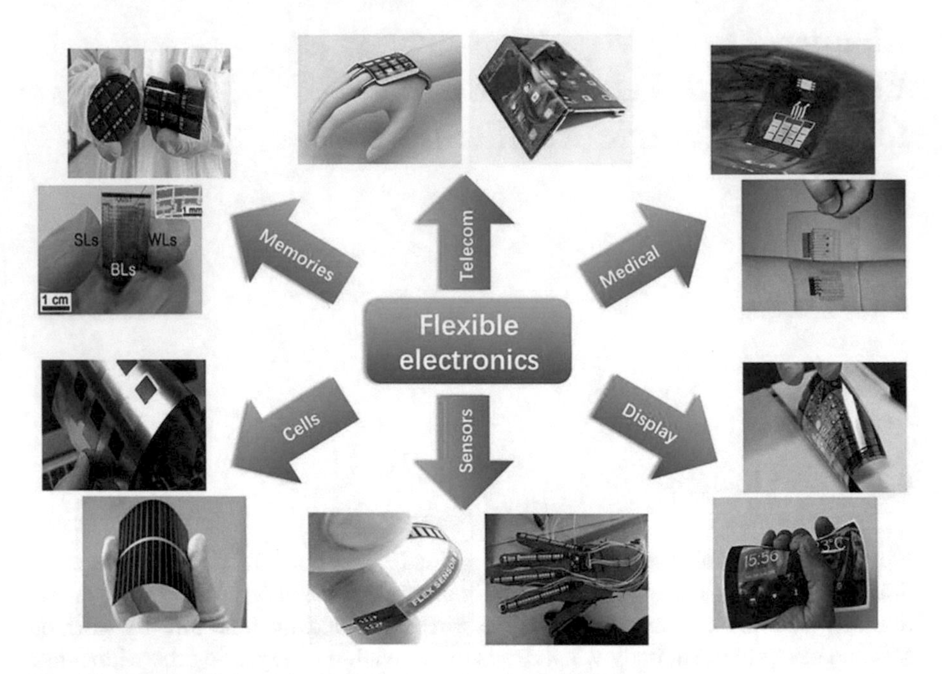

Fig. 4.1 Application areas for flexible electronics (After [1], Micromachines)

personal area network (WPAN) applications [12–14]. Traditionally, mmWave antennas are developed on ceramic type of substrates (LTCC, Alumina) due to the low loss presented in that frequency range. However, these materials and the associated processing are typically more expensive than PCB-line materials and cannot conform to various surfaces [13]. There is an increasing need for reconfigurable and deformable antennas for high-speed wireless communications and high-resolution sensing systems at millimetre waves [15, 16]. Millimetre-wave body area networks (BANs) is identified as a highly attractive solution for future wireless BANs with strong potential for healthcare, entertainment, identification systems, sports, smart home, space, and military applications.

In this chapter an overview of different substrate materials ranging from polymer based to textile have been presented. Various fabrication techniques such as lithography, screen printing, ink-jet printing have also been described. Simulation, design, and fabrication aspects of flexible wearable antennas suitable for UWB and mmWave range have been discussed. Electromagnetic and mechanical sustainability and durability has been considered while performing various flexibility tests. Measurement and study of return loss, radiation pattern, gain, efficiency, and various other performance parameters of flexible antennas in free space and on-body have been reported. Antenna performance and robustness under deformations (bending, crumpling, wrinkling, and wetting) are investigated and incorporated into the design of wearable antennas to meet conformal requirements.

4.2 Flexible Antenna Requirements

Wearable antennas are designed to work in the complicated body-centric environment which can lead to significant variation in the antenna performance and the influence of the antenna on the human body. A wearable antenna having advantages such as low cost, lightweight, flexible can be integrated with the clothing [2, 3]. Designing textile antenna requires the knowledge on electromagnetic properties such as permittivity, and loss tangent of the textile material. Most of the textile materials have less dielectric constant that reduces the surface wave losses and enhance the impedance bandwidth of the antenna.

Ultra-wide bandwidth implies substantial different antenna topologies and propagation aspects as compared to narrow band systems, which makes practical antenna design more challenging [8–11]. The antennas should have good performance in terms of impedance matching, efficient over the entire frequency band, independent of the wearer's morphology and of the antenna's position on the body. In order to obtain bandwidth enhancement, flexible planar UWB antennas do not have full ground plane on the back of the radiator but have partial/modified ground planes or coplanar waveguide topologies [11]. This in turn influences the antenna performance when placed on the human body as the high permittivity and lossy biological tissues will significantly affects the antennas performance.

The radiation from the wearable antenna should present the least specific absorption rate (SAR) [17–19] inside the human tissue to adhere to the health and safety requirements. Antenna designs are associated with back radiation, which, in turn, unavoidably increases the SAR inside the human body. To mitigate the problems, UWB antenna designs equipped with a full ground plane have recently been reported [17–19] using flexible substrates. Flexible antennas need to be bent or conformed to a specific surface during operation, and thus tests in various bending conditions are necessary to investigate bending-dependent characteristics [19, 20]. This is to ensure conformance to the users' body and ease of integration onto clothing and to guarantee comfort and safety to the users. Design parameters, such as the reflection coefficient, input impedance, radiation pattern, efficiency, gain, and fidelity behaviour, are gathered and analysed similarly as any conventional antenna. However, depending on the implemented materials, the prototype of such antennas is more challenging than conventional antennas.

When these antennas are placed close to a lossy medium, they experience strong radiation pattern distortion, shift in resonance frequency, and changes in the input impedance [21]. Developing textile antennas at millimetre waves is a technological challenge since high fabrication accuracy is required. To facilitate commercialization of the textile antennas, the choice of the substrate should be limited to commonly used fabrics. At millimetre waves, the textile substrate should be properly selected and characterized as its thickness, permittivity, and loss tangent are of great importance to design high-efficiency antennas with satisfactory performance [14–16].

4.3 State-of-the-Art Fabrication and Printing Techniques

Printed electronics has gained high attention due to low-cost, eco-friendly, and easy fabrication techniques. There are several widely adopted fabrication processes for flexible and wearable antennas. This section reviews the commercial methods in addition to techniques used by the research and development sector. Various manufacturing processes are available such as photolithography, conducting paint, screen printing, inkjet printing, embroidery and weaving using conductive threads [2, 22, 23]. Some of the commonly used techniques are described below.

A. ***Photolithography***
Photolithography is the process of producing metallic patterns using photoresist and etching agents to mill out a desired area corrosively. This technique has emerged in the 1960s targeting the printed circuit board industry. Photolithography has become very popular since it can produce complex patterns accurately [2]. Current practice in the fabrication of antennas and RF circuits based on photolithography utilizes positive photoresists predominantly since negative photoresists often yield to edge swelling phenomenon as the counterpart is dissolved, which compromises the pattern's resolution. Figure 4.2 illustrates the fabrication procedure of the photolithography process. To produce flexible electronics, flexible single or double-sided substrates are used where the desired pattern is obtained by etching parts of both/either side.

B. ***Inkjet Printing***
Inkjet printing of antennas and RF circuits using silver and gold nanoparticles-based conductive inks have become extremely popular in recent years. The function of inkjet material printers is performed by depositing ink droplets of a size down to a few picolitres; hence, these printers can produce patterns with extremely high resolution (Fig. 4.3a) [2, 22, 23]. The quality of printing mainly depends on the ink properties, such as viscosity, surface tension, and particle diameter. The surface characteristics of the substrate, the platform temperature, and the print head parameters are also crucial factors. Inkjet printing is an emerging technique attractive due to the development of direct-write technology without use of masks, which reduces manufacturing costs. Photograph of the Dimatix printer DMP 2800 suitable for inkjet printing is presented in Fig. 4.3b [24]. Compared to traditional etching techniques, the design patterns of direct write technology with no mask requirements reduces material usage and waste generated from wet processes.

C. ***Sewing and embroidering***
Sewing or embroidering machine is mainly employed in textile-based antennas which is preferred over direct adhesion of E-textile over the fabric since no adhesive materials are introduced which may affect the electrical properties of the material [2, 3, 25]. Wrinkling and crumpling should be minimized to maintain the material qualities. For wearable antennas on clothing application, this is the most preferred fabrication process. Computer aided sewing machines are shown in Fig. 4.4a, b [25, 26].

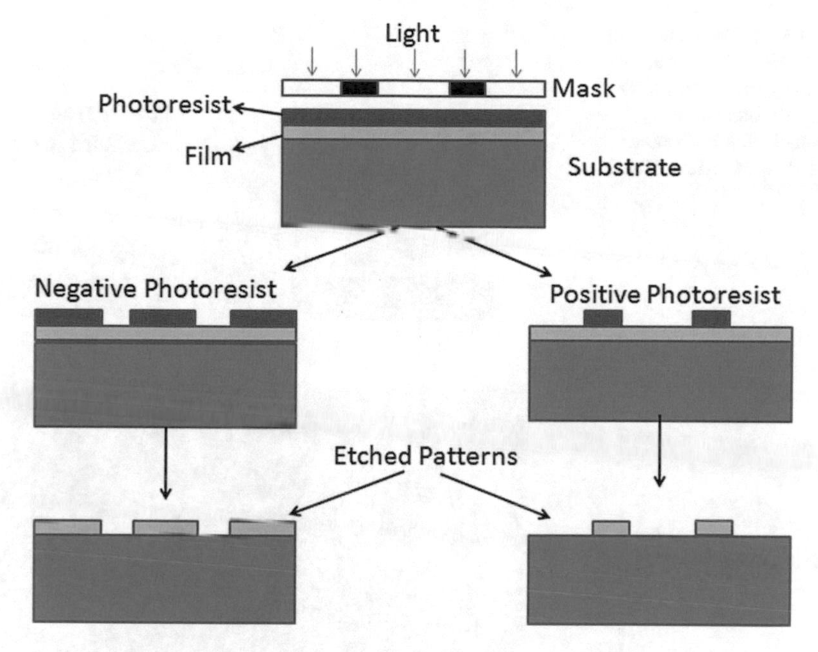

Fig. 4.2 Photolithography fabrication procedure (After [23], Intech Open)

4.4 Flexible Substrates Based UWB Antennas

4.4.1 Kapton

Kapton polyimide film is used as the antenna substrate due to its good balance of physical, chemical, and electrical properties with a low loss factor over a wide frequency range. It is known for its flexibility, robustness, low dielectric loss, and thermal endurance. Kapton polyimide substrate has a dielectric constant of 3.4 and a loss tangent of 0.002. Moreover, Kapton polyimide offers a very low profile (50.8 μm) yet is very robust with a tensile strength of 165 MPa at 73 °F, a dielectric strength of 3500–7000 V/mil, and a temperature rating of 65 °C to 150 °C [8].

A compact elliptical-shaped ultra-wideband (UWB) antenna printed on a 50.8 μm Kapton polyimide substrate is designed to work in the ISM and UWB band. The antenna is fed by a linearly tapered coplanar waveguide (CPW) that provides smooth transitional impedance for improved matching [8]. The antenna demonstrates a very low susceptibility to performance degradation due to bending effects in terms of impedance matching and far-field radiation patterns, making it suitable for integration with flexible electronic devices. The antenna maintains its omnidirectional radiation pattern despite being conformed on the foam cylinders. A conductive ink based on sliver nanoparticles is deposited over the substrate by a Dimatix DMP 2831 inkjet material printer followed by a thermal annealing at 100 °C for 9 h by an LPKF Proto flow industrial oven. Another flexible design with dimensions 30 × 33 mm^2 built on

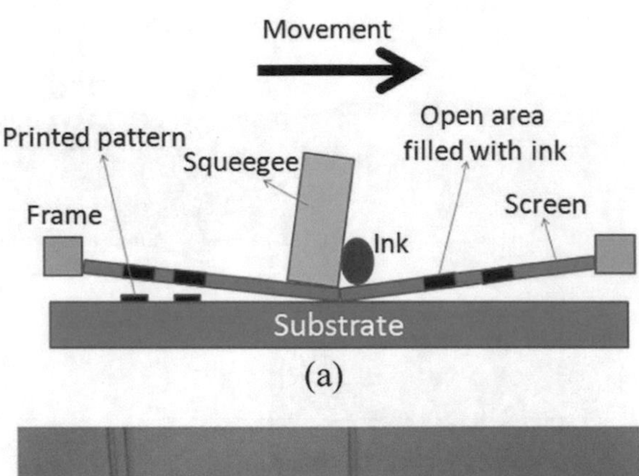

(a)

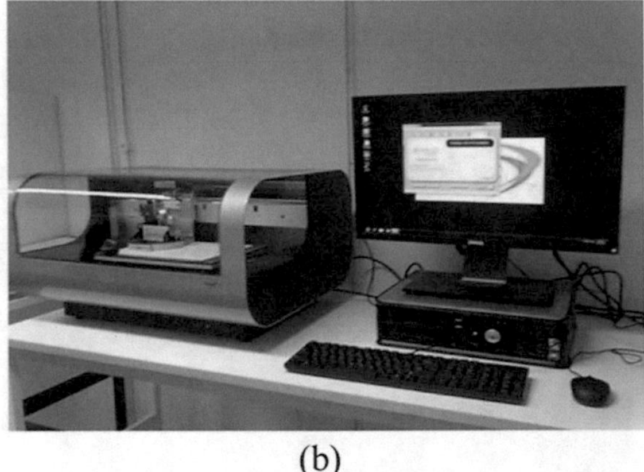

(b)

Fig. 4.3 **a** Overview of the inkjet printing process for antenna fabrication (After [23], Micromachines). **b** Dimatix Inkjet Printer (After [1], Micromachines)

Kapton polyimide substrate is proposed in [22]. The slot is confined by a rectangular frame and a semi elliptical pattern which has a major radius of 12 mm and a minor radius of 11.3 mm. The radiating element is fed by a linearly tapered CPW which provides smooth transitional impedance for improved matching.

Multiple-input multiple-output (MIMO) technique for the UWB systems has attracted considerable interest because MIMO technique can take full advantage of the rich diversity by abundant multipath to further enhance the spectral efficiency and break power limit in the UWB system. A miniaturized fully inkjet-printed flexible multiple-input-multiple-output (MIMO) antenna is proposed, occupying a compact size of $22 \times 31 \times 0.125\ \text{mm}^3$ [27]. The proposed antenna was inkjet-printed on a 0.125 mm flexible Kapton polyimide substrate using a Dimatix DMP 2800 printer as shown in Fig. 4.5a. A conductive silver nanoparticles ink was utilized. The measured impedance bandwidth of 2.9–12 GHz with $S_{11} < -10$ dB and the mutual coupling below -15 dB is reported.

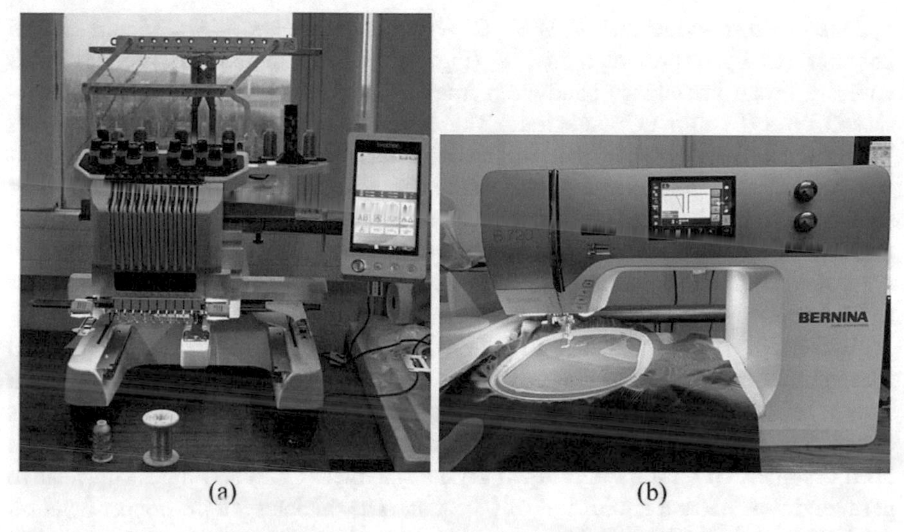

Fig. 4.4 **a** Computer aided embroidery machine (After [25], Electronics). **b** The Bernina 720 sewing machine tool used for sewing the radiating element structures in the form of embroidered pattern over the denim-cotton fabric (Reproduced from [26])

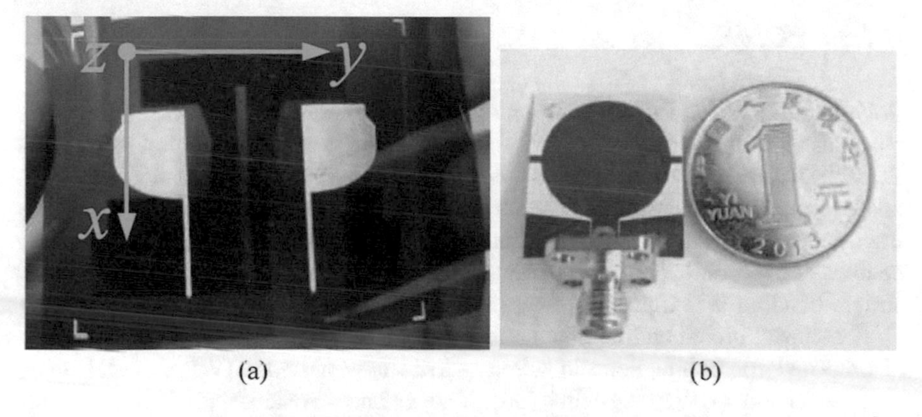

Fig. 4.5 **a** Fabricated MIMO UWB antenna prototype fabricated on Kapton polyamide film (Reprinted with permission from IEEE [27]). **b** A flexible ultrawideband CPW-fed antenna based on Liquid crystal polymer (LCP) (Reprinted with permission from IEEE [28])

4.4.2 LCP

Liquid Crystal Polymer (LCP) has emerged as an organic, lightweight, low cost, low loss factor and flexible material with excellent loss characteristics up to 110 GHz and excellent mechanical stability [2, 28, 29]. An additional advantage of LCP is its capability to create multilayer structures at low temperatures (~280 °C) allowing for embedding of active devices and circuits for complete RF front ends.

Flexible ultra-wideband (UWB) CPW-fed antenna based on Liquid crystal polymer (LCP) is presented in [28] (Fig. 4.5b). The monopole antenna which covers a broad impedance bandwidth from 2.7 to 13.9 GHz is designed and fabricated on a 0.1 mm LCP substrate. The dimension of antenna is 27 × 22 mm^2. The characteristics of the antenna are simulated and analysed under different bending radii. There are no significant differences for the gain and return loss of the antenna between bending and flat case from simulation results. The radiation pattern and the performances including return loss and gain of the proposed antenna are measured under both flat and bending cases. The resonant frequency shifts due to the bending effects in return loss curves are observed.

The design of an ultra-wideband (UWB) band-notched wearable antenna to avoid interferences from WLAN frequency bands is presented in [29]. Two curved tapered slots have been used as ground to allow smooth impedance transition and the notch is obtained by using an H-shaped slot. The coplanar waveguide fed UWB antenna has been designed on an ultra-thin liquid crystal polymer (LCP) substrate with relative permittivity of 2.9 and a thickness of 0.05 mm. The thickness of the copper layer on the LCP is only 0.018 mm. The presented return loss and radiation pattern results perform well in free space and on-body scenarios for flat and bending scenarios.

4.4.3 PDMS

Polydimethylsiloxane (PDMS) is an extremely flexible polymer with a very low Young's modulus that is compatible with many silicon micromachining techniques. PDMS exhibits many attractive features: It is low-cost, lightweight, biocompatible, and chemically resistant. It can be formed to any size or thickness down to a few micrometres by spin coating or replica moulding [30, 31]. In addition, its flexibility enables one to conform it to any shape and to fabricate multi-layered structures by simple bonding techniques.

A compact broadband flexible antenna with dimensions 32 × 18 × 0.127 mm^3 is proposed for applications in wireless local area network (WLAN) and upper ultra-wideband (UWB) systems [30]. The antenna coated with bio-compatible ultra-thin parylene-c is fabricated on a thin polyimide substrate with a thickness of 127 μm. The flexible antenna is composed of polyimide film with a dielectric constant of 3.5 and a loss tangent of 0.0008.

A flexible circular-patch UWB antenna with monopole like radiation patterns with fabrication based on PDMS conductive fabric composite technique is presented in [31]. The radiating structure comprises an annular-ring circular patch loaded with two rectangular slots, and two additional parasitic rings added concentrically around the annular-ring circular patch as shown in Fig. 4.6a. By combining multiple TM_{0n} modes: TM_{01}, TM_{02}, TM_{03}, and TM_{04}, of an annular ring loaded circular patch antenna, the antenna bandwidth is improved. To verify the conformability of the antenna, antenna bending experiments were carried out at 40 mm radius in the x-axis and y-axis direction as shown in Fig. 4.6b. In both cases, it is shown that the return loss is below −10 dB over the frequency band of operation, depicting the antenna's robustness against physical deformation (Fig. 4.6c, d). The far-field radiation characteristics results are presented in Fig. 4.6e for 3.3, 5.6, and 7.8 GHz. The system fidelity factor (SFF) values are higher than 86% for flat and bent condition using Gaussian signals which is suitable for accurate transmission.

A flexible antenna with dimensions 19 × 38 mm^2 based on polyamide flexible substrate is proposed in [32]. The antenna substrate consists of two layers of flexible

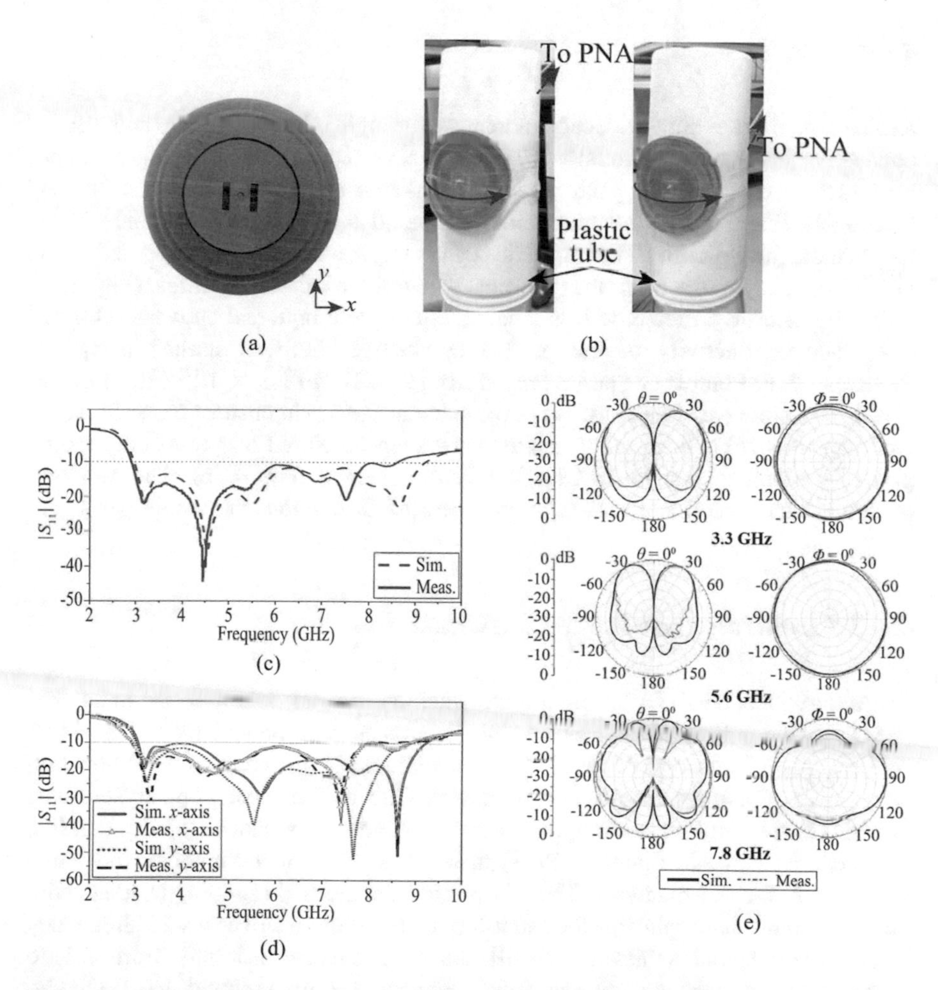

Fig. 4.6 **a** Photograph of the fabricated UWB antenna. **b** Bending experimental setups over a plastic tube having outer radius of 40 mm. Case 1 (left): *x*-axis; Case 2 (right): *y*-axis bending. Comparison between simulated and measured $|S_{11}|$: **c** in flat form. **d** Under *x*-axis and *y*-axis bendings (r_b = 40 mm). **e** Measured and simulated normalized radiation patterns in the x–*z* plane (left side) and *x*–*y* plane (right side) at: 3.3, 5.6 and 7.8 GHz (Reprinted with permission from IEEE [31])

polyimide ($\varepsilon_r = 3.5$, tan$\delta = 0.008$) with a thickness of 0.127 mm, and an air gap of 1.8 mm which benefits the impedance matching. An inverted-F, bow-tie shaped structure is introduced on the top layer to achieve wideband performance. The antenna is placed around a three-layer cylindrical model to mimic the human wrist at 3 mm for measurements related to wearable applications. The result shows that wide bandwidth can be achieved with $|S_{11}|$ less than -10 dB, covering from 3.09 to 11.96 GHz.

4.4.4 Paper

Realization of the ultra-wideband antennas through ink-jet printing and use of conductive inks on commercially available paper sheets is reported in [33]. Paper possesses several intriguing attributes that makes it suitable for low-cost "green" electronics. It is cellulose in nature, thus considered as a renewable resource and has low manufacturing costs. The properties of the paper substrates were studied in the UWB frequency range using the split-post dielectric resonator technique (Fig. 4.7a) [34]. The measured dielectric loss tangent values were bounded between 0.06 and 0.07. The conductivity of the printed conductive ink was studied using the Signatone Four Point Probe and it ranged from 9×10^6 to 1.1×10^7 S/m. A planar UWB monopole was fabricated on paper substrate with dimensions 58×58 mm^2. Good agreement between the simulated versus the measured responses at the input port of the antenna up to 16 GHz has been observed (Fig. 4.7b). The radiation pattern is omnidirectional with efficiency of 80% throughout the whole band.

4.4.5 Innovative Substrate Materials

An ultra-wideband (UWB) monopole antenna on an additive manufactured (AM) flexible substrate for footwear application is proposed in [35] (Fig. 4.8a). Additive manufacturing or 3D printing is a technology that enables the fabrication of complex structures from a digital model. One of the most popular and least expensive 3D printing techniques is fuse filament fabrication (FFF). Flexible polylactic acid plastic filament (PLA) material is used for the antenna and transparent PLA for the phantom. The 3D printed phantom (Fig. 4.8b) is filled with Indexsar liquid that replicates the inner human body tissues over a wide frequency range. The antenna achieves −10 dB input impedance matching from 3.1 to 10.6 GHz in free space, on the foot phantom and on the real human body. Simulated reflection coefficient of the planar and confromal UWB antenna with the human tissue model is presented in Fig. 4.8c.

A conductive polymer-based UWB antenna shows a reproducible non-destructive mechanical conformability and through adoption of a non-resonant efficiency-driven antenna design strategy, combined with use of poly

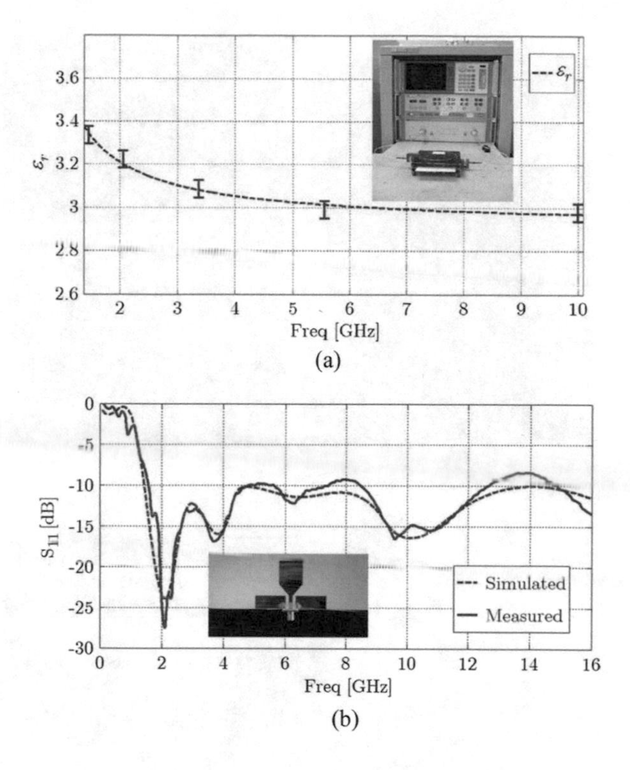

Fig. 4.7 **a** Characterization of the paper material through the split-ring resonator method. **b** Simulated and measured performance of the antenna fabricated on paper substrate (Reprinted with permission from IEEE [33])

(3,4-ethylenedioxythiophene) polystyrene sulfonate (PEDOT:PSS) materials featuring a substantial thickness (70 μm) and a relatively high measured DC conductivity (9532 S/m) (Fig. 4.9a) [36]. The antenna exhibits radiation efficiency of over 85% throughout its UWB operation band of 3–20 GHz. With increasing bending angle above 60°, the tapered slots between the trapezoidal elements and the CPW ground planes become narrower which shifts the operational range to higher frequencies as shown in Fig. 4.9b, c. Under bent condition, the cross-polarization reaches a similar or even higher level as the co-polarization.

Conductive polymer polyaniline (PANI) is selected for flexible antenna design, because of its very low cost, chemical, mechanical, and electrical stability. The proposed antenna is an elliptical monopole fed by a coplanar waveguide; it uses a Kapton substrate, and it is optimised to work from 1 to 8 GHz. The relatively low conductivity of PANI is overcome by addition of carbon nanotubes in the polymer matrix [37]. The flexibility of both the Kapton substrate and the nanocomposite (PANI/ Multi-walled carbon nanotubes (MWCNTs)) provides the ability to crumple the antenna paving the way to potential applications for body-worn wireless communications systems. The substrate used is 130 μm thick Kapton (polyimide). For both bending (along x/y-axis) and crumpling effects, a good impedance matching (return loss below −10 dB) is achieved.

A graphene-assembled film (GAF)-based compact and low-profile ultra-wide bandwidth (UWB) antenna fed by a CPW structure is presented in [38] (Fig. 4.10a).

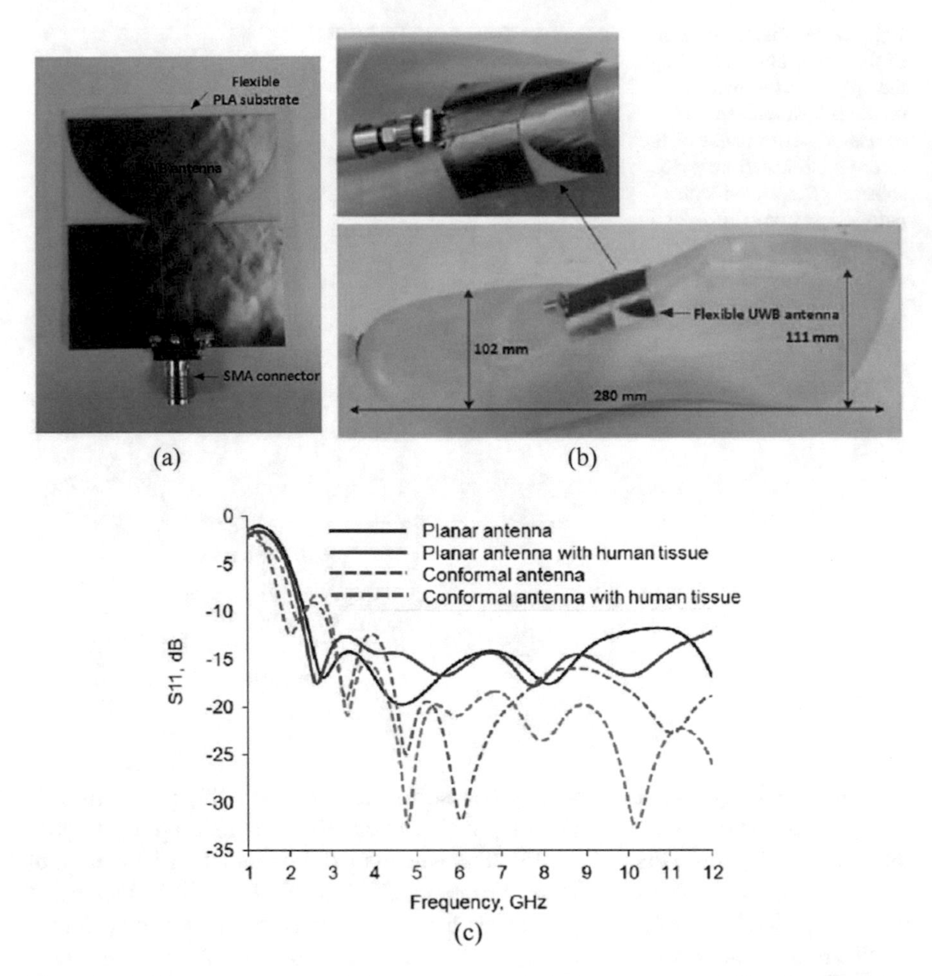

Fig. 4.8 **a** UWB antenna on flexible PLA substrate with the feed connector. **b** The 3D printed foot phantom with the flexible antenna. **c** Simulated reflection coefficient (S_{11}) of the planar and conformal UWB antennas with the human tissue model in free space (Reprinted with permission from IEEE [35])

Graphene-assembled film (GAF) provides high flexibility, light weight, and a conductivity of $\sim 10^6$ S/m, which is comparable to that of the traditional metal materials. The antenna is easy to fabricate with a onc-step laser-direct moulding engraving method. The thickness of the substrate dielectric material is 255 μm and using CPW structure, an antenna with very low profile of only 0.28 mm is reported. Two H-shaped slots are introduced on a coplanar-waveguide (CPW) feeding structure to adjust the current distribution and thus improve the antenna bandwidth. The GAF antenna with dimensions of $32 \times 52 \times 0.28$ mm^3 provides an impedance bandwidth of 60% (4.3–8.0 GHz) as shown in Fig. 4.10b.

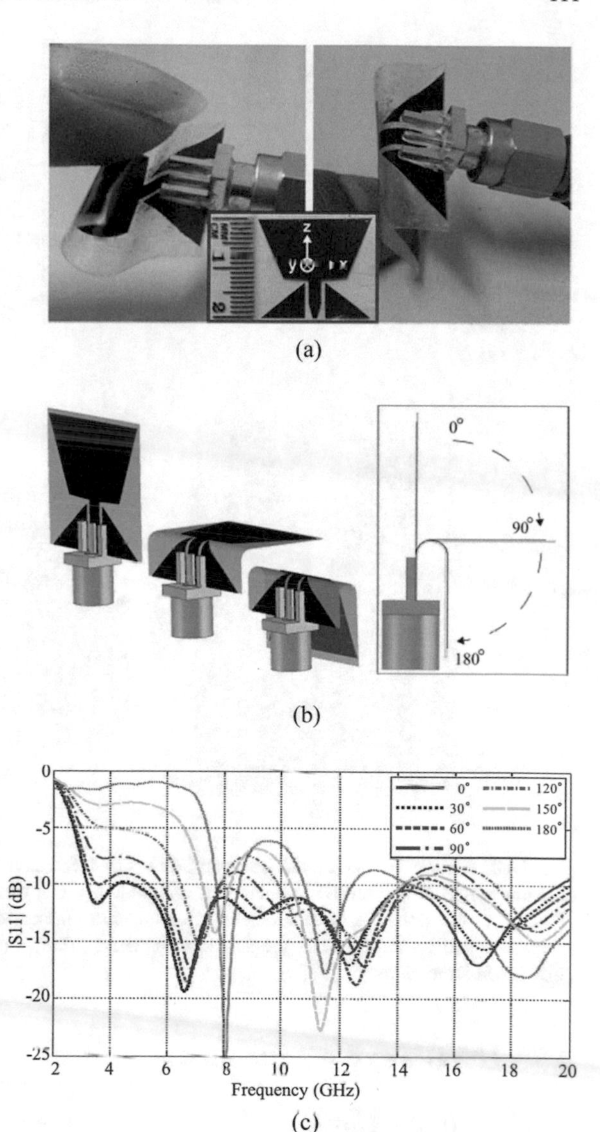

Fig. 4.9 Conductive polymer-based UWB antenna: **a** Antenna prototype front and side view. **b** The antenna and its 3D bending setup in CST Microwave Studio. Inset: Definition of the bending angle (side-view). **c** Simulated S_{11} for different bending angles from 0° to 180° (Reprinted with permission from IEEE [36])

The UWB characteristics are analysed through on-body measurements and show a bending insensitive bandwidth of ~67% (4.1–8.0 GHz), with the maximum gain of 3.9 dBi and 4.1 dBi in its flat state and bent state, respectively (Fig. 4.10c, d). When the antenna is placed on the wrist or integrated with clothes, the −10 dB bandwidth of the antenna widens, and the resonance point shifts left maintaining below −10 dB within the operational range.

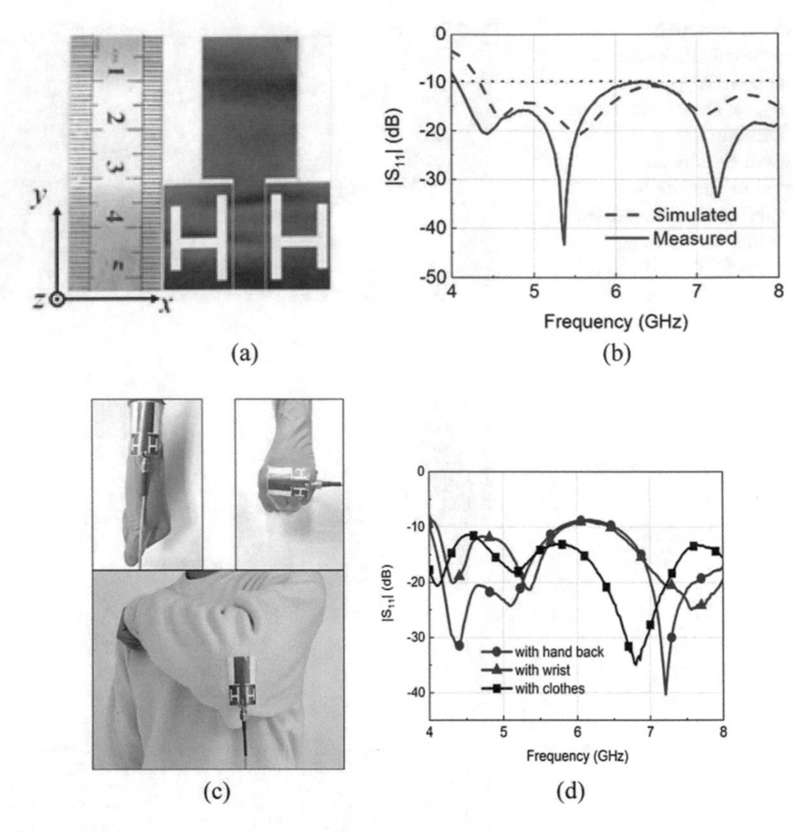

Fig. 4.10 **a** The fabricated GAF antenna prototype. **b** Simulated (dashed blue line) and measured (red line) $|S_{11}|$ curves of the GAF antenna. **c** Antennas under different application scenarios such as attached to the wrist, back of the hand, clipped onto clothes. **d** $|S_{11}|$ curves of the GAF antenna when attached to the back of the hand (red), wrist (blue), and clothes (black) (Reprinted with permission from Sensors [38])

4.5 UWB Textile Antennas

4.5.1 Cotton Cloth

A hexagonal shaped UWB wearable textile antenna with band-notched characteristic which can operate in the WBAN, ISM and UWB band has been presented in [39]. A C-shaped slot is designed to provide band notched characteristics from 2.5 to 3.1 GHz. Military cotton cloth with dielectric permittivity of 2.25, thickness 0.43 mm and loss tangent ($\tan\delta = 0.03$) is chosen as substrate to design the textile antenna. For the metallic components, ShieldIt conductive textile is used from LessEMF Inc. with the estimated conductivity of 1.18×10^5 S/m and thickness 0.17 mm. The antenna was connected to the SMA connector by soldering at maximum temperature of 250 °C to protect the textile conductor. The antenna has dimensions of 58 × 42 × 0.77 mm^3 and operates over a bandwidth from 1.76 to 17.78 GHz with good performance over the whole frequency range.

4.5.2 *Felt*

A wearable low profile textile octagonally shaped UWB antenna (OSUA) is proposed in [40] which is a multi-layered structure with full ground plane. The antenna topology is shown in Fig. 4.11a. It consists of two substrate layers and three metallic layers, which include a top octagonal patch, a parasitic circular monopole patch, and a full ground plane, respectively. The conductive components of the antenna (the patches, parasitic and reflector) are built using ShieldIt conductive textile and the substrate layers are formed using 2 mm thick felt. Felt is a thermally isolating material with an estimated relative permittivity (ε_r) of 1.45 and a loss tangent (tanδ) of 0.044. The overall dimensions of the OSUA are 80 × 61 × 4.51 mm^3.

SAR values are calculated using the three layer (skin–fat–muscle) human body model. The simulations show that the SAR values at the frequencies of 3, 6, and 10 GHz are all below the limit of 2 W/kg averaged over 10 g of tissue. The antenna is placed on various locations of the human subject as shown in Fig. 4.11b. Simulated and measured S_{11} in free space and antenna in flat orientation is presented in Fig. 4.11c. The on-body S_{11} is below -10 dB for the antennas placed on the human arm and chest indicating excellent on-body performance (Fig. 4.11d). High on-body fidelity is obtained for several types of deformations.

A fully textile microstrip antenna topology with felt as substrate and Shieldit Super as conductor (for radiators and ground plane) as shown in Fig. 4.12a is reported in [41]. The probe feeds a small transmission line, which feeds the three primary radiators A, B, and C. Patches B and C are fed in parallel, and their combination is fed in parallel with A. An additional radiator surrounds these primary radiators and is excited solely through capacitive coupling. The S_{11} of the final all-textile antenna with full ground plane (ATA-FGP) topology in free space is given in Fig. 4.12b. The simulated −10 dB bandwidth (BW) is from 3.4 (f_1') to 10.2 (f_2') GHz and the measured −10 dB bandwidth is from 3.6 (f_1) to 10.3 (f_2) GHz. For on-body simulations, a simplified two-third muscle equivalent homogenous body model with relative permittivity is 50.8, and conductivity of 3 S/m is used. On-body simulations were performed with the proposed ATA-FGP centred above the tissue model with an air gap of 10 mm. Experimental on-body evaluations were performed with the ATA-FGP placed on two locations: on the chest and back of two male human volunteers in an anechoic chamber.

A compact UWB flexible antenna with the radiator on top of the overall structure with a full ground plane on its reverse side is reported in [42]. The radiator is based on a microstrip patch combined with multiple miniaturization and broad banding methods, with dimensions of 39 × 42 × 3.34 mm^3. The full ground plane enables

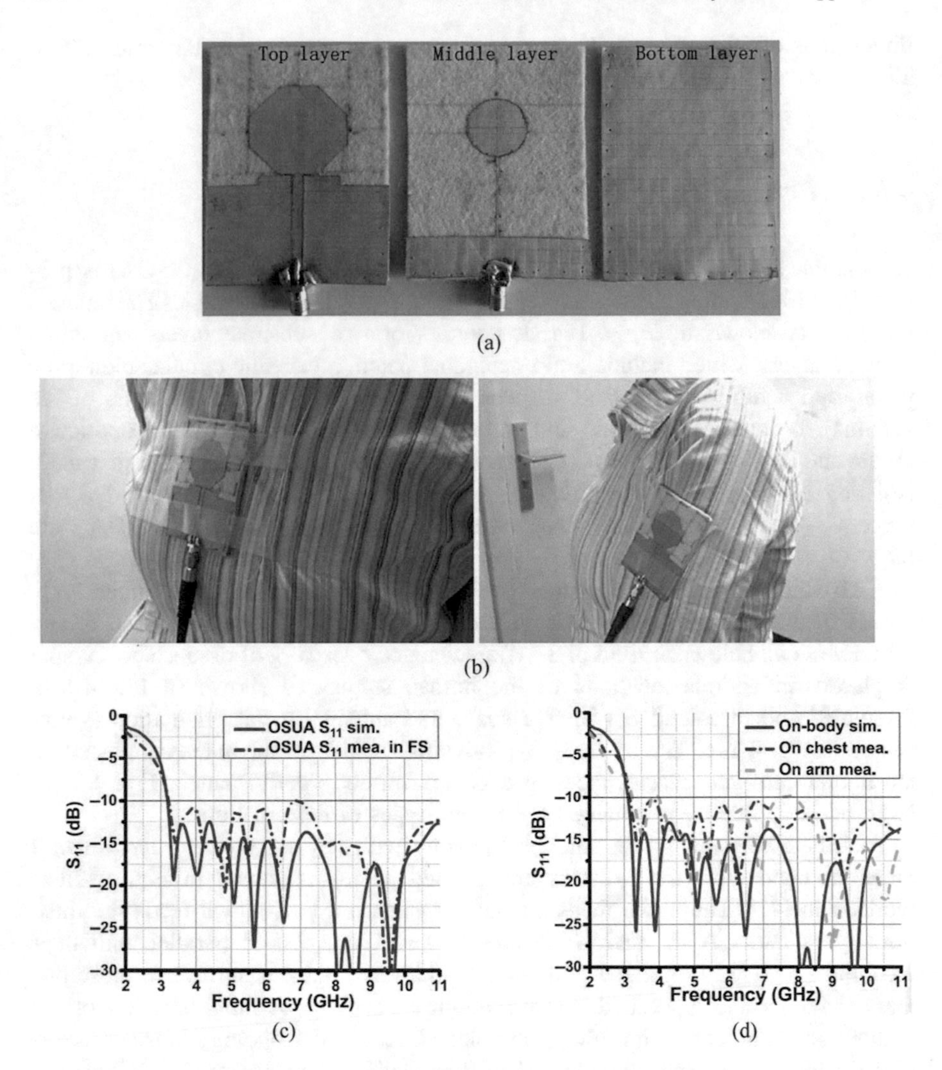

Fig. 4.11 Antenna design **a** From left to right: fabricated top, middle and bottom metallic layers. **b** OSUA on-body measurement setup with the antenna placed on the chest and on the arm, respectively. **c** Simulated and measured S_{11} in free space for the OSUA. **d** Comparison of the on-body simulated and measured S_{11} (Reprinted with permission from IEEE [40])

the safe antenna operation in the vicinity of the human body with minimal body coupling and radiation towards it.

The textile antenna substrate is made using a 3 mm-thick felt textile. Felt is chosen in this work based on its cost effectiveness, ease of accessibility, and ease of fabrication. The relative permittivity (ε_r) of felt is 1.45, and its loss tangent (tanδ) is 0.044. ShieldIt Super from LessEMF Inc. is used to form the ground plane and radiating elements. The middle layer Felt substrate is sandwiched between two conductive ShieldIt layers. The top layer is the radiating layer (with radiators A and B), and the full ground plane layer is at the bottom. To achieve wide bandwidth, multi resonance technique and the parasitic coupling technique between two radiators has been applied.

The antenna is also assessed under two bent conditions (30° and 60°) indicating the ability in preserving the performance when placed on human body. The SAR is evaluated using a detailed human body voxel model in CST at 4 and 8 GHz, with an input power of 0.5 W (rms). The calculated SAR are well under the (FCC regulated limit of 1.6 W/kg averaged over 1 g of tissue, with values of 0.335 and 0.491 W/kg respectively. The free space (FS) simulation of the fidelity factor was performed by placing far field probes at a distance of 10 cm from the antenna and Gaussian excitation signal was used. The maximum computed fidelity is 87% at $\theta = 140°$ in the x–z plane and 71.2% at $\theta = 100°$ in the y–z plane.

4.5.3 (PDMS)-Embedded Conductive-Fabric

A planar UWB antenna designed with suppressed backside radiation, suitable for wearable applications in the 3.7–10.3 GHz band [18] is presented in Fig. 4.13a. The antenna is fabricated via polydimethylsiloxane (PDMS)-embedded conductive-fabric technology, which enables conformity, flexibility, easy realization, robustness, water resistance, and thermal and chemical stability. A nickel–copper–silver-coated nylon ripstop from Marktek Inc., having a thickness of 0.13 mm, was used as the patch layer due to its high conductivity. Alongside, a nickel–copper-coated ripstop from Less EMF Inc., with thickness of 0.08 mm, was used as the ground layer. The latter choice was attributed to its higher porosity than the first fabric, which is beneficial for the PDMS-fabric adherence over the ground plane.

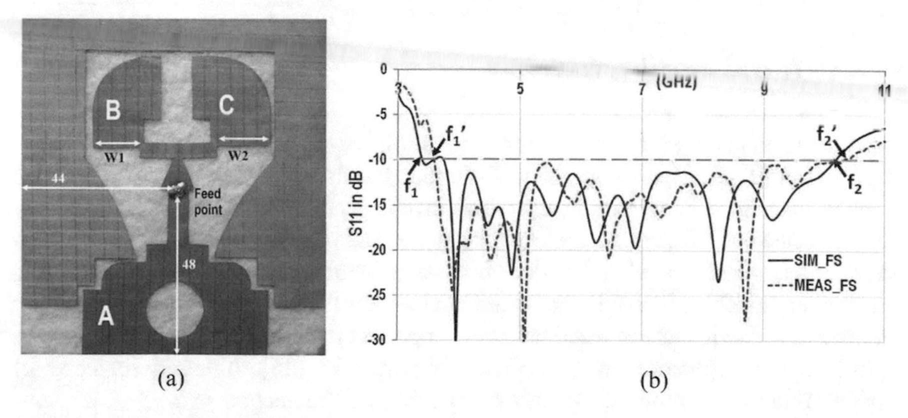

Fig. 4.12 UWB textile antenna fabricated on Felt substrate **a** All-textile antenna with full ground plane (ATA-FGP). **b** Simulated and measured S_{11} in free space (FS) for the prototyped ATA-FGP (Reprinted with permission from IEEE [41])

To achieve an ultra-wide bandwidth for thin patches with full ground plane, several bandwidth enhancement techniques, including the use of multiple resonators and slots, are applied.

Free-space and near human-body environments show good performance in both scenarios (Fig. 4.13b, c). The latter was performed via a flat canonical phantom and an anatomical human head/torso/arm phantom. Due to the isolation provided by the ground plane, the peak SAR results at 5, 7, and 9 GHz were found to be equal to 0.147, 0.174, and 0.09 W/kg, respectively, which conform to the SAR requirement of lower than 2 W/kg. The system fidelity factor (SFF) for the free-space for three sine modulated pulses I, II, and III were 0.88, 0.77, and 0.72, respectively, whereas on the flat phantom, the values obtained were 0.88, 0.79, and 0.68, respectively.

4.5.4 Denim Jean

Denim jean cloth is a strong and durable fabric which is very absorbent and resistant to tearing. The measured reflection coefficients in dry condition over 1.6–12 GHz is reported in [43, 44]. Denim jean with permittivity of 2.3 at 6 GHz, is used as the substrate, while the radiating elements (patch and ground) are composed of ShieldIt super conducting textile. The antenna operates from 1.6-12 GHz. The effect of rainwater and seawater on the permittivity of denim jean substrate and performance of an ultra-wideband (UWB) eye-shaped textile antenna is investigated in [43]. The thicknesses of ShieldIt conductor and denim jean substrate are 0.17 mm and 0.7 mm respectively. The antenna can operate over the entire range in dry conditions but cannot be used when immersed in rainwater and seawater. This is due to the existence of various particles and minerals in the absorbed water, which largely change the permittivity of the substrate.

4.5.5 Novel Textile Materials

Textile antennas based on automated embroidery of conductive E-threads are reported in [2, 3]. These composite E-threads consisted of 7 twisted silver-plated copper filaments. They exhibit excellent mechanical strength and flexibility, low DC resistance of 1.9 Ω/m, and allow for geometrical precision down to 0.1 mm. For fabrication, an automated embroidery process was employed using a programmable embroidery machine. Most E-thread antennas suffer from increased losses at higher frequencies, due to surface roughness and imperfect metallization. An Archimedean spiral antenna operating at 1–6 GHz is reported in [45]. It has a diameter of 160 mm and was "printed" on an organza fabric. The second is a 75 × 85 mm^2 patch antenna that operates from 3 to 11 GHz. The patch and ground planes are both "printed" using conductive textiles and placed on a flexible polydimethylsiloxane (PDMS) substrate.

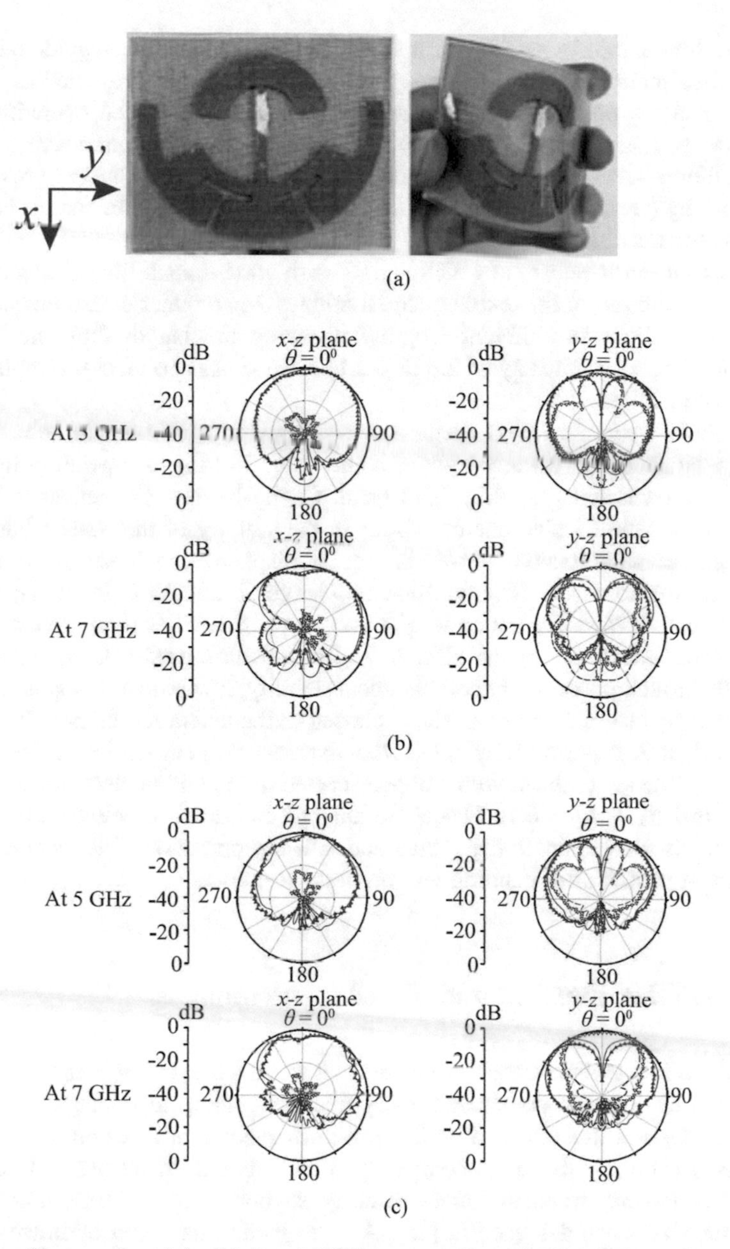

Fig. 4.13 **a** Photographs of the fabricated UWB antenna. Normalized radiation patterns of the antenna at 5 and 7 GHz. **b** In free space. **c** On the flat phantom (Reprinted with permission from IEEE [18])

Two different designs of UWB textile antennas: coplanar waveguide fed printed UWB disc monopole and UWB annular slot antenna are presented in [7]. The textile antennas possess excellent transient characteristics, when operating in free space as well as on the human body. As a conductor, high conductive metallized Nylon fabric—*Nora* is used. Its three metallized layers (Ni/Cu/Ag) provide high conductivity (surface resistivity of 0.03 Ω/square) and protection against corrosion, as well as extreme flexibility. For the textile antennas, acrylic fabric 0.5 mm thick is used as a dielectric substrate which is light with good drapability. The technique to extract permittivity of the textile material utilizes S-parameter measurements of two transmission lines of different lengths. Knowing the length difference and the S-parameters, a permittivity of 2.6 ± 0.1 between 3 and 10 GHz was obtained for the acrylic textile.

A robust wearable UWB textile cavity-backed slot antenna in substrate integrated waveguide (SIW) technology, is presented in [46] for operation in the low duty cycle restricted [3.4–4.8] GHz band. Cavity-backed slot antennas have the potential to exhibit stable characteristics in the vicinity of the human body, since they only transmit/receive power in one hemisphere and these demonstrate an excellent isolation from its environment. The configuration is based on a rectangular shaped SIW cavity which is split into two half-cavities (half-cavity A and B) by a non-resonant rectangular slot. A 3.94-mm-thick closed-cell expanded rubber protective foam, possessing good chemical stability, low moisture regain, and good resistance against oils and solvents, is chosen as the antenna substrate. It exhibits a relative dielectric permittivity $\varepsilon_r = 1.495$ and loss tangent $\tan\delta = 0.016$. The slot and feed plane are realized with a copper-coated nylon taffeta electro-textile, with a surface resistivity $R_S = 0.18$ Ω/sq. The antenna can achieve below −10 dB impedance bandwidth, maintaining robust and efficient operation. The antenna characteristics vary little after bending and on-body scenarios.

4.6 60 GHz Flexible and Textile Antennas

Bulk PDMS is demonstrated to be lossy at millimetre waves, hence membrane-supported devices are considered using micromachining technology. It is shown that transmission lines printed on 20- μm-thick membranes exhibit similar performances as bulk substrates commonly used at millimetre-wave frequencies. A 4 × 2-element microstrip antenna array supported by a 20-μm-thick PDMS membrane has been designed in [13]. A taper section has been optimized to minimize the return loss. The radiating elements are entirely supported by a large PDMS membrane (21.5 × 10.3 mm^2). The antenna is well matched at resonance and the relative shift between the predicted (56.2 GHz) and measured (56.3 GHz) resonance frequencies is less than 0.2%.

A compact end fire 60-GHz tapered slot antenna (TSA), which is printed on a low-permittivity liquid crystal polymer (LCP) substrate is reported in [47]. The antenna features a novel wideband collinear microstrip-to-slot transition, which is of smaller size compared to other transitions. The antenna also features a fork-shaped metallic carrier, which gives it good rigidity. In addition, the antenna features a metallic reflector, which increases its directive gain. The antenna performs well in the 53.3–69.6 GHz frequency range with directive gain varying between 6.8–9.9 dB and 94% radiation efficiency.

A textile Yagi–Uda antenna is fabricated on 0.2 mm thick cotton fabric [48]. The permittivity is extracted using the open stub methodology, and the loss tangent is assessed using two microstrip lines of different lengths. At millimetre waves, the cutting accuracy required for the radiating elements is not achievable using electro textiles, hence flexible copper foil with thickness of 0.07 mm is used for all conductive elements. It is glued to the textile and cut directly on the textile substrate using a laser machine (ProtoLaser S, LPKF, OR) with high accuracy.

Due to the high path loss at 60 GHz, high gain values and end fire radiation patterns are needed. The proposed structure is made of a driven dipole and 10 directors printed on the top layer of the textile substrate (Fig. 4.14a) [48]. The reflection coefficient of the free space and on-body scenario mounted on a skin-equivalent phantom is below −10 dB (Fig. 4.14b). The measured gain is 9.2 dBi and measured antenna efficiency is 78%. The textile antenna will be slightly subject to bending, crumpling, and twisting. The homogeneous phantom is placed and maintained inside a foam ($\varepsilon_r = 1.05$) support (Rohacell 51 HF). In the H-plane, a tilt of 15° is observed due to the reflections from the phantom surface. A maximum gain of 11.9 dBi and 48% efficiency is measured when the antenna is placed on the phantom.

A wearable hybrid textile antenna array with a copper foil has been presented at millimetre waves with good performance in terms of reflection coefficient, gain and efficiency [49]. The textile substrate is characterized in V-band using the open stub technique. Designed for short-range off body communications, it operates over the whole 57–64 GHz range. The measured gain in free space is 8 dBi. It is shown that bending and crumpling has a small impact on the reflection coefficient and antenna gain. The gain drops by 1 dB and the reflection coefficient remains below −10 dB over the 57–64 GHz frequency range.

An inkjet-printed millimetre wave antenna on Polyethylene Terephthalate (PET) substrate is presented in [50]. The proposed antenna is fed by a CPW line in the middle of the ground, where the centre transmission line has a width of 460 μm and a spacing of 40 μm from each lateral ground plane edge. A PET substrate was selected ($\varepsilon_r = 3.4$, $\tan\delta = 0.01$), which offers the unique advantages of being much cheaper, highly flexible, harmless to the human body. The measured impedance bandwidth is less than −10 dB from 60 to 65 GHz.

Design, fabrication, and measurement of a 60 GHz printed antenna with inkjet technology over a flexible substrate which is 125 μm-thick polyethylene naphthalate (PEN) is presented in [51] (Fig. 4.15a). DMP2800 inkjet table printer from DIMATIX, Inc. and the Dimatix 10pL cartridge (DMP 11610) was used for printing purposes. The printer head used a Cabot conductive ink CCI-300, which contains 20% mass of ultra-fine silver nanoparticles. The antenna is a coplanar square monopole with quasi-omnidirectional radiation characteristics due to the thin

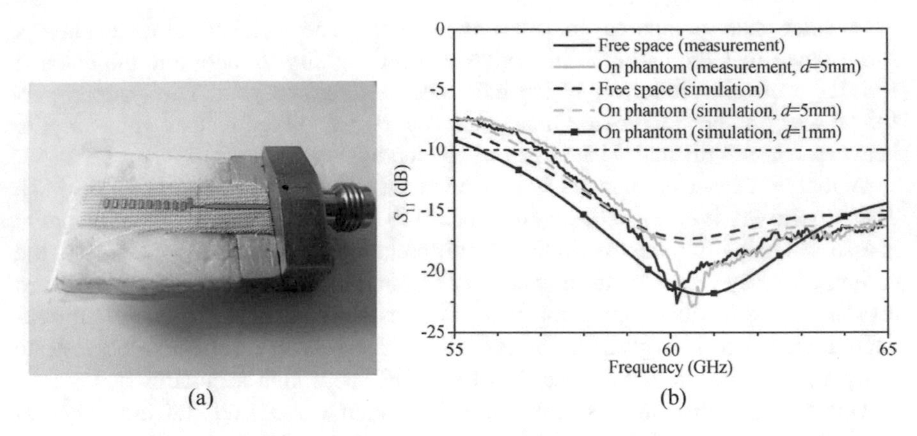

Fig. 4.14 **a** 60 GHz flexible antenna prototype placed on a foam support and fed by a V-connector for measurement purpose. **b** Computed and measured reflection coefficients of the textile antenna (Reprinted with permission from IEEE [48])

and low relative permittivity. Total efficiency of 68% and 1.86 dBi maximum realized gain were measured at 60 GHz. Environmental-friendly flexible materials such as polypropylene, which has lower losses than PEN, and aluminium inks could be used to develop prototypes for "green electronics."

The design of a low-profile antenna is presented for body-centric applications operating in the mmWave frequency band cantered at 60 GHz in [52] (Fig. 4.15b). The antenna has an overall size of $14 \times 10.5 \times 1.15$ mm^3 and is printed on a flexible printed circuit board. The S_{11} and radiation pattern results for free space and on-body scenarios are measured. The antenna has a bandwidth of 9.8 GHz and offers a gain of 10.6 dBi and 12.1 dBi in free space and on-body respectively. It also achieves an efficiency of 74% in free-space and 63% in on-body scenario. The antenna consists of two layers of flexible printed circuit boards (FPCB), with relative permittivity of 2.7, $\tan\delta = 0.005$ and thickness of the FPCB is 0.15 mm fed through a microstrip feed line. The radiating element is the combination of a rectangular loop and two U-shaped patches fed through a short microstrip feed line. The antenna has also successfully established a good wireless link for on–off-body LOS, on–off-body NLOS, and body-to-body communication scenarios with a path gain of −43, −55, and −52 dB, respectively.

Path loss analysis in free space and under skin-equivalent phantom of horn-to-horn and textile-to-textile communications is presented in [53]. The path loss differs considerably in the analysed situations; with horn-to-horn links showing a path loss exponent between 1.13 and 2.35, while it is reaching 4.7 for textile-to-textile links. The obtained results provide additional input for the design of communications links in the context of future 5G and IoT eco-systems. The antenna structure is made of a driven dipole and 10 directors printed on the top layer of the textile substrate. The available bandwidth under −10 dB is 2 GHz around 58.86 GHz, with a gain of 10 dBi at the center frequency.

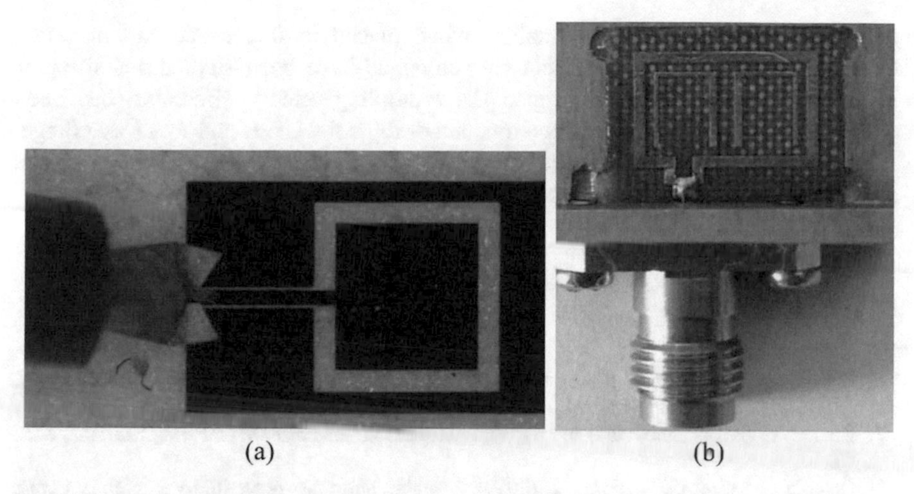

(a) (b)

Fig. 4.15 **a** Photograph of the monopole fed by a 400- μm GSG probe (Reprinted with permission from IEEE [51]). **b** Fabricated prototype of the proposed 60 GHz antenna fed by a V-connector (Reprinted with permission from IEEE [52])

A textile multiple input multiple output (MIMO) Yagi-Uda antenna at Millimetre Wave band (mmWave) has been designed for on-body communication in the 57–64 frequency range [54]. The antenna shows a wide bandwidth of 8.4 GHz with good efficiency and gain at the mmWave frequency band. Good results indicate the proposed antenna's great potential for current and future IoT applications and also in communication systems.

4.7 Conclusion

Flexible antennas can be used for a wide variety of applications and are most suited for body-centric communication due the conformal nature of the antennas which provide more flexibility and freedom to the user to perform daily activities. UWB antennas and mmWave have received much attention in many different fields for applications related to healthcare, smart home and living, sports, defence, and security purposes. Flexible antennas have been successfully applied to communication services such as wireless local area network and wireless personal area networks with the human subject being an integral part of the network. Various strategies have been elaborated to develop and use different kinds of materials for substrates and the conductive region of the antennas to realise flexible and textile printed antennas. Different fabrication methodologies have also been applied to find more simple and cost-effective solutions for flexible antenna fabrications without compromising on the performance of the antennas. A wide variety of designs have been elaborated and presented with details related to the electromagnetic performance and mechanical stability when placed in free space and on-body. Conformal nature of the flexible/textile antennas have been studied and the performance under various bending angle, crumpling effects, stretching has been evaluated for various materials and antenna designs for UWB and mmWave range. Environmental factors such as presence of rain or sea water and the influence on the antenna characteristics have also been considered.

References

1. Kirtania SG, Elger AW, Hasan MR, Wisniewska A, Sekhar K, Karacolak T, Sekhar PK (2020) Flexible antennas: a review. Micromachines 11:847
2. Khaleel H (2014) Innovation in wearable and flexible antennas. WIT Press, Billerica, MA, USA
3. Corchia L, Monti G, Tarricone L (2019) Wearable antennas: nontextile versus fully textile solutions. IEEE Antennas Propag Mag 61(2):71–83
4. Kiourti A, Lee C, Volakis JL (2016) Fabrication of textile antennas and circuits with 0.1 mm precision. IEEE Antennas Wirel Propag Lett 15:151–153
5. Castano L, Flatau AB (2014) Smart fabric sensors and e-textile technologies: a review. IOP Publishing Smart Mater Struct 23(5):1–27
6. Ramos-Silva JN, Ramírez-García E, Alcántara-Gavilan BA, Rodríguez-Méndez LM, Enciso-Aguilar MA (2019) Design of a compact ultra wide band flexible antenna for personal mobile communications. In: 2019 IEEE international fall meeting on communications and computing (ROC & C), Acapulco, Mexico, 2019, pp 13–17
7. Klemm M, Troester G (2006) Textile UWB antennas for wireless body area networks. IEEE Trans Antennas Propag 54(11):3192–3197
8. Khaleel HR, Al-Rizzo HM, Rucker DG, Mohan S (2012) A compact polyimide-based UWB antenna for flexible electronics. IEEE Antennas Wirel Propag Lett 11:564–567
9. Catherwood PA, Bukhari SS, Watt G, Whittow WG, McLaughlin J (2018) Body-centric wireless hospital patient monitoring networks using body-contoured flexible antennas. IET Microw Antennas Propag 12(2):203–210
10. Yadav A, Singh VK, Bhoi AK, Marques G, Garcia-Zapirain B, de la Torre Díez I (2020) Wireless body area networks: UWB wearable textile antenna for telemedicine and mobile health systems. Micromachines 11:558
11. Holland SA, Baiya D, Elkhouly E, Fathy AE (2013) Ultra wideband textile antenna development for indoor localization. In: IEEE MTT-S international microwave symposium digest (MTT), Seattle, WA, 2013, pp 1–4
12. Aun NFM, Soh PJ, Al-Hadi AA, Jamlos MF, Vandenbosch GAE, Schreurs D (2017) Revolutionizing wearables for 5G: 5G technologies: recent developments and future perspectives for wearable devices and antennas. IEEE Microw Mag 18(3):108–124
13. Hage-Ali S et al (2009) A millimeter-wave microstrip antenna array on ultra-flexible micromachined polydimethylsiloxane (PDMS) polymer. IEEE Antennas Wirel Propag Lett 8:1306–1309
14. Wang JC, Leach M, Wang Z, Lim EG, Ma KL, Huang Y (2015) State-of-the-art of 60 GHz antennas in wireless body area network. In: 2015 international SoC design conference (ISOCC), Gyungju, 2015, pp 171–172
15. Hettak K, Petosa A, James R (2014) Flexible plastic-based inkjet printed CPW fed dipole antenna for 60 GHz ISM applications. In: IEEE antennas and propagation society international symposium (APSURSI), Memphis, TN, 2014, pp 328–329

16. Pourahmadazar J, Denidni TA (2017) 60 GHz antenna array for millimeter-wave wireless sensor devices using silver nanoparticles ink mounted on a flexible polymer substrate. Microw Opt Technol Lett 59(11):2830–2835
17. Yan S, Poffelie LAY, Soh PJ, Zheng X, Vandenbosch GAE (2016) On-body performance of wearable UWB textile antenna with full ground plane. In: 2016 10th european conference on antennas and propagation (EuCAP), Davos, 2016, pp 1–4
18. Simorangkir RBVB, Kiourti A, Esselle KP (2018) UWB wearable antenna with a full ground plane based on PDMS-embedded conductive fabric. IEEE Antennas Wirel Propag Lett 17 (3):493–496
19. Cicchetti R, Miozzi E, Testa O (2017) Wideband and UWB antennas for wireless applications: a comprehensive review. Int J Antennas Propag
20. Mahmood SN, Ishak AJ, Saeidi T, Alsariera H, Alani S, Ismail A, Che Soh A (2020) Recent advances in wearable antenna technologies: a review. Prog Electromagnet Res B 89:1–27
21. Amadjikpe L, Vera A, Choudhury D, Papapolymerou J (2008) Study of a 60 GHz rectangular patch antenna on a flexible LCP substrate for mobile applications. In: 2008 IEEE antennas and propagation society international symposium, San Diego, CA, 2008, pp 1-4
22. Khaleel HR (2014) Design and fabrication of compact inkjet printed antennas for integration within flexible and wearable electronics. IEEE Trans Compon Packag Manuf Technol 4 (10):1722–1728
23. Khaleel HR, Al-Rizzo HM, Abbosh AI (2013) Design fabrication and testing of flexible antennas in advancement in microstrip antennas with recent applications, Austria, Vienna: InTech
24. Dimatix Inkjet Printer [Online]. http://www.dimatix.com, accessed 2020
25. Tsolis A, Whittow WG, Alexandridis AA, Vardaxoglou JC (2014) Embroidery and related manufacturing techniques for wearable antennas: Challenges and opportunities. Electronics 3:314–338
26. Varma S, Sharma S, John M, Bharadwaj R, Dhawan A, Koul SK (2021) Design and performance analysis of compact wearable textile antennas for IoT and body-centric communication applications. Int J Antennas Propag
27. Li W, Hei Y, Grubb PM, Shi X, Chen RT (2018) Compact inkjet-printed flexible MIMO antenna for UWB applications. IEEE Access 6:50290–50298
28. Yu Y, Jin P, Ding K, Zhang M (2015) A flexible UWB CPW-fed antenna on liquid crystal polymer substrate. In: Asia-Pacific microwave conference (APMC), 2015, pp 1–3
29. Abbasi QH, Rehman MU, Yang X, Alomainy A, Qaraqe K, Serpedin E (2013) Ultrawideband band-notched flexible antenna for wearable applications. IEEE Antennas Wirel Propag Lett 12:1606–1609
30. Qiu Y et al (2014) Compact parylene c-coated flexible antenna for WLAN and upper-band UWB applications. Electron Lett 50(24):1782–1784
31. Mohamadzade B, Simorangkir RBVB, Hashmi RM, Lalbakhsh A (2020) A conformal ultrawideband antenna with monopole-like radiation patterns. IEEE Trans Antennas Propag 68(8):6383–6388
32. Xu LJ, Wang H, Chang Y, Bo YA (2017) Flexible UWB inverted-F antenna for wearable application. Microw Opt Technol Lett 59:2514–2518
33. Shaker G, Safavi-Naeini S, Sangary N, Tentzeris MM (2011) Inkjet printing of ultrawideband (UWB) antennas on paper-based substrates. IEEE Antennas Wirel Propag Lett 10:111–114
34. Krupka J, Derzakowski K, Abramowicz A, Riddle B, Baker-Jarvis J, Clarke RN, Rochard OC (2002) Bounds on permittivity calculations using the $TE01_{\delta}$ dielectric resonator. In: Proceedings 14th international conference microwave radar wireless communication, 2002, pp 394–396
35. Jun S, Sanz-Izquierdo B, Summerfield M (2015) UWB antenna on 3D printed flexible substrate and foot phantom. In: Loughborough antennas & propagation conference (LAPC), Loughborough, 2015, pp 1–5
36. Chen SJ, Chivers B, Shepherd R, Fumeaux C (2015) Bending impact on a flexible ultra-wideband conductive polymer antenna. In: International conference on electromagnetics in advanced applications (ICEAA), Turin, 2015, pp 422–425

37. Hamouda Z, Wojkiewicz J, Pud AA, Kone L, Bergheul S, Lasri T (2018) Flexible UWB organic antenna for wearable technologies application. IET Microw Antennas Propag 12 (2):60–166
38. Fang R, Song R, Zhao X, Wang Z, Qian W, He D (2020) Compact and low-profile UWB antenna based on graphene-assembled films for wearable applications. Sensors 20:2552
39. Sopa P, Rakluea P (2020) The hexagonal shaped UWB wearable textile antenna with band-notched characteristics. In: 8th international electrical engineering congress (iEECON), Chiang Mai, Thailand, 2020, pp 1–4
40. Yimdjo Poffelie LA, Soh PJ, Yan S, Vandenbosch GAE (2016) A high-fidelity all-textile UWB antenna with low back radiation for off-body WBAN applications. IEEE Trans Antennas Propag 64(2):757–760
41. Samal PB, Soh PJ, Vandenbosch GAE (2014) UWB all-textile antenna with full ground plane for off-body WBAN communications. IEEE Trans Antennas Propag 62(1):102–108
42. Samal PB, Soh PJ, Zakaria Z (2019) Compact microstrip-based textile antenna for 802156 WBAN-UWB with full ground plane. Int J Antennas Propag
43. Shakhirul MS, Jusoh M, Sahadah A, Nor CM, Rahim HA (2014) Embroidered wearable textile antenna on bending and wet performances for uwb reception. Microw Opt Technol Lett 56(9):2158–2163
44. Yahya R, Kamarudin MR, Seman N (2014) Effect of rainwater and seawater on the permittivity of denim jean substrate and performance of UWB eye-shaped antenna. IEEE Antennas Wirel Propag Lett 13:806–809
45. Kiourti A, Volakis JL, Simorangkir RBVB, Abbas SM, Esselle KP (2016) UWB antennas on conductive textiles. In: IEEE international symposium on antennas and propagation (APSURSI), Fajardo, 2016, pp 1941–1942
46. Lemey S, Rogier H (2014) SIW textile antennas as a novel technology for UWB RFID tags. In: IEEE RFID technology and applications conference (RFID-TA), Tampere, 2014, pp 256–260
47. Pazin L, Leviatan Y (2010) A compact 60-GHz tapered slot antenna printed on LCP substrate for WPAN applications. IEEE Antennas Wirel Propag Lett 9:272–275
48. Chahat N, Zhadobov M, Le Coq L, Sauleau R (2012) Wearable endfire textile antenna for on-body communications at 60 GHz. IEEE Antennas Wirel Propag Lett 11:799–802
49. Chahat N, Zhadobov M, Muhammad SA, Le Coq L, Sauleau R (2013) 60-GHz textile antenna array for body-centric communications. IEEE Trans Antennas Propag 61(4):1816–1824
50. Hettak K, Ross T, James R, Momciu A, Wight J (2013) Flexible polyethylene terephthalate-based inkjet printed CPW-fed monopole antenna for 60 GHz ISM applications. In: 2013 european microwave integrated circuit conference, Nuremberg, 2013, pp 476–479
51. Bisognin et al (2014) Inkjet coplanar square monopole on flexible substrate for 60-GHz applications. IEEE Antennas Wirel Propag Lett 13:435–438
52. Ur-Rehman M, Malik NA, Yang X, Abbasi QH, Zhang Z, Zhao N (2017) A low profile antenna for millimeter-wave body-centric applications. IEEE Trans Antennas Propag 65 (12):6329–6337
53. Ghandi M, Tanghe E, Joseph W, Benjillali M, Guennoun Z (2016) Path loss characterization of horn-to-horn and textile-to-textile on-body mmWave channels at 60 GHz. In: International conference on wireless networks and mobile communications (WINCOM), Fez, 2016, pp 235–239
54. Abu Shaweesh YI, Himat ZFM (2017) Textile multiple-input multiple-output (MIMO) antennas at the millimeter-wave band (mmW). In: 4th IEEE international conference on engineering technologies and applied sciences (ICETAS), Salmabad, 2017, pp 1–4

Chapter 5
Implantable Antennas for WBANs

5.1 Introduction

With the increase in the elderly population, lifestyle changes and the growing demands for next generation medical and healthcare system, body-centric wireless body area network (WBAN) has become a promising solution which enables the real-time physiological data collection and transfer, continuous monitoring, remote telemetry, etc. [1, 2]. The ability to establish a link between outside and inside the human body will increase the opportunity to detect diseases or abnormalities suffered by patients [3]. Wearable smart devices and sensors in WBANs are generally limited to monitoring specific types of physiological parameters that are accessible from outside the human body. Recently, research and development are being carried out in biomedical implantable devices which have paved a path to a new realm of applications such as transplanted organ monitoring, deep brain neural recording system, real-time capsule endoscope and localization system etc. These require high data rates up to tens of Mbit/s to obtain high quality of system communication [3–5].

For implantable communication, UWB techniques are becoming more promising for novel implant applications due to several benefits over narrow band systems such as high transmission speed, low power consumption, highly integrated systems with easy miniaturized antenna [6–8]. UWB has advantages like large bandwidth, low profile, high data rate, highly integrated systems featuring smaller antenna size, low power consumption and so on which make it a potential candidate to be used for in-body communication [8]. The lower part of the UWB frequency range is the best choice to overcome the problem in which the attenuation through the human body tissues increases dramatically as the frequency increases [9, 10].

In this chapter in-body and on-body channel characteristics of an implantable UWB system are presented giving insights into the propagation aspects of the wireless links. The path loss and shadow fading models are presented and analysed based on the numerical techniques using human and animal anatomical models, phantom-based measurements and in-vivo living animal experiments. Then the

S. K. Koul and R. Bharadwaj, *Wearable Antennas and Body Centric Communication*, Lecture Notes in Electrical Engineering 787,
https://doi.org/10.1007/978-981-16-3973-9_5

path-loss models for in-body Tx (e.g., brain, heart, small intestine, inside the arm) to on-body Rx are developed to provide useful insight in the physical layer design. The analysis of various channel parameters and specific absorption ratio (SAR) and the effects of the human body will facilitate the development of robust implantable systems suitable for various biomedical applications in the healthcare domain.

5.2 UWB Implantable Antennas

5.2.1 Antenna Design Considerations

On-body and in-body antenna design face various design challenges to ensure a proper communication link between them. To perform a reliable UWB in-body measurement campaign, the antennas play a crucial role. They should have a compact structure to be implanted inside the human body in different locations as well as to be located over the body surface. Besides, an omnidirectional radiation pattern to communicate with a sensor array located around or inside the body should be achieved. The size of the on-body antenna needs to be small enough to be suitable to be placed on the human body. The antenna matching, achievement of the desired bandwidth are important aspects that need to be taken into consideration. The propagation of radio waves through the human body tissues is found as a challenge for such antenna [11].

Human body is a combination of different tissues with different electrical properties that are highly frequency dependent. Due to direct relationship between the implanted device and body tissues, the design of implanted system is critical [11, 12]. The design of the implanted antenna requires deep knowledge of the electrical properties of the human tissues. To design a reliable antenna for wireless capsule endoscopic (WCE), electrical properties of the tissues must be taken into account at the time of antenna design. An antenna surrounded by biological tissues in its near-field acts as a new effective antenna with new propagation behavior and return loss which is different from the actual antenna. As a result, an antenna designed for one part of the body (i.e., designed according to the dielectric properties of that part of the body) might not operate as expected in another part of the body [13]. Various antenna designs and state-of-the art techniques are presented below for various healthcare related applications.

5.2.2 Antenna Design Examples for Various Applications

Planar antennas are presenting satisfactory characteristics such as low-profile, low-cost, ease of manufacturing and conformability. These antennas are strong candidates for capsule endoscopy, neural recording where their dimensions become manageable for most implanted medical applications. More recently, antenna designers have been opting for the 3–5 GHz frequency range for capsule localization and implantable communication purposes. This frequency selection is supported by the achievement of high image resolution and low path losses.

Compact planar elliptical ring implanted in-body antennas working at the lower part of UWB frequency band is proposed in [14]. The antennas are compact and can be placed in a wireless capsule which can be easily swallowed by the patients. Rogers TMM10 substrate with 75 mil thickness is used for fabrication of the antennas having radius of 5.85 mm with the fabricated antenna shown in Fig. 5.1a. The simulation and measurement of S_{11} results are presented in Fig. 5.1b inside the phantom with reflection coefficients results below—10 dB for the whole required frequency band (3.1–5.1 GHz). For the simulation and the measurements, the electrical properties of the phantom are similar with human muscle tissue with relative permittivity ε_r of 52.2 and conductivity of 3.3 S/m.

An on-body antenna is designed to measure the in-body to on-body transmission is shown in Fig. 5.1c [14]. The antenna is designed on RO3203 substrate with the relative permittivity of 3.02 and dimensions of 40 × 40 mm^2. The top semi-circular

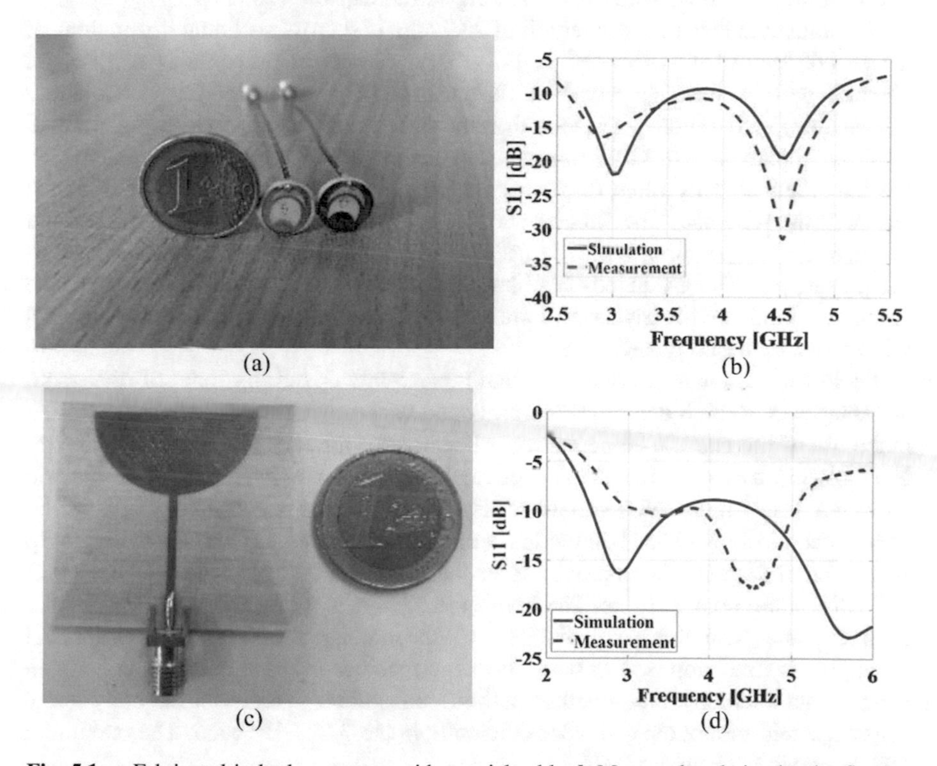

Fig. 5.1 **a** Fabricated in-body antennas with coaxial cable. **b** Measured and simulated reflection coefficient results of in-body antenna inside the muscle phantom. **c** Fabricated on-body antennas. **d** Measured and simulated reflection coefficient results of on-body antenna on the muscle phantom (Reprinted with permission from IEEE [14])

part is the radiating element of the antenna which has a wideband characteristic. The trapezoidal shape is used to tune the input impedance to 50 Ω. The fabricated antenna is well matched ($S_{11} < -10$ dB) within the lower part of the UWB band with some deviation between the measured and simulated results due to the thickness of the container of the phantom as shown in Fig. 5.1d.

5.2.2.1 Implantable Antennas for Wireless Capsule Endoscopy

Wireless capsule endoscopy (WCE) is a medical device used to detect diseases and capture images in the gastrointestinal (GI) tract without causing any painful experience to the patient. It cannot be controlled from the outside in terms of retracing the path and localizing the WCE. To overcome such limitations research on antenna performance to detect the location of the WCE and to have communication inside the body have been carried out and reported in [15]. A UWB antenna can be used for capsule endoscopic for the possibilities of high data rate transmission, allowing streaming of transmission of high-quality images or videos, and enabling the localization and tracking of the capsule [16, 17].

A compact antenna having length of 23.7 mm and cross-sectional dimensions of a size 000 capsule is proposed in [18]. The antenna in principle is a wide-slot antenna with a fork shaped feed. It consists of a grounded-CPW (Coplanar Waveguide), printed on a TMMi substrate with the dielectric constant of 9.8 and 1.27 mm thickness. The CPW has two grounds on both sides of the substrate which are interconnected by three 0.75 mm vias, symmetrically placed 2 mm from the centre of the feed pad. The circular pad at the feed end of the CPW is a broadband probe-to-CPW transition, which is needed for SMA connector feeding of the prototype antenna. The grounded-CPW has the characteristic impedance of 50 Ω. The antenna is loaded with Glycerine solution to achieve impedance matching across a wide range of tissue types.

Dielectric resonance antenna (DRA) is characterized by one of the main advantages of its high degree of flexibility and versatility. Hemispherical DRAs can make use of the capsule dome volume allocated for the antenna. A conformal strip is employed to excite the hemispherical DRA which allows the whole electric current to flow on the DRA surface leading to better energy coupling [19]. In order to broaden the impedance bandwidth, a conformal tapered copper conducting strip with 10 mm diameter is employed to excite the DRA and connected with a 50 Ω RG405 feeding coaxial cable. The hemisphere DRA is mounted on a finite circular ground plane with the same diameter as the hemisphere. The UWB trapezoid monopole is first proposed to be used as the receiver antenna outside the body to form a data telemetry link together with the transmitter antenna in the body torso. This trapezoid monopole can work efficiently in the 3–5 GHz band. The maximum gain reported is around 5.8 dB.

Morimoto et al. [20] proposed two implantable UWB (3.4–4.8 GHz band) hemispherical capsule-conformal antennas: helical and loop. The hemispherical antennas match the shape of capsule endoscopes. A simplified planar design of the hemispherical loop antenna was further proposed for easy of fabrication. Figure 5.2a shows the design of the planar loop antenna. The microstrip line with a width of 1 mm is built on the substrate with a relative permittivity of 4 and a thickness of 1.6 mm. The loop has a radius of 4 mm. The simulated S_{11} results performed in a tissue box model was compared with the measured result in a fluid phantom which is presented in Fig. 5.2 (b). The simulated directivity is presented in Fig. 5.2c.

To receive the transmitted signal from the implanted antenna, a planar unbalanced dipole on-body receiving antenna is designed [20]. The antenna is built on the substance with a relative permittivity of 4 and a thickness of 1.6 mm as presented in Fig. 5.2d. This antenna has two different shaped elements to widen the bandwidth. The S_{11} performance is measured with the antenna placed near the fluid phantom as shown in Fig 5.2e. The spacing between the antenna and the fluid phantom is 1 cm. As shown in Fig. 5.2e, the antenna exhibits a S_{11} performance smaller than −9.5 dB in the entire UWB low-band.

A wide-slot antenna design having dimensions (11.85 × 9 × 1.27 mm^3) with a U-shaped feed fabricated on Rogers TMM10i high frequency material is proposed for WCE applications [21]. It forms a magnetic dipole, which is less susceptible to variations in the near-field propagation environment. The antenna operates at 4 GHz center frequency with a bandwidth of approximately 1 GHz. Using an insulating material between the dispersive tissue medium and the radiating element of the antenna improves the impedance-matching characteristics of the implantable antenna [22, 23]. The radiating element of the antenna used in the simulations, which occupies the lower half of the antenna, is inserted in glycerine for this purpose as shown in Fig. 5.3c along with the simulated human model (Fig. 5.3a, b). The fabricated antenna is presented in Fig. 5.3d. Dimensions of the UWB antenna top side and bottom side are presented in Fig. 5.3e, f respectively. Glycerine has a relative permittivity of 50, which is close to the relative permittivity of the surrounding tissue material, and hence, allows minimal reflections of the electromagnetic wave near the transitional boundaries between the tissue medium and the capsule.

An antenna at UWB frequency with very small form factor is proposed in [24] with overall size of 8.4 × 6 × 1.036 mm^3 that could be fitted in the commercial WCEs that have proper sizing as swallowable pills. The antenna is a single metallic layer and printed on one side of Rogers RT Duroid 6100 (TM) substrate with dielectric constant of ε_r = 10.2 and thickness of 0.8 mm. Copper of 0.036 mm thickness has been used as a metallic layer.

An ultra-wideband conformal capsule slot antenna, which has a simple configuration and a stable impedance matching characteristic, is described in [25]. The antenna is first printed on a 10 mil-thick single-sided Rogers 5880 substrate, which is flexible, and then wrapped and inserted into the capsule (Fig. 5.4a). The antenna provides a wide impedance bandwidth ranging from 1.64 to 5.95 GHz (113.6%). A typical application scenario a capsule will move through the digestive system, and thus experience varying environments, hence, a wide bandwidth and stable performance are both desirable attributes of the antenna. The inner patch is connected to the

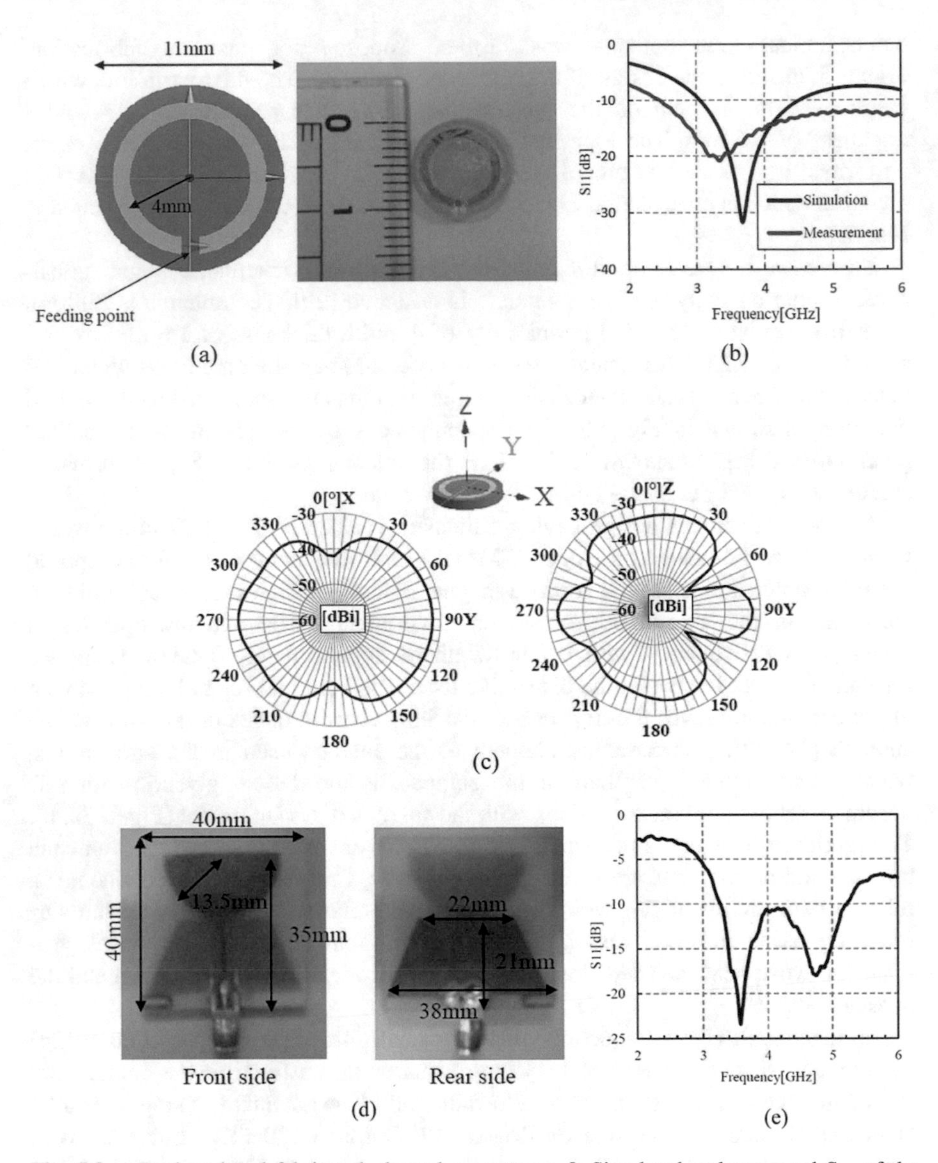

Fig. 5.2 **a** Designed and fabricated planar loop antenna. **b** Simulated and measured S_{11} of the planar loop antenna. **c** Simulated directivity of the planar loop antenna at 4.1 GHz in the tissue box model. **d** Fabricated planar unbalanced dipole. **e** Measured S_{11} of the planar unbalance dipole when it is near the fluid phantom (Reprinted with permission from IEEE [20])

center core of a coax through a via and the outer patch is connected to the metallic shield of the coax through another via. The capsule is placed at the centre of a muscle phantom simulated using ANSYS/HFSS. The size of the small phantom is $60 \times 60 \times 70$ mm^3 and that of the large phantom is $100 \times 100 \times 110$ mm^3.

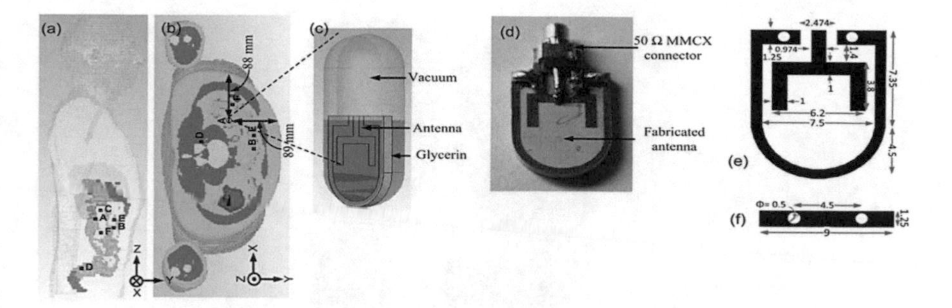

Fig. 5.3 Simulated WCE device positions. **a** Side view. **b** Top view. **c** UWB antenna model in glycerine insertion. **d** Dimensions of the fabricated UWB antenna. **e** Top side. **f** Bottom side (Reprinted with permission from IEEE [21])

Figure 5.4b shows that the impedance matching of the antenna remains stable when the phantom size is changed. The curve "constant" refers to the case where the small phantom is used, and the material properties of the phantom are set to those at 2.4 GHz (frequency independent). And the curve "constant flat" refers to the case for which the "constant" case antenna is changed to flat based [25]. The fabricated prototype of antenna is presented in Fig. 5.4c.

5.2.2.2 Implantable Antennas for Wireless Brain-Machine-Interface

Wireless Brain-Machine-Interface (B-MI) is a technology that allows communication between brain and a machine outside a body by mean of wireless connectivity, where neural activities of brain will transform into action [26]. This rapidly growing field has potential to provide handling for patients suffering from neurological injuries and diseases. The B-MI should be designed for short-range applications with relatively low power with possible frequency band candidate such as ultra-wideband (UWB) with frequency band of 3.1–10.6 GHz. Neural activity monitoring in the brain requires high data rate (800 kb/s per neural sensor), and should support large number of sensors.

The work presented in [28] focuses on the Tx and Rx antenna design for a reliable wireless link for neural recording system. A 12 × 12 mm^2 monopole antenna on FR4 PCB board of thickness 0.8 mm and dielectric constant 4.4 is used to design the antenna. 50 Ω sub-miniature (SMA) connector was used to connect the feeding strip; its outer side was grounded to the ground plane. A 27 × 30 mm^2 monopole receiver antenna was fabricated which acts as the Rx. The purpose of the work proposed in [27] is the design of an ultra-wideband antenna to transmit medical data from an implant inside the head of a patient to a nearby receiver. To protect the antenna and the circuits from moisture, casing made of aluminium oxide is necessary and must be considered in the design process. For the design,

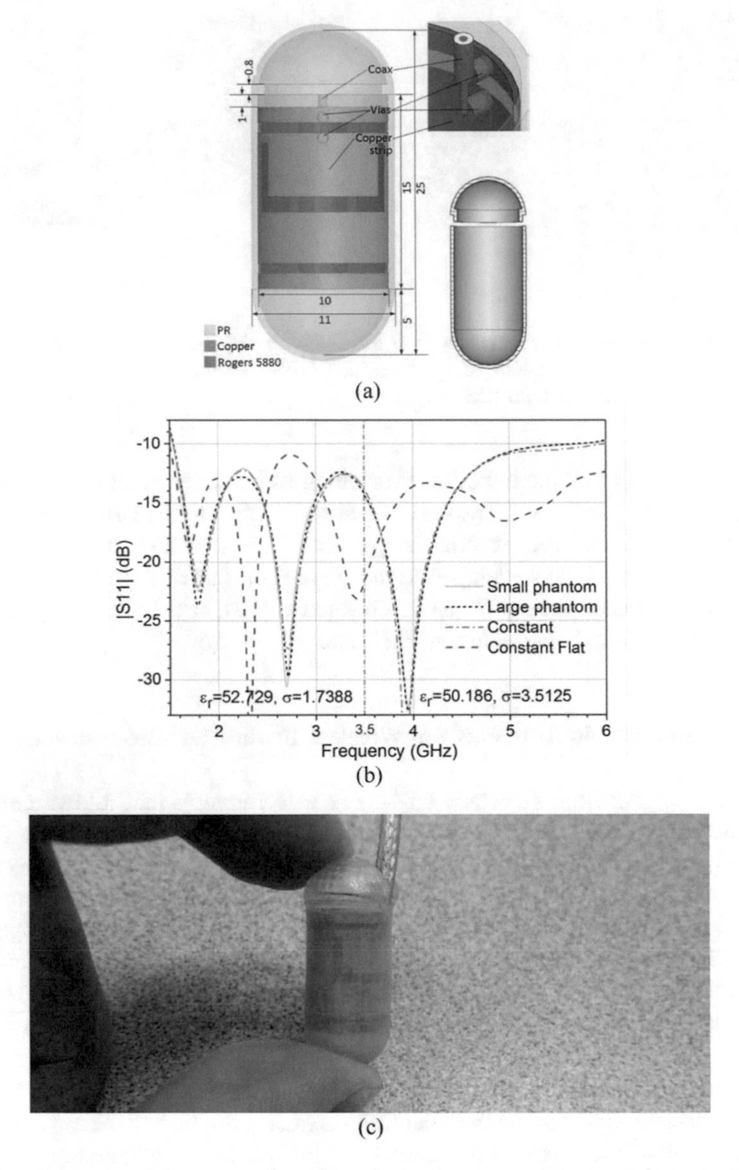

Fig. 5.4 **a** Geometry of the proposed conformal capsule antenna. The top right inset shows the feeding structure. The bottom right inset shows the cross section of the capsule shell. **b** Effects of the size and material properties of the phantom on S_{11} of the proposed antenna. **c** Fabricated capsule antenna fed by a coaxial cable (Reprinted with permission from IEEE [25])

0.254 mm thick RO4350B laminate is used. To minimize the profile of the antenna due to the limited space inside the casing, a two-layer design is utilized with the second layer mainly serving as ground plane underneath the antenna structure.

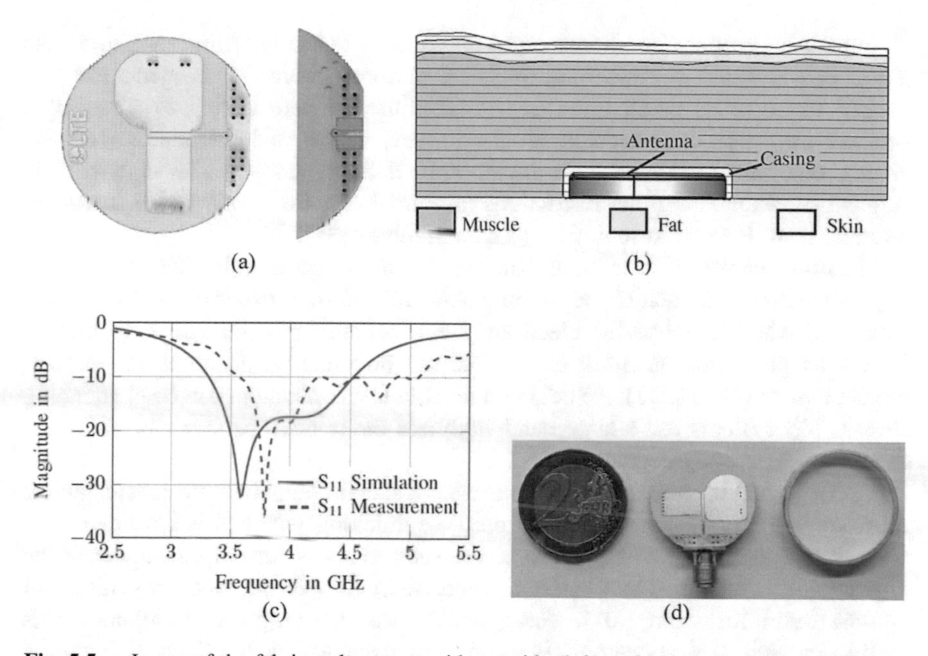

Fig. 5.5 **a** Image of the fabricated antenna with top side (left) and enlarged view of the bottom side feeding point (right). **b** Cross section of the simulated model of the antenna. **c** Simulated and measured input matching for the antenna PCB with casing and loading with human tissue (simulation) and porcine tissue (measurement). **d** Fabricated antenna with SMA connector, aluminium oxide casing and 2 Euro coin for size comparison (Reprinted with permission from IEEE [27])

The layout of the antenna design is shown in Fig. 5.5a and the schematic of the setup is shown in Fig. 5.5b. The S_{11} parameters are presented in Fig. 5.5c and the antenna with SMA connector and packaging is shown in Fig. 5.5d.

5.3 UWB Phantoms for Implantable Communication

Software-based simulations tools such as CST Microwave Studio (MWS), ANSYS HFSS are easily accessible and help to get an idea of the propagation channel characteristics [22, 29, 30]. The results may not directly correlate with realistic experiments especially for in-Body and implantable communications due to the various biological effects such as breathing, heart beat rate, blood flow etc. The high complexity of the simulations and computation time makes it a challenge to get the results in short duration and consume large memory to properly characterize a WBAN channel.

Implanting devices into human subjects to carry out propagation measurements is not possible due to ethical and physical reasons. *In-vivo* experiments are performed in hospitals or laboratories with facilities and are similar to a real case because they take into account all the tissues, blood, and internal movements among other factors that appear in the animal living bodies. The high cost of surgical procedures and the restrictions in animal experimentations due to ethical reasons, there is need to look for other alternatives [31].

Phantom based measurements are becoming a good substitute to in-vivo experimentation, reducing the requirement for animal experiments. Chemical solutions, which emulate the electromagnetic behaviour of human body tissues, known as phantoms are used to recreate the propagation phenomenon through human tissues [23, 31, 32]. Achieving a reliable UWB phantom can be challenging since UWB systems use a large bandwidth and the relative permittivity of human tissues are frequency dependent.

For implantable communication, muscle is the main human tissue considered for most wireless applications, hence a muscle-mimicking phantom is considered for such experiments which is liquid in consistency to allow ease of movement of the implantable antennas. The phantom proposed in [33] is an aqueous solution of 54.98% acetonitrile and 1.07% NaCl, which owns the targeted conditions and is easily prepared. This phantom (Fig. 5.6) was designed at 24 °C as the solution reproduces the dielectric values of the human muscle that were characterized at body temperature, as reported in [31]. The main component of this phantom is the use of acetonitrile, a polar molecule that gets dielectric constant values like those of the muscle within a wide frequency band. The addition of salt increases the loss factor so the tissue's behavior is better imitated in both parts of the complex relative permittivity [33].

$$\varepsilon_r = \varepsilon_r' - j\varepsilon_r'' \tag{5.1}$$

where ε_r' is the dielectric constant and ε_r'' is the loss factor. In Fig. 5.7, the complex relative permittivity of the muscle phantom is compared with the reported measured values for a real muscle tissue [33]. Various other phantom models based on liquid solutions which mimic the human tissues are presented in [21, 31, 34]. Biological equivalent liquid phantom proposed in [31] is shown in Fig. 5.7a, b having dielectric properties like 2/3 muscle's value. A multilayer phantom container is shown in Fig. 5.8 which is made of polyethylene terephthalate (PET) and has two layers for the muscle and fat phantom with dimensions of [$25 \times 25 \times 25$ cm^3], [$23 \times 25 \times 25$ cm^3] respectively. Between the layers, there is a divider sheet which is made of PET of 1.5 mm [34]. All the proposed phantoms electrical properties correlate well with that of those reported for the human tissues which is mainly muscle for the current implantable communication links study and analysis.

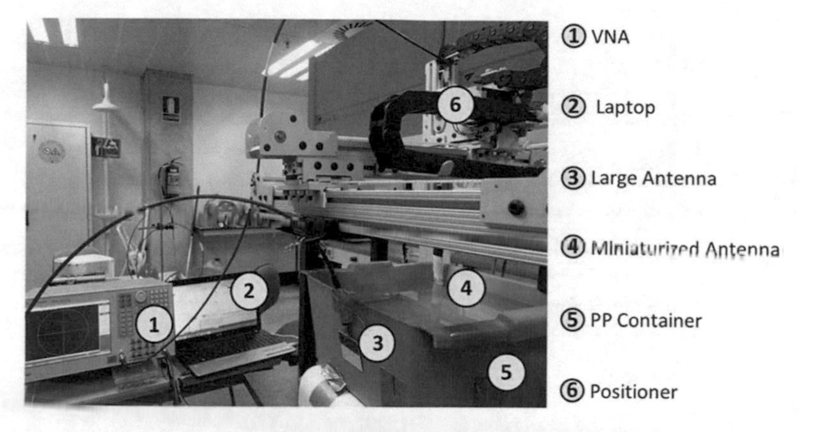

Fig. 5.6 Experimental measurement setup particularized for the IB2OB scenario using liquid phantom (Reprinted with permission from IEEE [31])

5.4 Channel Characterization for Implantable Communication

5.4.1 Phantom-Based Channel Characterization

UWB in-body channel characterization within 3.1–8.5 GHz using a new realistic UWB phantom is reported in [31], which emulates the electromagnetic behavior of the human muscle, with high accuracy in the whole frequency band considered. Two in-body scenarios are considered depending on the location of the receiving antenna: in-body to in-body (IB2IB) and in-body to on-body (IB2OB) scenarios.

1. IB2IB, where both antennas are located inside the human body.
2. IB2OB, where the in-body antenna is located inside the body, whereas the on-body antenna is placed over the human body surface.

In Fig. 5.9, the two UWB monopole antennas used in the measurement campaign are depicted. One antenna is approximately 2.2. times larger than the other. The small antenna, which is located at the center of Fig. 5.9, is an UWB monopole with microstrip feeding. The overall size is 23×20 mm^2. This miniaturized antenna was designed considering the human muscle permittivity values [33]. Thus, antenna matching as well as a quasi-omnidirectional radiation pattern in the x–z plane within the UWB frequency range was achieved. The large antenna also presents a quasi-omnidirectional radiation pattern in the xz-plane in the UWB frequency band.

During the measurements, the forward transmission coefficient (S_{21}) was measured and processed for both scenarios. Measurements were performed from 3.1 to 8.5 GHz with $N = 1601$ resolution points resulting in a resolution frequency of $\Delta f = 1.875$ MHz using a VNA. The antenna was covered with a protective latex

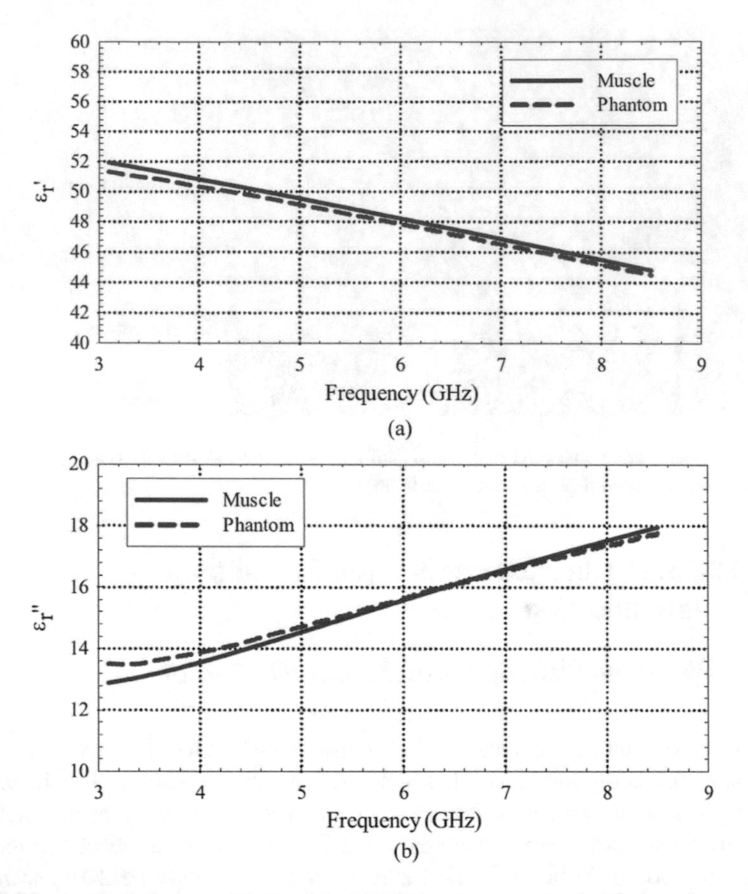

Fig. 5.7 Measured relative permittivity of the UWB phantom compared with the values of human muscle tissue reported in [33]. **a** Dielectric constant. **b** Loss factor (Reprinted with permission from IEEE [32])

layer and submerged into the liquid phantom present into a polypropylene container having dimensions of 30 × 30 × 15 cm^3 with a negligible wall thickness. The size of the box was chosen to mimic a cross section of the human torso and voxel models. The miniaturized antenna was mounted on a robotic arm and moved to several (x, y, z) locations inside the liquid phantom using a high resolution three-axis positioner while VNA measurements were recorded. Two different setups were considered depending on the antenna locations which is described below:

1. *Setup IB2OB:* The on-body antenna [31] (large antenna) was fixed in the center of the external container's wall, whereas the in-body antenna [31] (miniaturized antenna) was in a *xyz* mesh of 19 × 19 × 3 measurement points.
2. *Setup IB2OB:* The on-body antenna was in an initial position and moved away λ, 2λ, 4λ, and 8λ in the *y*-axis, where λ is the wavelength corresponding to the central frequency of the frequency band under analysis considering the propagation speed inside the phantom.

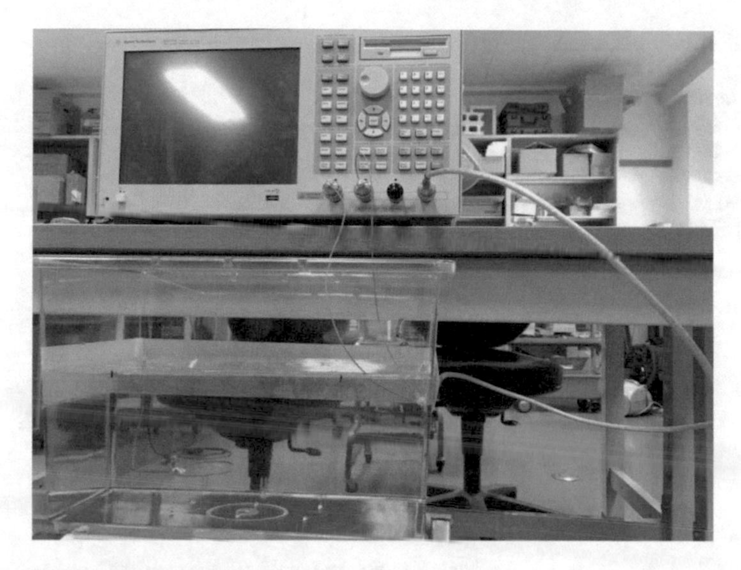

Fig. 5.8 UWB biological-equivalent liquid phantom for implantable communications measurements (Reprinted with permission from IEEE [23])

Fig. 5.9 UWB monopoles antennas used in both experimental scenarios (Reprinted with permission from IEEE [31])

5.4.1.1 IB2IB Scenario

In Fig. 5.10a, the path loss obtained from the measurements for the IB2IB scenario is shown. One can observe that the best fit is achieved with a linear approximation model as:

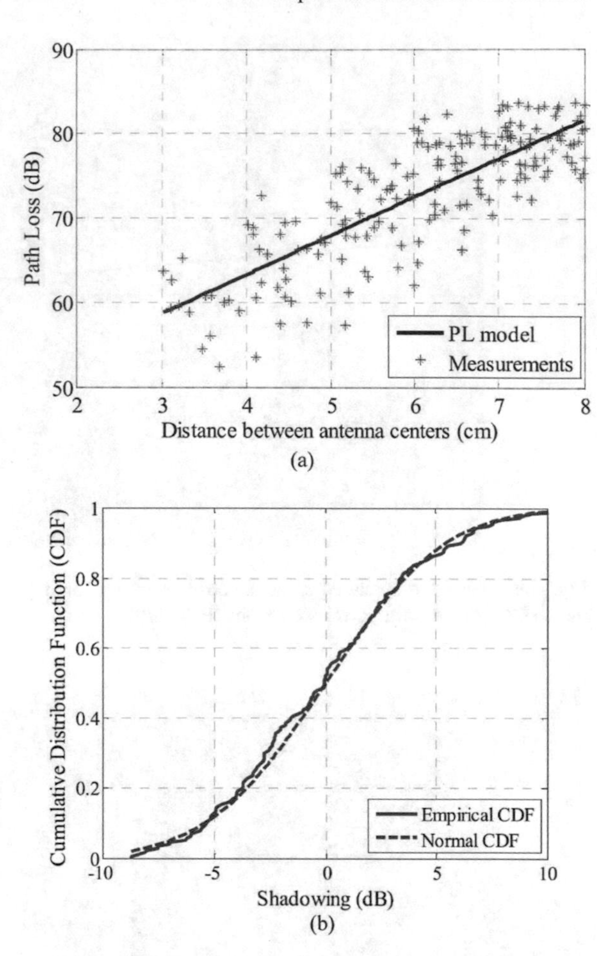

Fig. 5.10 **a** Path loss as a function of distance between antenna centres in the experimental IB2IB scenario. **b** Shadowing in the experimental IB2IB scenario (Reprinted with permission from IEEE [31])

$$PL(dB) = PL_0 + \alpha.d(cm) + X_{\sigma IB2IB} \quad (5.2)$$

where PL_0 is the value of the path loss when the distance between antenna centres d tends to 0, α is a fitting parameter, and $X_{\sigma IB2IB}$ is the shadowing term for IB2IB scenario. In this case, $PL_0(dB) = 45.03$ dB and $\alpha = 4.57$ dB/cm. Figure 5.10b shows the cumulative distribution function (CDF) of the measurements. It can be noted that the empirical model of shadowing can be fitted by means of a normal distribution with zero mean and standard deviation $\sigma = 4.26$ (N (0, σ)). It is important to note that the path loss is given from $d = 3$ cm to $d = 8$ cm because most of the measurements above 8 cm were at noise level.

5.4.1.2 IB2OB Scenario

The path loss values obtained for the IB2OB case are shown in Fig. 5.11a. The log-distance equation is given by the following expression [35]:

$$PL(dB) + PL_{0,d_{ref}} + 10\gamma\log_{10}\left(\frac{d}{d_{ref}}\right) + X_{\sigma IB2OB} \quad (5.3)$$

being $PL_{0,d_{ref}}$ the path loss value in a reference distance d_{ref} equal to 1 cm, γ the path loss exponent, d the distance between antenna centres, and $X_{\sigma IB2OB}$ the shadowing term for the IB2OB scenario. In this case, the parameters of the log-distance equation were $PL_{0,d_{ref}} = 47.84$ dB and $\gamma = 1.98$. From Fig. 5.11b, one can also observe that the shadowing also follows a normal distribution with zero mean and $\sigma = 1.2$. The path loss model obtained from the measurements follows a logarithmic trend as the distance between antennas increases.

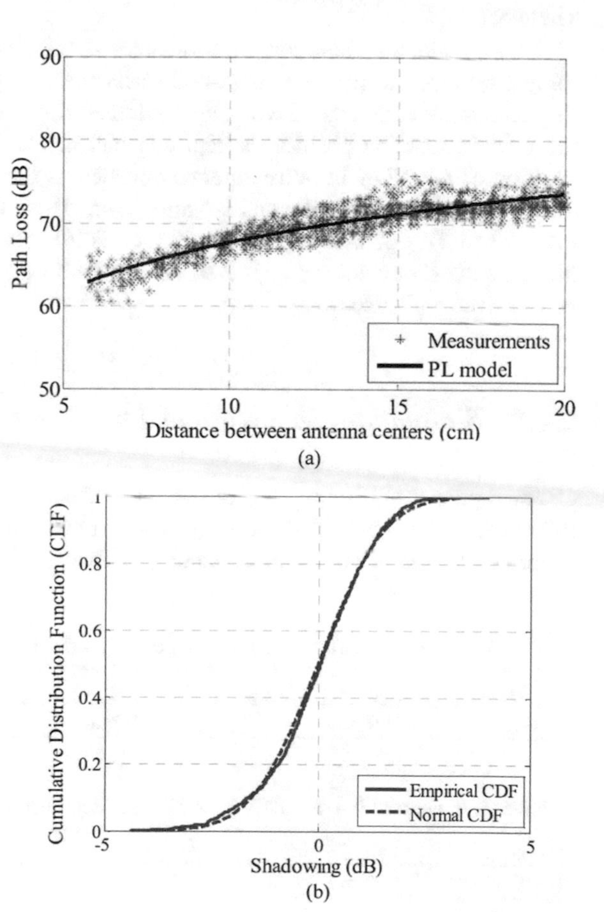

Fig. 5.11 **a** Path loss as a function of distance between antenna centres in the experimental IB2OB scenario. **b** Shadowing in the experimental IB2OB scenario (Reprinted with permission from IEEE [31])

5.4.2 Channel Modeling and Communication Link Analysis

In-body to on-body and in-body to in-body channel modelling inside a UWB liquid phantom for wireless capsule endoscopy based on a planar elliptical in-body antenna and semicircular monopole on-body antenna is presented in [14].

The channel model measurements are carried out inside a small anechoic chamber with its boundaries covered with absorbing materials, like the setup in [31]. The measurement setup consists of 250 × 250 × 250 mm^3 box containing the UWB liquid phantom which is developed on the basis of sugar and salt recipe composition in order to mimic the electrical properties of the human muscle. A robotic arm mounted with in-body antenna is used for the automatic changing of the position of the in-body antenna precisely inside the liquid phantom.

The in-body to on-body path loss model is combination of both the statistical variance of path loss and the fitted mean path loss. The standard deviation is fitted at 2.2 dB, which implies the variance of the path loss values because of the reflection and diffraction. All the extracted factors of the path loss model are summarized in Table 5.1.

To evaluate the radiation performance of the designed implanted antenna in lossy medium, the in-body to in-body transmission performance of the elliptical ring antenna is measured. Two planar elliptical ring antennas are used where one of the in-body antenna position is kept constant inside the liquid phantom while the position of the other in-body antenna attached to the robotic arm is changed. The distance between two antennas is varied from 18 to 61 mm with path loss values from 42 to 72 dB. The reference distance is 20 mm with $PL_0(dB)$ = 48 dB. The fitted standard deviation σ_{dB} is 2.9. All the parameters of the path loss model are summarized in Table 5.2.

5.4.3 Simulation, in Vivo and Phantom Based Comparison

Channel path loss (PL) model for the UWB frequency band in the gastrointestinal (GI) scenario is presented in [32]. Three methodologies are generally used to characterize the propagation channel, software simulations, and experimental

Table 5.1 In-body to on-body path loss models (Reprinted with permission from IEEE [14])

$PL_{0,dB}$	d_0	D	n	σ
59 dB	60 mm	[57,110] mm	1.7	2.2

Table 5.2 In-body to In-body Path Loss Models (Reprinted with permission from IEEE [14])

$PL_{0,dB}$	d_0	D	n	σ
48 dB	20 mm	[18,61] mm	5.2	2.9

measurements either in phantom or in in vivo animals. The details of the implantable and on-body antennas used in this study are given in Fig. 5.9 and Sect. 5.4.1.

Phantom

The transmitting antenna is located inside the intestine of the human body and the receiving antenna is placed over the abdominal region of the human body. In the GI scenario, the main issues involved in the signal transmission are: large or small bowel, muscle, fat, skin, and blood. As the muscle, colon, and small bowel have similar dielectric constant and conductivity in the low UWB frequency band, a muscle phantom tissue is considered for simulations and phantoms measurements. The set up for the measurements is like that presented in Sect. 5.3.

Experiment

The in vivo experiment was performed in a living porcine model in the facilities of the Hospital Universitari I Politècnic la Fe, Valencia, Spain. This experiment was conducted for digestive surgeons and the method used to perform the surgery in the animal was a laparoscopy. The set up in the in-vivo measurements consisted of the measurements based on the VNA, in-body, on-body antennas, and the 3D magnetic tracker like that in the phantom experiment.

Simulations

Software simulations were performed with the commercial software CST MW using time-domain transient solver. Two types of simulations were performed:

1. The experimental test bed is replicated in CST.
2. The abdominal part of a human female computer-aided design (CAD) model (Nelly) was chosen to confirm whether the measurements in a living pig were well performed.

To reduce the complexity of the simulation, only the skin, fat, and muscle of the human body was considered. In case B simulations, the in-body and on-body antennas were located in different positions along the *x*-, *y*-, and z-axes with total range of distances between the in-body and on-body antennas ranging from d = 4 to 7.95 cm.

Path Loss Analysis

The PL values for the numerical simulations are a few decibels below those deduced from the in vivo measurements due to alignment issues in the in-vivo experiments and the lack of blood flow, respiration, reflections, and minor movements modelled in the simulated version of the analysis. The data values obtained from the multilayer phantom container have a shift of two centimetres in the *x*-axis. The experiments performed in phantom measurements considered muscle and a 2 cm width of the fat phantom layer. Therefore, the measurements performed with fat phantom, can be shifted up to 2 cm to the left to replicate as much as possible the *in-vivo* results, where the pig has less than 2 cm of the abdominal fat. Figure 5.12, presents the *in-vivo* measurements, phantom measurements, shifted

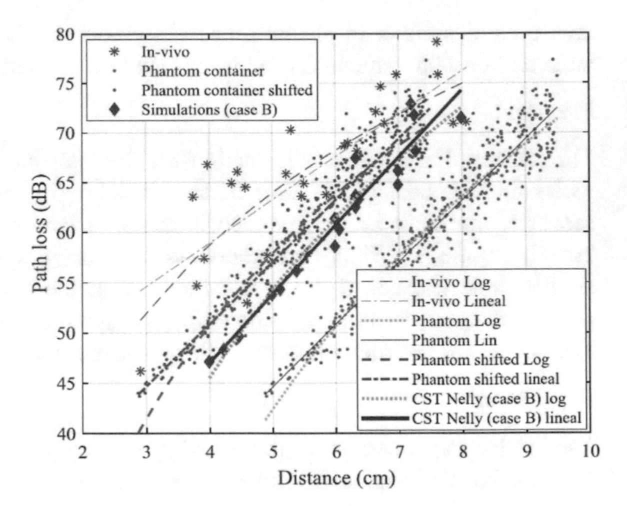

Fig. 5.12 Path loss measured and simulated data along with their fitted linear and logarithmic models (Reprinted with permission from IEEE [32])

phantom measurements, and software simulations (Nelly model, case B) are shown with the logarithmic and linear fitting models. In this situation, the agreement between results coming from the three methodologies is clearly shown leading to a more general model.

Path Loss Models

Equations (5.4) and (5.5) describe the logarithmic and linear PL models. The logarithmic PL model is given by:

$$PL(dB) = 10\gamma \log_{10}\left(\frac{d}{d_0}\right) + PL_0(dB) + X(\sigma, \mu) \tag{5.4}$$

where d is the separation between antennas, d_0 is the reference distance, $PL_0(dB)$ is the reference PL value at the reference distance d_0, γ is the logarithmic PL exponent, and $X(\sigma, \mu)$ is the statistical distribution modeling the shadowing term of the signal. The statistical distribution follows a Gaussian model with mean, μ, and standard deviation of σ. The linear model is given by:

$$PL(dB) = \alpha\left(\frac{d}{d_0}\right) + PL_0(dB) + X(\sigma, \mu) \tag{5.5}$$

where α is the slope of the line. The maximum distance achieved for simulations, phantom measurements, and in vivo measurements is not the same. In the case of the *in-vivo* experiments, the maximum distance measured is d_{max} = 8.067 cm, and for the multilayer phantom model, it is d_{max} = 9.5 cm or d_{max} = 7.5 cm (if the samples are shifted 2 cm due to the fat layer). In software-based simulations, the maximum distance d = 7.95 cm.

Table 5.3 Path loss models (Reprinted with permission from IEEE [32])

	Logarithmic	Linear
In vivo	PL_0 = 26.226 dB d_0 = 1 cm γ = 5.3967 $\mu \approx 0$ σ = 4.4972 RMSE = 28.1	PL_0 = 41.5635 dB d_0 = 1 cm α = 4.337 $\mu \approx 0$ σ = 4.6007 RMSE = 28.8
CST Nelly model (Case B)	PL_0 = -8.3767 dB d_0 = 1 cm γ = 8.9681 $\mu \approx 0$ σ = 1.9795 RMSE = 7.5	PL_0 = 19.7445 dB d_0 = 1 cm α = 6.8262 $\mu \approx 0$ σ = 1.9489 RMSE = 7.37
Phantom	PL_0 = -29.7593 dB d_0 = 1 cm γ = 10.3395 $\mu \approx 0$ σ = 2.3069 RMSE = 3.64	PL_0 = 13.8201 dB d_0 = 1 cm α = 6.1719 $\mu \approx 0$ σ = 2.2724 RMSE = 3.54
Phantom shifted	PL_0 = 6.29 dB d_0 = 1 cm γ = 7.3824 $\mu \approx 0$ σ = 2.3772 RMSE = 3.88	PL_0 = 26.1639 dB d_0 = 1 cm α = 6.1719 $\mu \approx 0$ σ = 2.2724 RMSE = 3.51

For the GI scenario with a distance range from 3 to 8 cm and a frequency band of 3.1–5.1 GHz, the PL exponent is in between 4.3 and 6.8 and 5.4 and 8.9 for linear and logarithmic models, respectively. Results are given in Table 5.3 where the root mean squared estimator (RMSE) is computed for both the linear and logarithmic methods, for the three methodologies.

5.4.4 Diversity Experimental Results

Spatial diversity techniques have been proposed to reduce the attenuation level in the UWB frequency range to more robust implant communication links [23]. In this work, UWB transmitter diversity antenna is presented, and its performance is evaluated numerically and experimentally.

The antennas for the polarization diversity scheme require low reflection coefficient and low coupling effect between each one in the UWB low band. Figure 5.13a, b shows the dual-polarized diversity antenna, which is composed of two planar elliptical loop antennas fabricated on dielectric substrate with permittivity 4 and thickness 1.6 mm. The elliptical loop is made of copper whose thickness, major axis, and minor axis are 0.1, 4.8, and 3.4 mm, respectively.

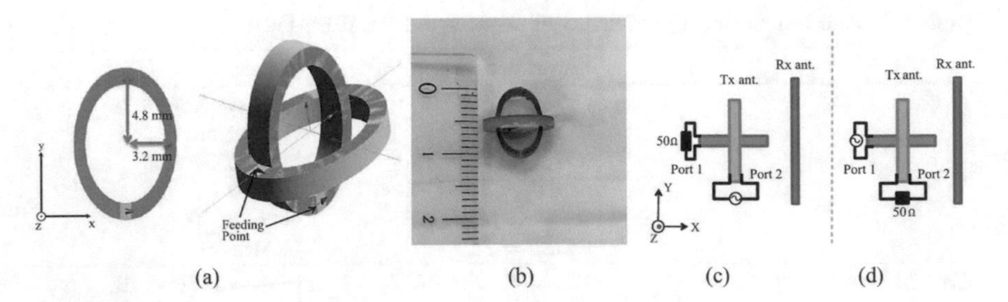

Fig. 5.13 Polarization implant side diversity antenna for UWB communication. **a** Part of dual-polarized planar elliptical loop antenna. 3-D rendering of the dual-polarized elliptical loop antenna. **b** Photograph of developed implant side diversity antenna. Antenna arrangement for performance evaluation. Arrangement of **c** "parallel." **d** "vertical" (Reprinted with permission from IEEE [23])

The volume of this antenna is 11.3 × 11.3 × 8.5 mm^3 which can be easily contained in a commercial capsule endoscope of volume packaged with same UWB pulse generator chip with dimensions 0.4 mm^2, circuits and battery [36].

Figure 5.13c, d shows the antenna arrangement for the performance evaluation. The first antenna arrangement is shown in Fig. 5.13c, where the transmit antenna was placed parallel to the receive antenna. In the other case, as shown in Fig. 5.13d, the transmit and receive antennas are set vertically to each other. In this work all the antenna arrangements are addressed as "parallel" and "vertical," respectively. In "parallel" arrangement, port 1 is excited, while port 2 is excited in "vertical" arrangement.

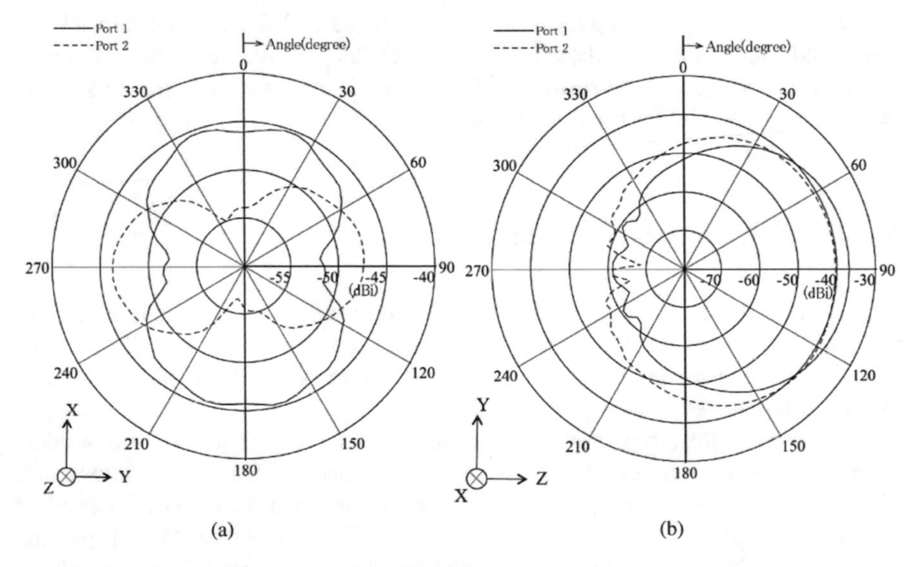

Fig. 5.14 Simulated antenna radiation pattern. **a** Pattern of *X*–*Y* plane. **b** Pattern of *Y*–*Z* plane (Reprinted with permission from IEEE [23])

The measured performance using liquid phantom and simulated coupling performance is carried out using FDTD. S_{11} and S_{21} measured and simulated results are less than −10 dB and −20 dB respectively in the UWB low band. The antenna performance is analysed using a finite-difference time-domain simulation and physical experiment in a liquid phantom. Path loss performance is computed in an implant communication link using a liquid phantom and a living porcine subject. In the first antenna arrangement, the transmit antenna was placed parallel to the receive antenna. In the other case, the transmit and receive antennas are positioned vertically to each other. In "parallel" arrangement, port 1 is excited, while port 2 is excited in "vertical" arrangement. Figure 5.14 demonstrates the antenna radiation pattern when excited at ports 1 and 2. From Fig. 5.14, both radiation patterns can be accepted in the used frequency band. The result suggests that the developed diversity antenna can also operate as polarization diversity antenna.

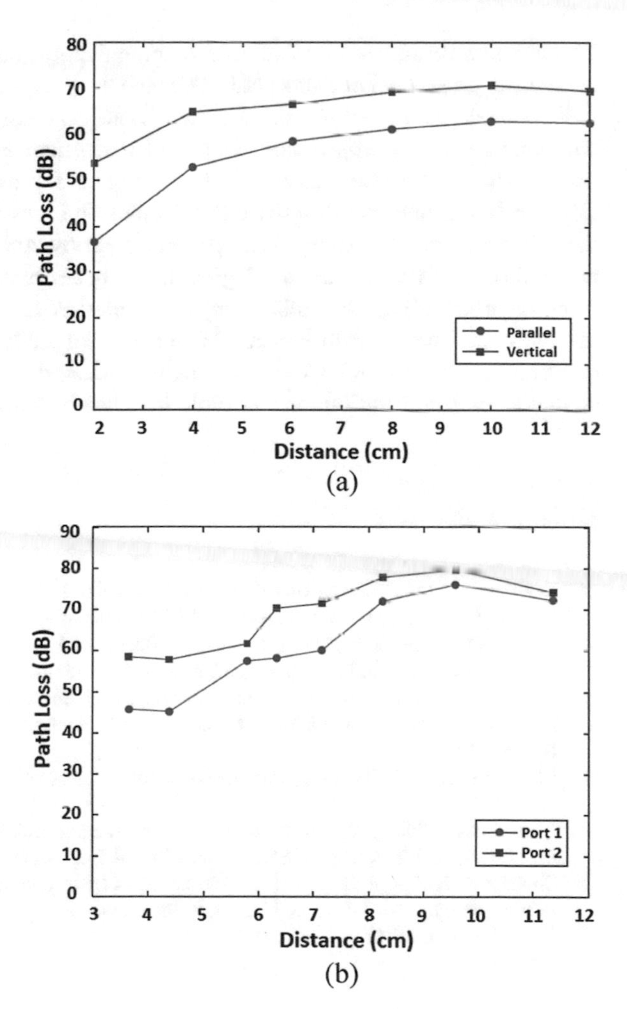

Fig. 5.15 **a** Measured path loss performance in the liquid phantom experiment. **b** Measured path loss performance in the living animal experiment (Reprinted with permission from IEEE [23])

The results of the average path loss from 3.4 to 4.8 GHz with respect to 8 the communication distance is presented in Fig. 5.15a, b. The path loss for the "parallel" case is smaller than that for the "vertical" case. The difference of around 10 dB at all communication distances means the developed diversity antenna can perform as a polarization diversity antenna. Furthermore, at a communication distance of 10 cm, both situations of "parallel" and "vertical" have accomplished the acceptable path loss levels of around 60 and 70 dB, respectively. The measured average path loss results for the living animal measurements are also carried out. The path loss at the distance of more than 10 cm is reduced which is due to the difference of the biological tissue between the transmit and receive antennas.

5.5 Conclusion

The antenna design for implantable communication has far more challenges and need more novel design concepts to enhance the performance of the antenna which must operate in a biological medium. Biological tissues influence the antenna parameters such as impedance, gain and performances to a great extend which further affects the communication link suitable for implantable communication. Capsule based antenna designs and designs for brain neural-function monitoring have been presented along with general in-body antennas suitable for various applications. Various methodologies have been presented to characterize the in-body propagation channels such as simulation based, phantoms and realistic in-vivo experiments. Path loss models have been analysed for IN2OB and IN2IN communication channel which will help in design of a reliable communication device suitable for various implantable healthcare monitoring applications.

References

1. Dominique MM-P (2015) Schreurs Wearable wireless health monitoring: current developments, challenges, and future trends. IEEE Microw Mag 16(4):55–70
2. Bharadwaj R, Swaisaenyakorn S, Parini C, Batchelor J, Alomainy A (2014) Motion tracking of a human subject in healthcare applications using compact ultra wideband antennas. In: 4th international conference on wireless mobile communication and healthcare—transforming healthcare through innovations in mobile and wireless technologies (MOBIHEALTH), Athens, 2014
3. Montón E et al (2008) Body area network for wireless patient monitoring. IET Commun 2 (2):215–222
4. Kiourti A, Nikita KS (2017) A review of in-body biotelemetry devices: implantables, ingestibles, and injectables. IEEE Trans Biomed Eng 64(7):1422–1430
5. Shubair RM, Elayan H (2015) In vivo wireless body communications: State-of-the-art and future directions. In: Loughborough antennas & propagation conference (LAPC), Loughborough, 2015, pp 1–5

6. Bharadwaj R, Swaisaenyakorn S, Parini CG, Batchelor JC, Alomainy A (2017) Impulse radio ultra-wideband communications for localization and tracking of human body and limbs movement for healthcare applications. IEEE Trans Antennas Propag 65(12):7298–7309
7. Floor PA, Chávez-Santiago R, Kim AN, Kansanen K, Ramstad TA, Balasingham I (2019) Communication aspects for a measurement based UWB in-body to on-body channel. IEEE Access 7:29425–29440
8. Garcia-Pardo C et al (2018) Ultrawideband technology for medical in-body sensor networks: an overview of the human body as a propagation medium, phantoms, and approaches for propagation analysis. IEEE Antennas Propag Mag 60(3):19–33
9. Brumm J, Bauch G (2020) Verification of a simplified channel modeling technique for ultra wideband in-body communication with simulations. In: 14th european conference on antennas and propagation (EuCAP), Copenhagen, Denmark, 2020, pp 1–5
10. Chavez-Santiage R et al (2013) Propagation models for IEEE 802156 standardization of implant communication in body area networks. IEEE Wirel Commun Mag 51(8):80–87
11. Kissi C et al (2019) Directive low-band UWB antenna for in-body medical communications. IEEE Access 7:149026–149038
12. Haas M, Schweizer B, Anders J, Ortmanns M (2017) A miniaturized UWB antenna for implantable data telemetry. In: 39th annual international conference of the IEEE engineering in medicine and biology society (EMBC), Seogwipo, 2017, pp 1086–1089
13. Kamaruddin NA, Azemi SN, Ibrahim SZ, Azremi AAH, Kahar NF (2019) Antenna for in-body communications. In: IEEE Asia-Pacific conference on applied electromagnetics (APACE), Melacca, Malaysia, 2019, pp 1–6
14. Fang X et al (2019) Experimental in-body to on-body and in-body to in-body path loss models of planar elliptical ring implanted antenna in the ultra-wide band. In: 13th international symposium on medical information and communication technology (ISMICT), Oslo, Norway, 2019, pp 1–5
15. Särestöniemi M, Kissi C, Raez CP, Sonkki M, Hämäläinen M, Iinatti J (2019) Propagation and UWB channel characteristics on the human abdomen area. In: 13th european conference on antennas and propagation (EuCAP), Krakow, Poland, 2019, pp 1–5
16. Särestöniemi M, Pomalaza-Raez C, Berg M, Kissi C, Hämäläinen M, Iinatti J (2019) In-body power distribution for abdominal monitoring and implant communications systems. In: 16th international symposium on wireless communication systems (ISWCS), Oulu, Finland, 2019, pp 457–462
17. Barbi M, Garcia-Pardo C, Nevárez A, Pons Beltrán V, Cardona N (2019) UWB RSS-based localization for capsule endoscopy using a multilayer phantom and in vivo measurements. IEEE Trans Antennas Propag 67(8):5035–5043
18. Dissanayake T, Esselle KP, Yuce M (2009) UWB antenna impedance matching in biomedical implants. In: 3rd european conference on antennas and propagation, Berlin, 2009, pp 3523–3526
19. Wang Q, Wolf K, Plettemeier D (2010) An UWB capsule endoscope antenna design for biomedical communications. In: 3rd international symposium on applied sciences in biomedical and communication technologies (ISABEL 2010), Rome, 2010, pp 1–6
20. Morimoto Y, Anzai D, Wang J (2013) Design of ultra wide-band low-band implant antennas for capsule endoscope application,. In: 7th international symposium on medical information and communication technology (ISMICT), Tokyo, 2013, pp 61–65
21. Thotahewa KMS, Redoutè J, Yuce MR (2015) Propagation, power absorption, and temperature analysis of UWB wireless capsule endoscopy devices operating in the human body. IEEE Trans Microw Theory Tech 63(11):3823–3833
22. Perez-Simbor S, Garcia-Pardo C, Cardona N (2019) Initial delay domain UWB channel characterization for in-body area networks. In: 13th international symposium on medical information and communication technology (ISMICT), Oslo, Norway, 2019, pp 1–5
23. Shimizu Y, Anzai D, Chavez-Santiago R, Floor PA, Balasingham I, Wang J (2017) Performance evaluation of an ultra-wideband transmit diversity in a living animal experiment. IEEE Trans Microw Theory Tech 65(7):2596–2606

24. Yazdandoost KY (2016) Antenna for wireless capsule endoscopy at ultra wideband frequency. In: IEEE 27th annual international symposium on personal, indoor, and mobile radio communications (PIMRC), Valencia, 2016, pp 1–5
25. Bao Z, Guo Y, Mittra R (2017) An ultrawideband conformal capsule antenna with stable impedance matching. IEEE Trans Antennas Propag 65(10):5086–5094
26. Yazdandoost, Miura R (2015) Miniaturized UWB implantable antenna for brain-machine-interface. In: 9th European conference on antennas and propagation (EuCAP), Lisbon, 2015, pp 1–5
27. Frank M, Lurz F, Kempf M, Röber J, Weigel R, Koelpin A (2020) Miniaturized ultra-wideband antenna design for human implants. In: IEEE radio and wireless symposium (RWS), San Antonio, TX, USA, 2020, pp 48–51
28. Bahrami H, Mirbozorgi SA, Rusch LA, Gosselin B (2015) Biological channel modeling and implantable UWB antenna design for neural recording systems. IEEE Trans Biomed Eng 62 (1):88–98
29. Brumm J, Strohm H, Bauch G (2019) A stochastic channel model for ultra wideband in-body communication. In: 41st annual international conference of the IEEE engineering in medicine and biology society (EMBC), Berlin, Germany, 2019, pp 4032–4035
30. Brumm J, Bauch G (2017) On the shadowing distribution for ultra wideband in-body communication path loss modelling. In: IEEE international symposium on antennas and propagation & USNC/URSI national radio science meeting, San Diego, CA, 2017, pp 805–806
31. Andreu C, Castelló-Palacios S, Garcia-Pardo C, Fornes-Leal A, Vallés-Lluch A, Cardona N (2016) Spatial in-body channel characterization using an accurate UWB phantom. IEEE Trans Microw Theory Tech 64(11):3995–4002
32. Perez-Simbor S, Andreu C, Garcia-Pardo C, Frasson M, Cardona N (2019) UWB path loss models for ingestible devices. IEEE Trans Antennas Propag 67(8):5025–5034
33. Gabriel C (1996) Compilation of the dielectric properties of body tissues at RF and microwave frequencies, Brooks Air Force, San Antonio, TX, USA, Tech. Rep. NAL/OE-TR-1996-0037
34. Perez-Simbor S, Barbi M, Garcia-Pardo C, Castelló-Palacios S, Cardona N (2018) Initial UWB in-body channel characterization using a novel multilayer phantom measurement setup. In: IEEE wireless communications and networking conference workshops (WCNCW), Barcelona, 2018, pp 384–389
35. Rappaport TS (1996) Wireless communications: principles and practice. Prentice-Hall, Upper Saddle River, NJ, USA
36. Yuce MR, Dissanayake T (2012) Easy-to-swallow wireless telemetry. IEEE Microw Mag 13 (6):90–101

Chapter 6
Body Centric Localization and Tracking Using Compact Wearable Antennas

6.1 Introduction

In recent years, the indoor localisation and tracking of human subjects has received significant research and development interest for several applications in wireless body area networks (WBANs). Wearable wireless systems are attracting significant interest to access and monitor human activity for sports, healthcare, medical monitoring, entertainment, military applications, and day-to-day activities [1–4]. The desired key performance criteria in such scenarios are the estimation of the target location with high accuracy and precision. There are various technologies such as infrared, inertial, ultrasound, optical and radio frequency (RF) based systems [4, 5] that can be applied for body-centric localization and tracking. Infrared (IR) signals are low power and inexpensive, but they cannot penetrate through obstructions. The systems based on ultrasound technology are relatively cheap; however, they have lower precision in comparison with IR systems and are suitable for short range only [5, 6]. Inertial sensors such as accelerometers suffer from a fluctuating offset and complex calibration procedure [3, 4]. Optical-based motion capture systems provide high accuracy but are expensive, require long calibration procedures, suffer from occlusions, and are mostly confined to laboratory based measurements [4–6].

Among radio frequency technologies, the ultra-short pulse ultra-wideband (UWB) (3.1–10.6 GHz) based systems enables high localization accuracy and have several advantages such as low cost, low power, high data rate, portability, integration with other technologies (such as MEMS, inertial sensors) and can carry signals through many obstacles in comparison to narrowband systems [7–9]. Due to its signal bandwidth of 500 MHz and above, UWB technology promises interesting perspectives for position location in the short-range environment [10–12]. The fine time resolution of an UWB system allows employing accurate time dependent techniques for location estimation, e.g., time-of-flight (TOF), time of arrival (TOA) or time-difference of arrival (TDOA) [2, 6, 12–15]. These techniques rely on

S. K. Koul and R. Bharadwaj, *Wearable Antennas and Body Centric Communication*, Lecture Notes in Electrical Engineering 787,
https://doi.org/10.1007/978-981-16-3973-9_6

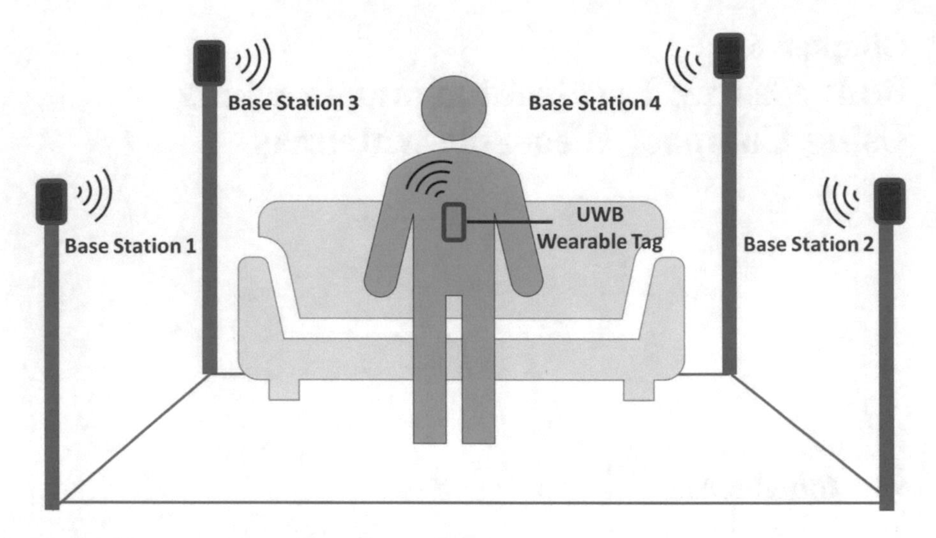

Fig. 6.1 Human indoor localization using ultra-wideband wearable technology

the high precision of the timing information exchanged through short pulses. A layout of a typical indoor system for tracking human subject is presented in Fig. 6.1 which consist of a mobile tag placed on the user and four base stations/ anchors mounted on the wall of the room. The positioning information can be collected in real time and further communicated with other devices [16].

This chapter gives an insight into the propagation characteristics when antennas are placed on various body locations and the factors affecting the accuracy achieved while localizing/tracking the antennas worn by the user in a potential wireless body sensor networks setup within the body/personal area networks. The focus of the work is to accurately determine and track a person's position and motion activity in an indoor environment using UWB channel characterization information and time of arrival positioning techniques. Several issues and challenges prevail for accurate localization of the human subject in an indoor environment before the system can be deployed for commercial applications. These include effect of human body in the localization area, portable and cost-effective localization system, multi-user interference, multipath effects and mitigation of non-line-of-sight (NLOS) propagation. Simple and effective techniques based on channel information and time of arrival localization techniques are presented in this chapter for localization accuracy enhancement. Indoor localization of human body and tracking of limb movements is studied and analyzed using wearable compact UWB antennas placed on different locations/joints of the body. UWB channel characterization is studied in detail for various activities and the information obtained is used to minimize the error obtained in the range and localization results, hence enabling accurate localization and motion tracking of the wearable nodes. The applicability of machine learning algorithms for enhancing localization accuracy by accurately classifying the channel links is also discussed.

6.1.1 State-of-the-Art Localization Techniques

UWB technology has widespread advantages in positioning systems making it a natural choice for localisation using body worn antennas [17]. Several commercial products and research work has been carried out in this domain which is detailed in this section. Commercial UWB based localization systems such as Ubisense, Eliko, BeSpoon, DecaWave, Time Domain PulsON, Open RTLS, Pozyx, have accuracy of 10–15 cm [18–21]. The systems are suitable for simultaneous real-time tracking of hundreds of objects and is robust enough for industrial environments [22]. The localization algorithms used in these products are generally based on time of arrival, time difference of arrival and angle of arrival techniques. 3D motion tracking products based upon miniature Micro-Electro-Mechanical Systems (MEMS) inertial sensors and UWB technology enables 5–8 cm positioning accuracy in an area of 20 × 20 m^2 [23].

Research has been carried out in various domains such as patient tracking for healthcare applications [24, 25], robotics [26], asset tracking [22], monitoring sportsmen (Fig. 6.2a) [27], indoor motion gaming [28] and localisation of the human subjects [8, 29, 30]. Various state-of-the art techniques such as ray tracing [26], Kalman and extended finite impulse response (FIR) filter (Fig. 6.2b, c) [31, 32], integrated UWB and inertial tracking [32], propagation path information [33], array antenna [34], carrier-based UWB technology [35] for localization is reported in the open literature.

High accuracy in the range of 1–5 cm has been reported in the open literature for short-range indoor UWB positioning systems [25–31] and in [35], sub-millimetre accuracy is achieved using a carrier based UWB localisation system. In [30], time of arrival (TOA) estimation using IR-UWB devices mainly for on-body arm tracking through round trip-time of flight (RT-TOF) is presented obtaining an accuracy of 20 cm. TOA ranging error for indoor human tracking applications is investigated in [17] through experimental measurements for different bandwidths ranging from 500 MHz to 5 GHz with ranging accuracy varying from several metres for low bandwidth to 0.19 m. Investigations related to the feasibility of UWB technology for indoor track cycling is evaluated in [27]. Optimal position to mount the UWB hardware for application specific use is carried out by placing the UWB tag on various locations of the human subject such as lower back, wrist etc. during the cycling activity Fig. 6.2a. Different positions on the bicycle and cyclist were also evaluated based on accuracy, received power level, line-of-sight, maximum communication range, and comfort. An 8 base station set up is chosen for accurate localization of the human subject while cycling. It was found that the optimal hardware position was the lower back, with a median ranging error of 22 cm (infrastructure hardware placed at 2.3 m).

Hybrid motion tracking based on inertial measurement unit (IMU)/UWB technology and Kalman filter fusion algorithm has been studied in [31] which uses inertial motion capture system and commercial UWB localisation system (Fig. 6.2b). The proposed sensor fusion algorithm accurately tracks the spatial

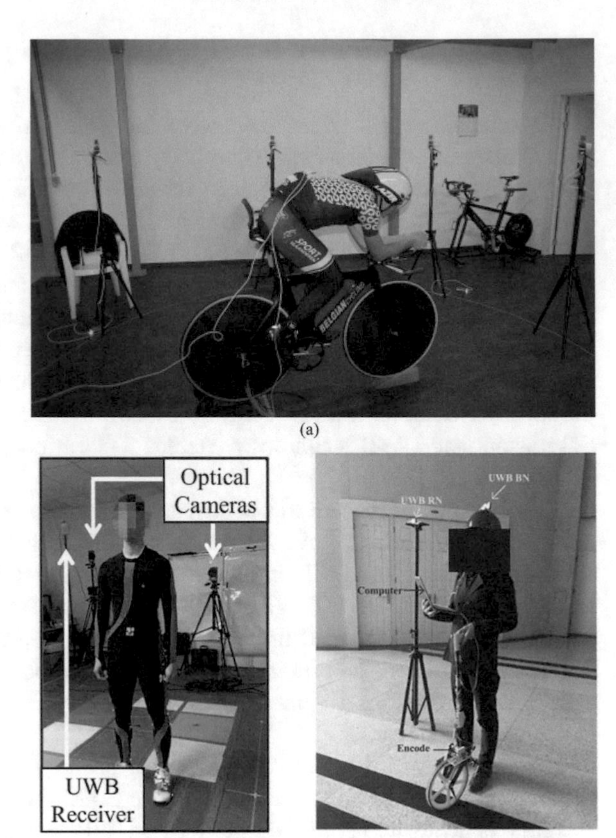

Fig. 6.2 **a** Bike Static measurement with Anchor layout and Tag placed on the lower back (After [27], Sensors). **b** Experimental setup with the test subject, UWB receiver, tags and optical cameras and markers (Reprinted with permission from IEEE [31]). **c** The indoor experimental environment for the human localization system based on least squares algorithm and by the time-delay model employing the Extended Kalman Filter (EKF) and EFIR estimators for the Group 1 of parameters (Reprinted with permission from IEEE [32])

location and motion of a human under various dynamic activities, such as walking, running, and jumping with an accuracy of 13 cm. An ultra-wideband (UWB)-based time-delay indoor human localization scheme is proposed to provide indoor human localization with time-delay measurements [32]. The experimental set up related to the tracking of the human subject is presented in Fig. 6.2c with an accuracy ranging between 20–50 cm. The human position is localized using the UWB-based distance data and the extended finite impulse response (EFIR) estimator employing the time delay localization model. High accuracy localization results have been reported in the literature using compact UWB body worn antennas in laboratory based [3, 8] and cluttered indoor environments [29]. The localization algorithms are based on the channel information and time of arrival localization techniques and the results are comparable with the optical motion capture system proving the suitability of the UWB technology for localization and tracking applications [3, 8].

6.2 Body Worn Antenna Localization

Measurements were performed in the motion capture studio at the University of Kent, UK [3, 8]. A human test subject 1.68 m tall and of average male build with a BMI of 21.3 was chosen for localizing antennas on the body (Fig. 6.3a). The antenna locations chosen for limb movement analysis are the three joints, (shoulder, elbow, and wrist) for the arm and (thigh, knee and ankle) for the leg. The arm movement was measured in nine different positions (0° to 120°) with intervals of 15° forward and sideways. The leg was moved in six different positions (−30° to +45°) in total (forward and backward) by 15° intervals during the motion capture measurements. A digital protractor was used to measure the angles with respect to the shoulder/thigh joint which is considered as reference for the arm/leg movement. The localization of the whole body is performed using 21 antenna positions placed on the (arms and legs) limbs and torso when the subject is standing at the center of the localization area. The four base stations (BS) with BS1 as reference zero co-ordinate (x_0, y_0, z_0) are positioned near the vertices of the cuboid in an area of 1.8×1.8 m^2 to obtain high accuracy positioning in three dimensions.

Compact and low cost tapered slot coplanar waveguide fed UWB antennas (TSA) (Fig. 6.3b, [36]) were used as transmitters placed on the body and as receivers in the base stations (BS). The TSA antenna has dimensions of 27×16 mm^2 and offers a return loss below −10 dB with good radiation performance and relatively constant gain across the UWB band when off, or on a human body.

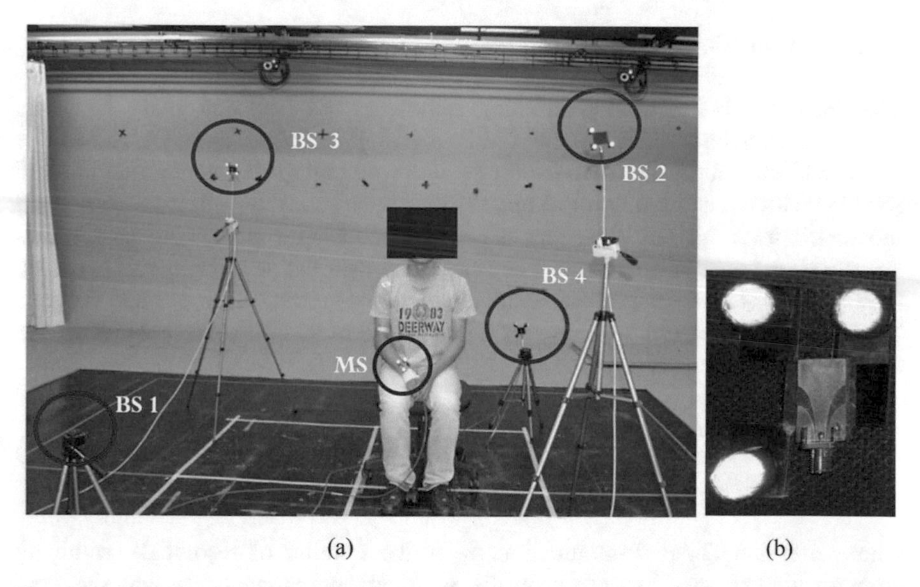

Fig. 6.3 **a** Measurement set up and venue in an area of 1.8×1.8 m^2. The base stations (red circles) and the body worn antenna (blue circle) were connected to a four-port vector network analyser. **b** Tapered slot UWB antenna placed on a plastic frame with reflective markers (Reprinted with permission from IEEE [3])

The antennas were mounted on plastic frames with 3 reflective markers on each (Fig. 6.3b) to provide benchmark 3D position information through a VICON motion capture system [3]. The 8 cameras motion capture system which gives high accuracy results (better than 1 cm) was used to compare with the UWB localization results and to obtain reference coordinates of the base stations (receiver antennas) [37].

Frequency domain measurements were performed in the 3–10 GHz band using a 4-port vector network analyzer (VNA R&S ZVA-50) to capture S_{21} (channel transfer function) parameters between each transmitter antenna location on the body and the BS (receiver) antenna. For each BS and mobile station (MS) antenna position, the channel impulse response (CIR) was obtained from the S_{21} collected from the VNA set to 6400 data samples, which is adequate to obtain the required channel information. For the case of limb movement tracking, measurements were taken for each of the wearable antenna location and the data is further combined to get the overall limb localization in three dimensions. The subject is made to hold to a particular position before proceeding to the next position with the VNA scanning time of around 800 ms to mimic the limb motion, hence obtaining a pseudo-dynamic motion for the limbs. For the case of the whole body, the antenna is positioned at different locations and readings are taken from the VNA.

6.2.1 Body-Worn Antenna Localization Techniques

6.2.1.1 Algorithm for Localization

The time of arrival (TOA) positioning techniques are used due to the fine time resolution/high bandwidth of the UWB signal [38, 39]. Figure 6.4 shows the proposed localization algorithm [3] for dealing with multipath and non-line-of-sight (NLOS) situations mainly developed for human body localization and limb movement tracking. The algorithm is based on channel impulse response characterization and time of arrival peak detection and data fusion techniques [3, 8, 29]. Firstly, the CIR and the received signal are obtained between each MS and BS pair. To obtain the CIR, an inverse fast fourier transform (IFFT) is applied to the measured S_{21}. The CIR is given by [40]:

$$h(\tau, t) = \sum_{k=1}^{K} a_k(t)\delta(\tau - \tau_k)e^{j\theta_k(t)} \tag{6.1}$$

where δ is the Dirac delta function, K is the number of resolvable multipath components, τ_k are the delays of the multipath components, a_k are the path amplitude values and θ_k are the path phase values. As seen from Fig. 6.5a, b there is significant difference in the channel type in terms various channel parameters such as number of multipath components and varies for different threshold levels.

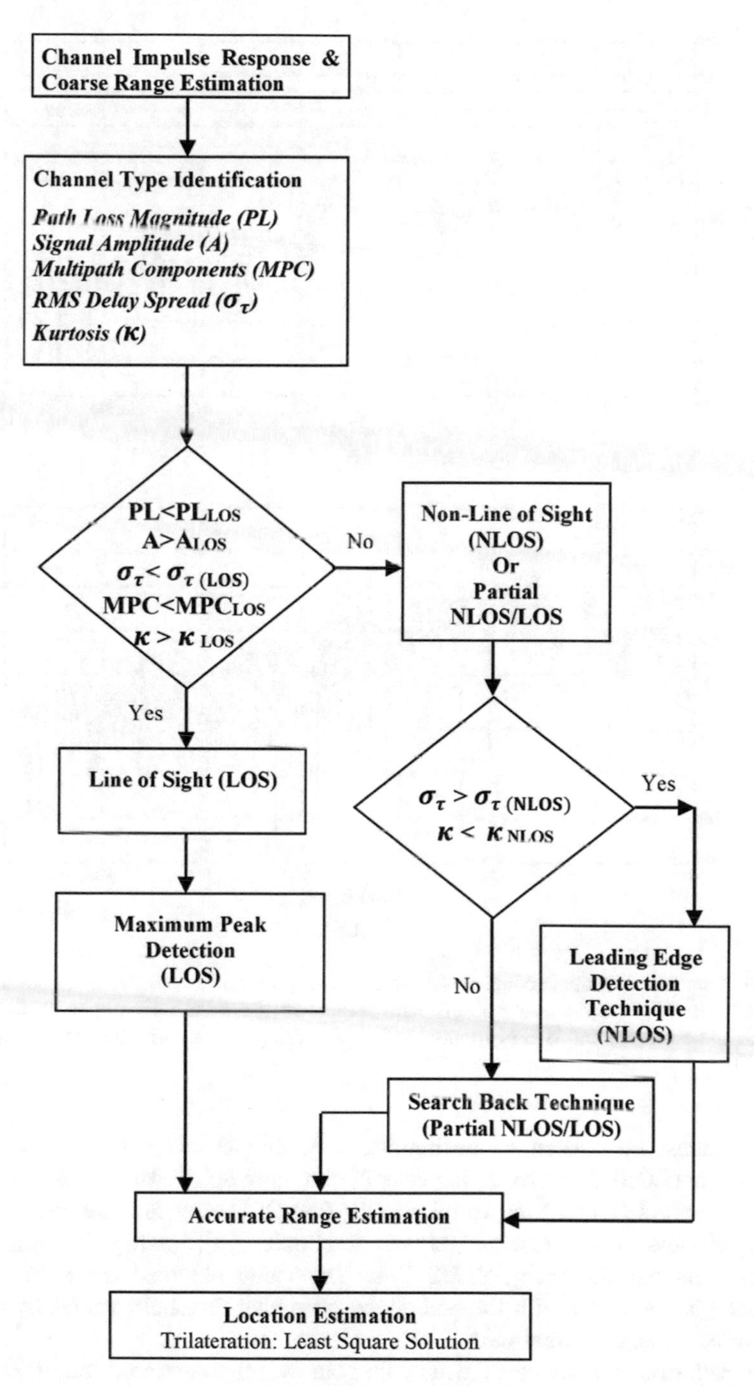

Fig. 6.4 Proposed localisation algorithm for body centric UWB localization and motion tracking applications (Reprinted with permission from IEEE [3])

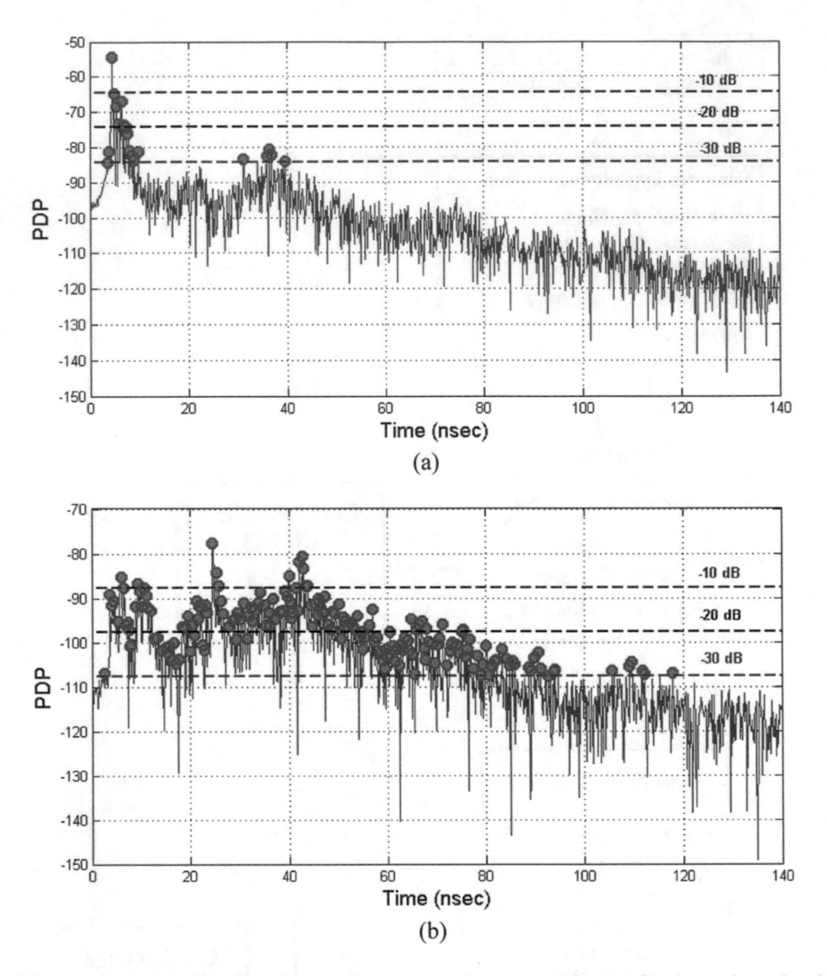

Fig. 6.5 Number of multipath components present in the power delay profile for various threshold levels of −10 dB, −20 dB and −30 dB with respect to the maximum amplitude for **a** BS1 (line-of-sight scenario) and **b** BS4 (non-line-of-sight scenario) (Reprinted with permission from IEEE [3])

Such features are chosen to distinguish between various channel links. Total line-of-sight (LOS) links are distinguished (with more than 95% accuracy) from the non-line-of-sight (NLOS) or partial NLOS (PNLOS) using the measured channel data [path loss, (PL); received signal amplitude, (A); multipath components, (MPC)] obtained for each MS-BS link. The value obtained for each channel parameter is compared with the predefined parameter threshold values by considering prior statistical information.

The path loss is obtained as mean path gain over the measured frequency band, as given by the following equation [41]:

$$PL(d(p)) = -20.\log_{10}\left\{\frac{1}{10}\frac{1}{N_f}\sum_{j=1}^{10}\sum_{n=1}^{N_f}\left|H_j^p(n)\right|\right\} \tag{6.2}$$

where *PL*(*d*(*p*)) is the path loss at the position of *p*, at which the distance between the Tx and Rx is a function of the position *p*, thus the distance is denoted by *d*(*p*). N_f is the number of frequency samples of the VNA. $H_j^p(n)$ is the measured S_{21} for the position *p*, *j*th snapshot, and *n*th frequency sample. High PL is observed for obstructed links leading to attenuation of the signal whereas lower magnitude of PL is observed for direct path links.

The power delay profile (PDP) is the squared magnitude of the impulse response [40].

$$P(\tau, t) = h(\tau, t)h^*(\tau, t) = \sum_{k=1}^{K} a_k^2\delta(\tau - \tau_k) \tag{6.3}$$

The PDP gives the intensity of a signal received through a multipath channel as a function of time delay where the time delay is the difference in the travel time between multipath arrivals.

Once, total LOS scenarios are distinguished, the remaining NLOS or partial NLOS/LOS links, are further classified with 95% accuracy using two parameters: rms delay spread σ_τ and kurtosis (κ). The rms delay spread is defined as [40]:

$$\sigma_\tau = \sqrt{\frac{\sum_k (\tau_k - \tau_m)^2.\ |h(\tau_k; d)|^2}{\sum_k |h(\tau_k; d)|^2}} \tag{6.4}$$

where τ_k are the multipath delays relative to the first arriving multipath component and *d* is the separating distance between the Tx and Rx. This parameter helps to distinguish between LOS and NLOS links as the rms delay spread values for NLOS scenarios are much larger.

Due to the dense multipath components, NLOS CIR has a wider distribution in comparison to LOS, which has a clearly distinguishable maximum peak with multipath components of very low magnitude. The kurtosis is a statistical parameter that indicates the fourth order moment of the received signal amplitude [3, 42]. Kurtosis κ is defined as follows:

$$\kappa(x) = \frac{1}{\sigma^4}\frac{\sum_i (x_i - \bar{x})^4}{N} \tag{6.5}$$

where σ is the standard deviation of the variable x and $\overline{x}$ is the mean value of x. N is the number of samples of x. The kurtosis index κ is supposed to be much lower for NLOS scenarios in comparison to LOS.

The maximum peak detection algorithm provides an estimate of the time of arrival of the UWB signal between Tx and Rx for line-of-sight (LOS) situations [3, 39]. The strongest peak of the CIR is considered an estimate of TOA and gives accurate results for scenarios with low multipath components and interference levels. In NLOS situations, the direct signal between Tx and Rx is significantly attenuated due to the presence of obstructions. The strongest path in such scenarios does not give the direct path estimate, leading to large ranging errors. In addition, the human body acts like an obstacle, hence causing NLOS and high multipath situations [3, 8]. In order to mitigate NLOS errors and accurately estimate the range, threshold-based algorithms are used [42, 43]. Threshold-based search algorithms [43] compare individual signal samples with a certain threshold to identify the first arriving signal and obtain the range information [43, 44]. For total NLOS scenarios, threshold based leading edge detection algorithms [45] can be used with selected threshold generally around 10–20% of the peak CIR amplitude depending on the magnitude of the CIR peaks prior to the peak value. Search back technique can be used to estimate the range for partial LOS/NLOS scenarios that fall under the intermediate channel classification parameter values. The search back method first finds the strongest path (SP), and then looks for a peak arriving before the strongest path, which has greater power than a detection threshold level. A few iterations are required to obtain the peak value nearest the expected one or within the localization range based on the selected threshold level.

After acquiring accurate range estimates for each MS-BS link, TOA-based mobile location algorithm is applied based on trilateration and least square solution to determine the unknown position of the antenna [46–48].

6.2.1.2 Time of Arrival Data Fusion Method

The time of arrival data fusion method is based on combining estimates of the TOA of the MS signal that arrives at four different BSs [46, 47]. Since the wireless signal travels at the speed of light (c = 3 × 10^8 m/s), the distance between the MS and BSi is given by

$$r_i = (t_i - t_0)c \tag{6.6}$$

where t_0 is the time instant at which the MS begins transmission and t_i is the TOA of the MS signal at BSi. The distances r_1, r_2, r_3, r_4 can be used to estimate (x_m, y_m, z_m) by solving the following set of equations:

$$r_1^2 = x_m^2 + y_m^2 + z_m^2 \tag{6.7}$$

$$r_2^2 = (x_2 - x_m)^2 + (y_2 - y_m)^2 + (z_2 - z_m)^2 \tag{6.8}$$

$$r_3^2 = (x_3 - x_m)^2 + (y_3 - y_m)^2 + (z_3 - z_m)^2 \tag{6.9}$$

$$r_4^2 = (x_4 - x_m)^2 + (y_4 - y_m)^2 + (z_4 - z_m)^2 \tag{6.10}$$

These equations can be solved by using least square solution. Subtracting (6.7) from (6.8), (6.9), (6.10) and rearranging the terms in matrix form, we get:

$$\begin{bmatrix} x_2 & y_2 & z_2 \\ x_3 & y_3 & z_3 \\ x_4 & y_4 & z_4 \end{bmatrix} \begin{bmatrix} x_m \\ y_m \\ z_m \end{bmatrix} = \frac{1}{2} \begin{bmatrix} K_2^2 - r_2^2 + r_1^2 \\ K_3^2 - r_3^2 + r_1^2 \\ K_4^2 - r_4^2 + r_1^2 \end{bmatrix} \tag{6.11}$$

where

$$K_i^2 = x_i^2 + y_i^2 + z_i^2$$

It can be rewritten as

$$Hx = b \tag{6.12}$$

$$H = \begin{bmatrix} x_2 & y_2 & z_2 \\ x_3 & y_3 & z_3 \\ x_4 & y_4 & z_4 \end{bmatrix} \quad x = \begin{bmatrix} x_m \\ y_m \\ z_m \end{bmatrix} \quad b = \frac{1}{2} \begin{bmatrix} K_2^2 - r_2^2 + r_1^2 \\ K_3^2 - r_3^2 + r_1^2 \\ K_4^2 - r_4^2 + r_1^2 \end{bmatrix}$$

The target coordinates can be found by rearranging the matrix equation through the following equation:

$$\hat{x} = \left(H^T H\right)^{-1} H^T b \tag{6.13}$$

6.2.1.3 GDOP Analysis

The positioning precision depends significantly on the geometry of the base station's distribution. Geometric dilution of precision (GDOP) [49–51] indicates the effectiveness of a geometric configuration. It is an indicator of three-dimensional positioning accuracy as consequence of relative position of the BSs with respect to a MS.

$$GDOP = \sqrt{tr(H^T H)^{-1}} \tag{6.14}$$

where $H = \begin{bmatrix} a_{x1} & a_{y1} & a_{z1} & 1 \\ a_{x2} & a_{y2} & a_{z2} & 1 \\ a_{x3} & a_{y3} & a_{z3} & 1 \\ a_{x4} & a_{y4} & a_{z4} & 1 \end{bmatrix}$.

And $a_{xi} = \frac{(x_i - x_m)}{r_i}$, $a_{yi} = \frac{(y_i - y_m)}{r_i}$, $a_{zi} = \frac{(z_i - z_m)}{r_i}$.

where x_i, y_i, z_i correspond to the base station positions, x_m, y_m, z_m is the estimated mobile station position and r_i is the estimated range between the mobile and each base station.

$$HDOP = \sqrt{\left((H^T H)^{-1}\right)_{1,1} + \left((H^T H)^{-1}\right)_{2,2}} \tag{6.15}$$

$$VDOP = \sqrt{\left((H^T H)^{-1}\right)_{3,3}} \tag{6.16}$$

where VDOP is the vertical dilution of precision which is a measure of localization accuracy in 1D (height (z axis)) and HDOP is the horizontal dilution of precision which is the measure of localization accuracy in 2D (x and y axis).

6.2.2 *Limbs Channel Classification:*

6.2.2.1 Path Loss Analysis

The path loss describes the fluctuations of the received signal with respect to the distance and signifies the local average received signal power (P_r) relative to the transmitted power (P_t) [40, 41]. It is directly calculated from the measured data by averaging the measured frequency transfers at each frequency point [41, 52].

The path loss magnitude for various positions of the arm is shown in Fig. 6.6 with respect to BS1 and BS4 during sideways arm movement activity. Figure 6.6a clearly shows the increase in path loss for the body worn antennas placed on the wrist as the arm moves from 0° to 120°. The variation in PL is maximum for the wrist (e.g. BS1: 55 to 70 dB) and ankle (e.g. BS2: 60 to 70 dB) as larger displacement is taking place during limb motion in comparison to the elbow/shoulder and knee/thigh location respectively. For BS2 and BS3, the magnitude of PL is high (e.g. for BS3: 70 dB) for 0° position and decreases as the arm moves to 120° position (e.g. for BS3: 55 dB) as the distance will reduce between the BS and MS. As observed in Fig. 6.6b, all NLOS links are formed with respect to BS4 due to obstruction of the direct path between the Tx and Rx.

6.2.2.2 Amplitude of Received Signal

Different levels of magnitude are observed for each wearable antenna position that is dependent on the distance between the MS-BS link, the channel type i.e. (LOS or NLOS) and orientation of the antenna with respect to each other (Fig. 6.7). BS1 is

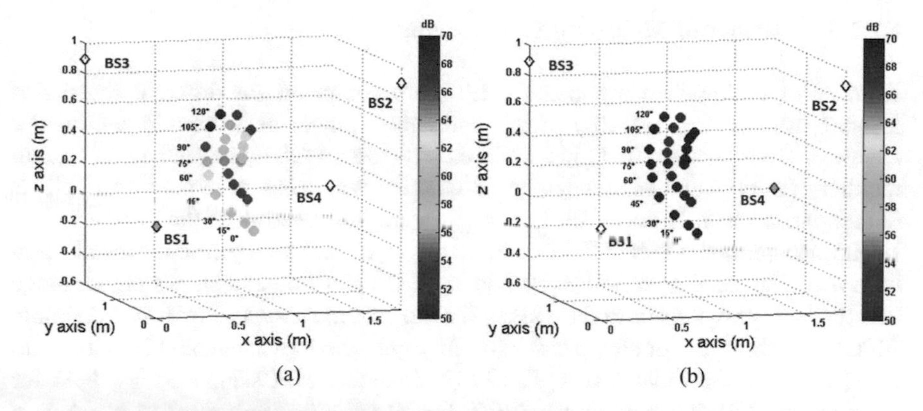

Fig. 6.6 Path loss magnitude colour graph for various wearable node positions (S1-S3) with respect to different base station location **a** BS1 and **b** BS4 for sideways right arm movement (Reprinted with permission from IEEE [3])

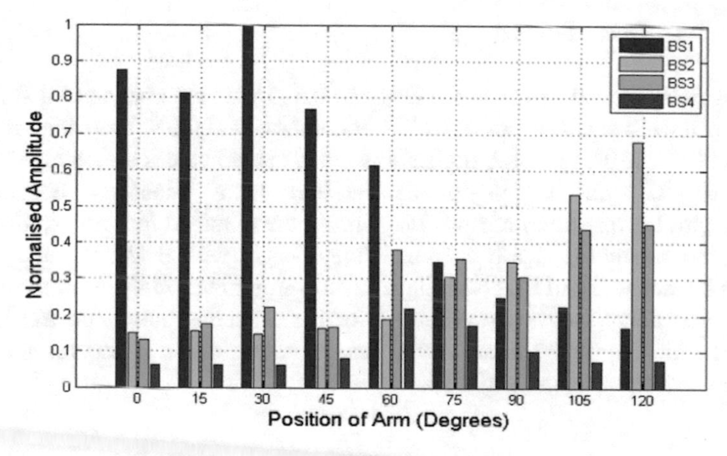

Fig. 6.7 Variation of amplitude of the received signal with respect to different base stations when the antenna is placed on the right wrist (S1) (Reprinted with permission from IEEE [3])

generally in LOS situation for nearly all antenna positions with highest signal amplitude in comparison to BS4, which is in NLOS situation due to shadowing by the human subject. For BS1, the signal magnitude decreases from 30° to 120° as MS-BS distance is increasing. It can be observed that the magnitude is generally lower for 0° to 60° orientations and then rises from 75° to 120° for BS2 and BS3. There is a drastic reduction in magnitude for BS4 as it is in NLOS situation with maximum percentage decrease as high as 90% of the magnitude level of BS1. Substantial amplitude variations are observed for antennas placed on the wrist/ankle due to higher displacements of positions caused by the arm/leg movements.

6.2.2.3 Number of Multipath Components

Number of multipath components (MPC) are computed for different thresholds levels (–10, –20 and –30 dB) of the normalised power delay profile (PDP). For sideways limb motion BS1 and BS3 mostly form LOS links and have similar number of multipath components. For BS2, there is an increase in number of MPCs in comparison to BS1 and BS3. For the forward limb motion of the right arm less MPC's are present for BS1,2 in comparison to BS3. The leg motion results also have a similar trend with BS1,3. BS2 is in NLOS situation and shows higher range of MPC's in comparison to the values for the arm movement. For BS4 maximum MPC's are observed for all the cases for the right limb movement and increase with the decreasing threshold levels. For LOS situations MPCs increase by 10% for decrease in PDP threshold and by 30% for NLOS situations. Highest number of MPCs are observed for the shoulder and thigh region as they are totally obstructed due to the torso of the body.

6.2.2.4 Rms Delay Spread

Low rms delay spread values in the range of 0.5–3 nsec are obtained for BS1 when the arm is moved sideways for 0° to 75° orientation, and the values increase up to 9 nsec for 90° to 120°. For BS2, higher rms delay spread values are obtained for arm motion (0°–45°) and leg motion as the body-worn antenna is not in direct line-of-sight. Lower values of rms delay spread are obtained for arm positions 90°–120° as the antenna is clearly in direct line-of-sight with BS2,3 and the distance between the antenna and BS also reduces as the arm moves in the upward direction. Highest rms delay values are obtained for the antennas placed on the leg with values reaching up to 20 nsec clearly demonstrating dense multipath and NLOS links between the antenna and BS4.

6.2.2.5 Kurtosis

The kurtosis values are in the range of 45–60 for BS1,2,3 showing higher probability of direct path and low multipath situation between the antenna and BS. For BS4, most of the antenna locations are in NLOS with the receiver, hence low kurtosis index values (8–20) are observed. Highest values are observed for antennas placed on the wrist with respect to BS1 during sideways movements. Lowest kurtosis values are observed for the legs especially for the thigh region, as maximum shadowing from the body occurs for this region as BS4 is placed behind the subject.

6.2.3 Human Body Localization

6.2.3.1 Path Loss Analysis

BS1,3 and BS1,2 have lower magnitude of PL as LOS links are formed for right limbs (S1-S6) and torso region (S16-S18) respectively. The overall variation in PL values obtained with respect to BS1 and BS4 is shown in Fig. 6.8a, b respectively. It is observed that, for BS1, the right limbs and torso have a PL magnitude ranging

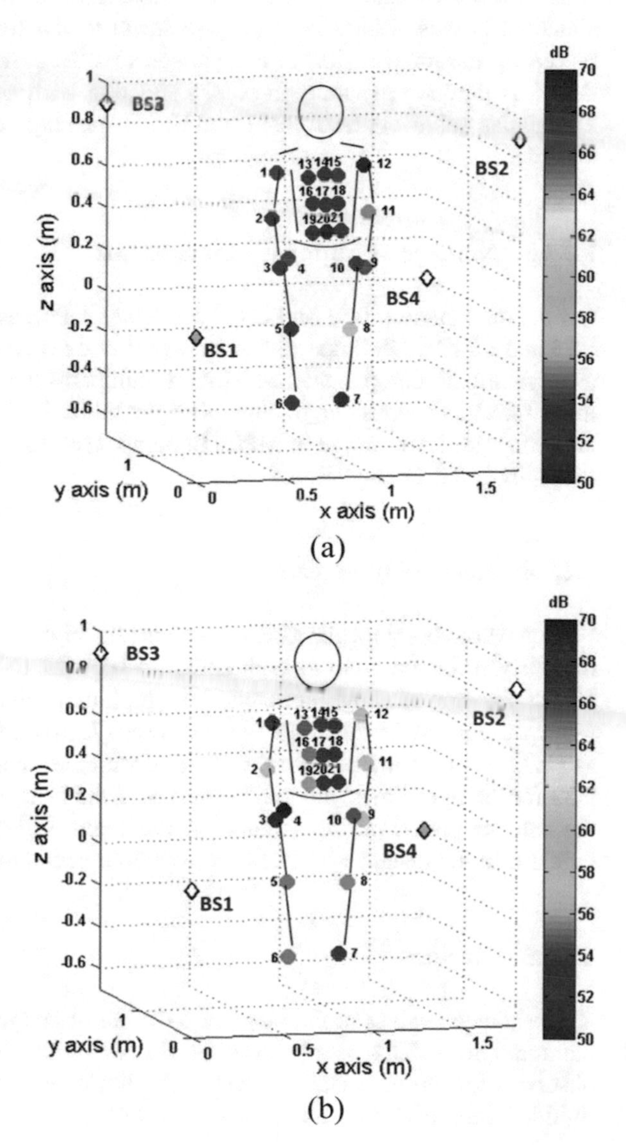

Fig. 6.8 Path loss magnitude colour graph for various wearable node positions (S1-S21) with respect to different base station locations **a** BS1 and **b** BS4 for the human body localisation (Reprinted with permission from IEEE [3])

from 50 to 55 dB whereas the left limbs have PL from 63 to 70 dB. The PL magnitude for the left shoulder is highest for BS1 case as it is at maximum distance from BS1 showing distance dependency. For BS4 higher magnitude of path loss is obtained for right limbs and torso in comparison to the left limbs.

6.2.3.2 Amplitude of Received Signal

The amplitude of the received signal is higher for BS2-left elbow and shoulder link as there is least distance between the BS and the antennas and is also in direct LOS situation. Lowest values of amplitude are observed for BS4 when the antenna is placed on the torso region in comparison to BS1. Lower values are observed for BS3,4 as they are generally in NLOS situation leading to distortion of the signal. The human subject is facing BS1 and BS2, hence high levels of signal amplitude is observed.

6.2.3.3 Number of Multipath Components

For the torso region, BS1 and BS2 have lower MPCs as LOS links are formed. For BS3 and BS4 NLOS links are formed as the body is obstructing the path leading to more multipath components. Maximum multipaths are for the BS4 link for all the three thresholds considered. For left/right limbs BS2,4/BS1,3 links are in LOS scenario with lower range of MPCs in comparison to BS1,3/BS2,4 links which are generally in NLOS situation.

6.2.3.4 Rms Delay Spread

For the whole-body localization measurements, higher rms delay spread (Fig. 6.9a) is observed for the torso as both BS3 and BS4 (12 to 17 nsec) at the back are in NLOS situation causing interference and high multipath scenarios. Generally lower rms spread values in the range of 0.1–5 nsec (75% of the total values) are observed as BS1,2 is in direct line-of-sight situation with the wearable antennas. Considering BS3 for the right limbs, the trend observed is the increase in rms delay spread from the antenna placed on the shoulder to the antenna placed on the wrist as there is increase in the distance between the wearable node and BS.

6.2.3.5 Kurtosis

Kurtosis index values in the range of 5–25 are observed for the antennas placed on the torso for BS3,4 as observed in Fig. 6.9b The highest kurtosis values are observed for the base stations facing the limbs, e.g., for antennas placed on the right/left limb BS1,3/BS2,4 show higher kurtosis index values. Very high kurtosis

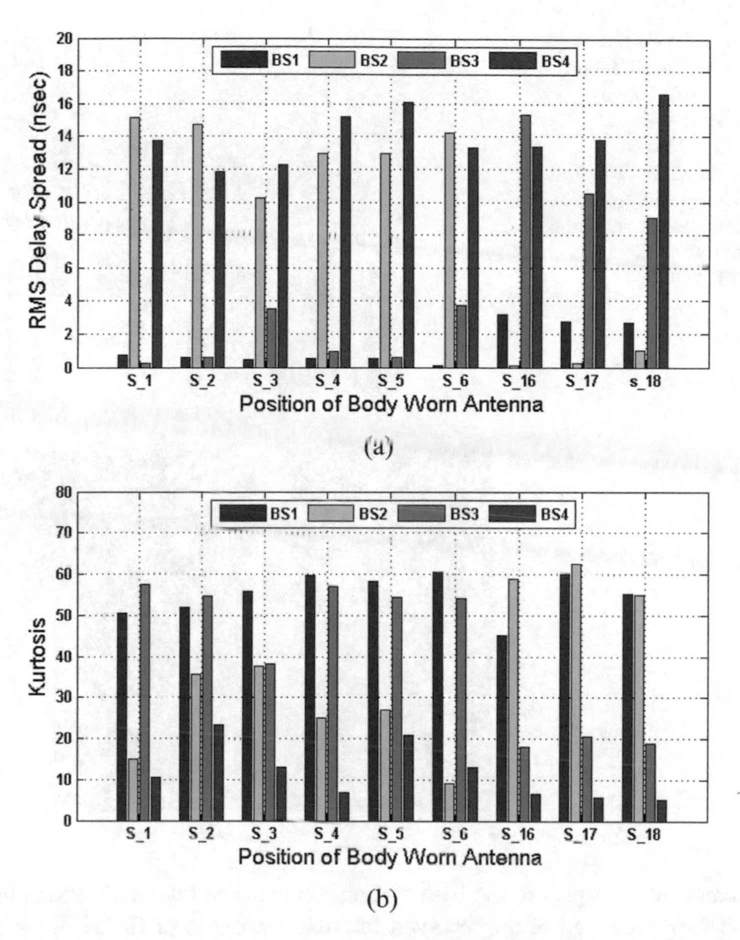

Fig. 6.9 Variation of the **a** Rms delay spread and **b** Kurtosis of the received signal with respect to the base stations (BS1-BS4) and the wearable antenna. Antenna positions: S_1-S_3; S_4-S_6; S_16-S_18 (Right arm, Right leg, Torso centre) (Reprinted with permission from IEEE [3])

index values (45–65) are observed for BS1 and BS3 for antennas (S_1-S_6) and for antennas (S_7-S_12) BS2 and BS4 show high kurtosis values, the magnitude of the values vary depending upon the location of the antenna with respect to the BS.

6.2.4 Localization Results for Various Activities

High centimeter range accuracy is obtained using time of arrival UWB localization techniques. The localization accuracy depends on various factors such as base station configuration, propagation environment, signal to noise ratio, presence of objects and obstacles specifically dynamic ones such as the human body, location of

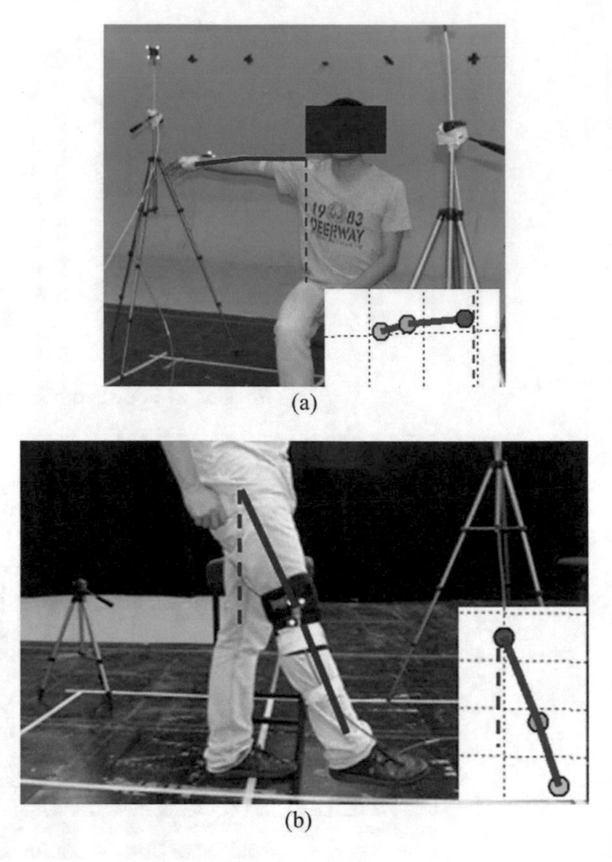

Fig. 6.10 **a** UWB Sideways Arm Motion Localisation Pink (shoulder), Green (Elbow), Blue (Wrist). **b** UWB Leg Motion Localisation: Blue (Thigh), Green (Knee), Pink (Ankle). All dimensions are in metres (Reprinted with permission from IEEE [3])

the antennas with respect to the base station, accuracy of the localization algorithm and sampling precision of the received impulse responses of IR UWB system.

The captured and analyzed location estimation results are compared with the results obtained by a standard optical motion capture system, which is used as a reference. Various positions taken by the arms and legs are depicted in Figs. 6.10 and 6.11, respectively along with the tracking results obtained by the approach proposed. The average displacement error between subsequent limb motion captures is around 1–2 cm with standard deviation in the range of 0.9–1.4 cm. The results obtained show high accuracy in capturing the position of the limbs showing the suitability of the UWB technology for localization and tracking.

Results for the localization of the body worn antennas for the whole-body localization are shown in Fig. 6.12. 3D localization error obtained is in the range of 0.5–2.5 cm with standard deviation of 0.7–1.3 cm for overall positioning of the body worn antennas. Localization accuracy results are best for the antennas placed on the arm in comparison to antennas placed on the legs as units placed on the legs incur more NLOS situations due to greater diameter than that of the arms. Average localization accuracy as small as 0.5–2.5 cm has been achieved for 90% of the scenarios, which is comparable to common commercial optical systems.

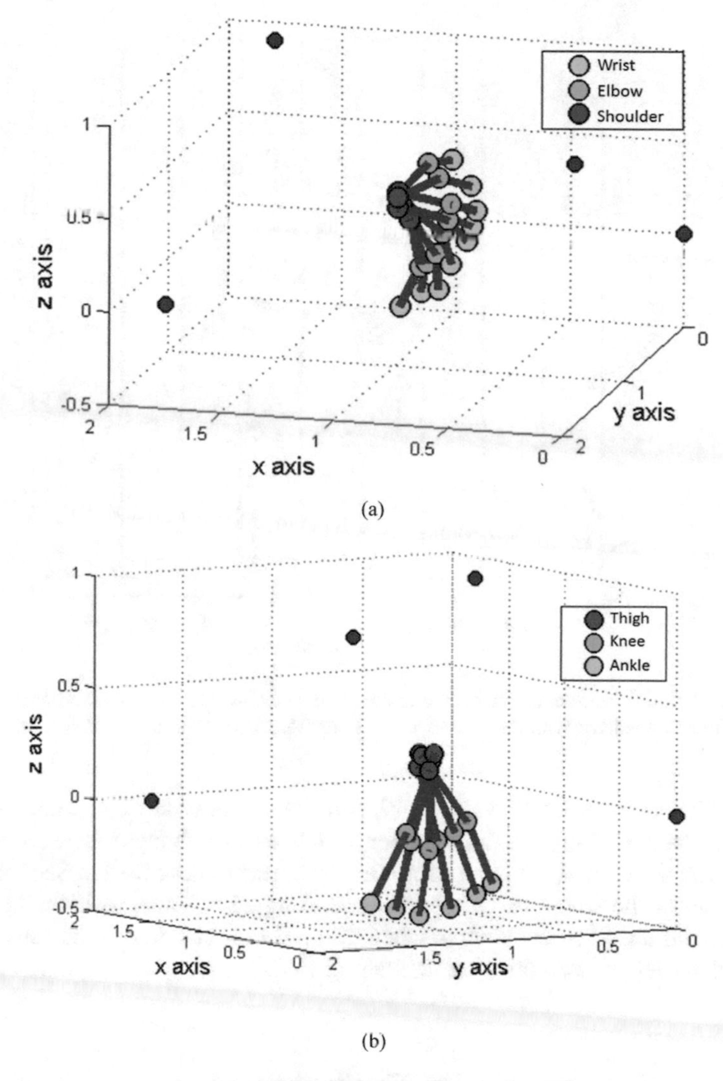

Fig. 6.11 Motion capture of UWB localisation for arm movement **a** 90° sideways motion and leg movement **b** 30° (forward motion) Inset: Pictorial representation of localisation of body worn antennas on the arm/leg using UWB technology (Reprinted with permission from IEEE [3])

6.3 Random Base Station Placement

Analytical and theoretical investigations have been carried out related to random placement of base station configurations and compared with optimised BS configurations used for body centric localization [53]. Localization accuracy analysis and the effectiveness of the BS geometrical configuration is performed using

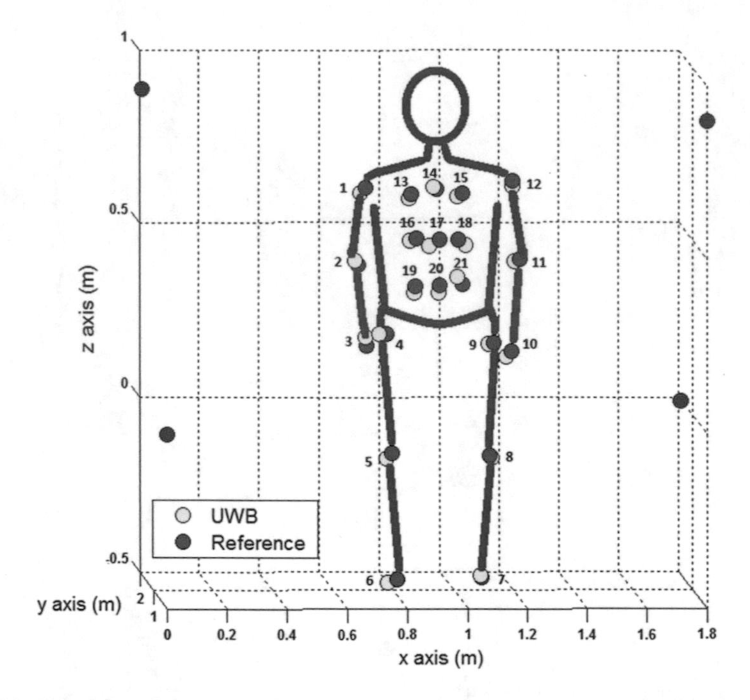

Fig. 6.12 Pictorial representation of the body worn antenna localisation of comparing UWB based estimated positions with the reference positions (Reprinted with permission from IEEE [3])

GDOP, VDOP and HDOP values [49, 50]. The results are compared with the optimized base station configuration for which realistic body-centric localization experiments are carried out in an indoor environment as described in Sect. 6.2. The localization of the whole body is performed using 21 antenna positions placed on the (arms and legs) limbs and torso when the subject is standing at the center of the localization area as with the measurement procedure given in Sect. 6.2.

6.3.1 Random Base Station Configurations

To obtain accurate position of the target, an appropriate geometrical distribution of base-stations is necessary. The optimized base station configuration chosen is a cuboid shape configuration with four base stations each at a corner of a square with a predefined area depending upon the application (Fig. 6.13a). Random placement of base stations, on the other hand can lead to degradation of the performance of the localization system, leading to inaccurate localization results of the target [50, 51, 53].

Six different random x–y positions of the BSs are considered and for each set of x–y positions, four different random z-axis positions are also considered. An example related to BS configuration analysis is shown in Fig. 6.13b for various

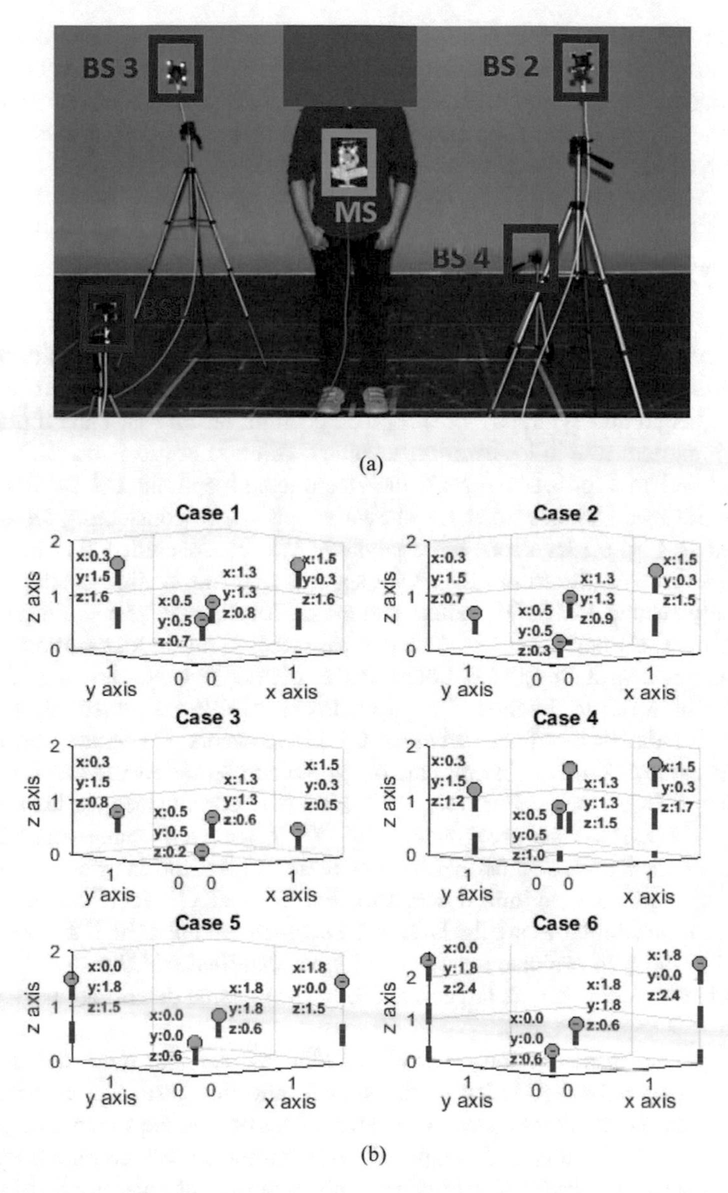

Fig. 6.13 **a** Localization of wearable antenna placed on the human subject in an indoor environment with 4 BSs placed in optimum position to obtain high accuracy localization. **b** Schematic of various BSs configurations. Case 1–4: (Random placement of BSs), Case 5: Optimized BS configuration, Case 6: Ideal BS configuration. All dimensions are in meters (Reprinted with permission from IEEE [53])

schematics of the BS station configuration. Fixed values for random *x*- and *y*-positions for the four base stations and variable *z*-axis co-ordinate values for the base stations are considered for cases 1–4 (C1-C4). These results are compared with the case 5 (C5) (optimized) and case 6 (C6) (ideal scenario, where distance between all the BS in *x*, *y*, and *z* axis direction is the same).

6.3.2 Localization Accuracy Analysis

It is observed that trilateration and least square algorithm gives best results when the antennas are placed at the corners of a square (C5 and C6) with 1–2 cm accuracy. Also, the height and symmetry of the *z*-axis position of the base station placement play an important role in localization accuracy. Dilution of precision (DOP) values are presented in Fig. 6.14a for all the random and optimal BS configurations. Average 3D localization results are shown in Fig. 6.14b considering all six cases presented in Fig. 6.13b. From the analysis it can be concluded that localization results are directly related to GDOP values. *x–y* axis localization results are more specifically related to HDOP values and *z*-axis localization results are related to VDOP values. C1 and C2 show some decrease (0.5–1 cm) in localization accuracy when compared with C5 and C6. For such scenarios DOP values are like C5 and C6 and fall within 0.5 to 5 range. C1 shows lower VDOP values leading to higher accuracy for the *z*-co-ordinate whereas C2 is vice-versa. C3 shows lower localization accuracy (2–6 cm) in comparison to the configuration used in C1 and C2 due to larger DOP values. For C3 and C4 *z*-axis direction values are least accurate as several VDOP values are greater than 20. When DOP values are >20 poor localization results are obtained with an increase in localization error by 2–4 times when compared with optimized scenario. For C5 and C6, localization accuracy results are quite similar as are the DOP values which fall under high accuracy range of 1–3. For C6 *z*-co-ordinate accuracy is higher than that of C5 as the distance of the BSs in the *z*- direction is larger for C6, but with some deterioration in *x–y* axis co-ordinate localization results.

Varying the base stations *z*-axis coordinates and keeping the *x–y* axis co-ordinates optimized (C5) led to different localization results in comparison to varying all the *xyz* co-coordinates. Random *z* values chosen are shown in Fig. 6.13b (C1–C4) and the results are compared with optimized BS configuration (C5). Results indicate significant deterioration in position estimation accuracy by 2–5 cm for the *z*-axis co-ordinate for the randomly positioned base stations in the *z*-axis only. The *x*- and *y*- co-ordinates estimation show only slight variation (0.5–1 cm) when compared with the optimized scenario. The *z*-axis error is high for this scenario as the nodes considered on the body are placed at different positions starting from shoulder to the ankle. Hence more randomly positioned BSs in the *z*-axis direction will play an important role in the *z*-co-ordinate accuracy of the localization results. Varying the base stations in *x–y* axis coordinates and keeping the *z*-axis co-ordinates optimized has less variation in the localization results. As the

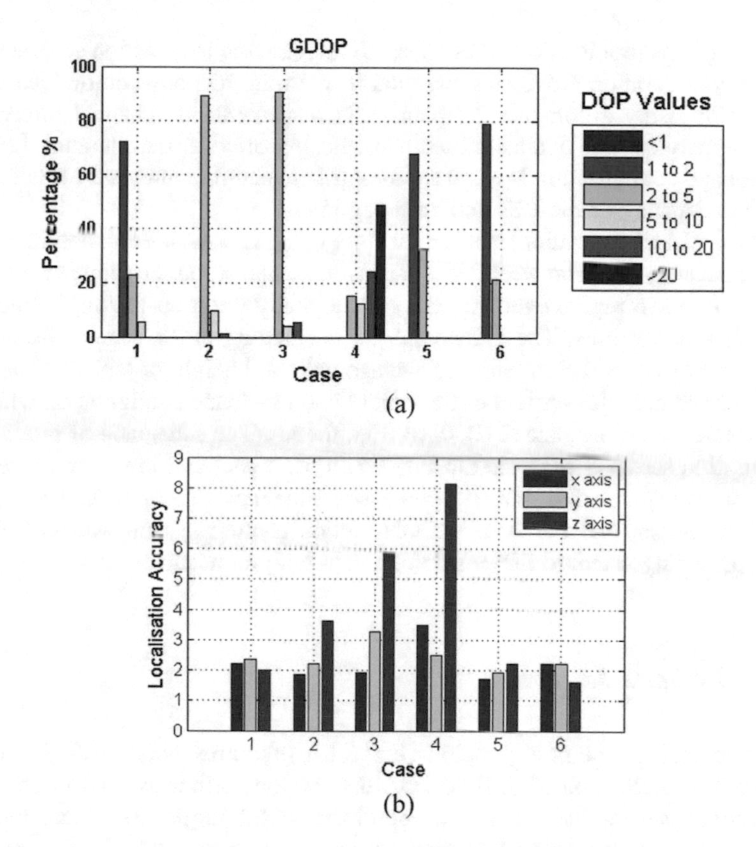

Fig. 6.14 **a** DOP values classified in different range for various base station configurations (C1-C6). **b** Localization accuracy for various base station configurations (C1-C6) (Reprinted with permission from IEEE [53])

human subject is standing in the center of the localization area, hence randomly placed *x*–*y* axis base stations do not have very significant effect on the localization accuracy for *x*- and *y*- axis co-ordinates estimation. Also, as *z*-coordinate values are optimized, overall 3D localization results are good with an average error of 2–3 cm.

6.4 L-shape Base Station Configuration Measurements

For applications that have more space constraints and still aim for very high tracking accuracy, more compact, scalable, and easily portable BSs configurations need to be investigated. This work presents a compact 2D plane L-shape configuration placed in various orientations in the localization domain and an in-depth channel analysis to mitigate NLOS effects to increase the localization accuracy. The L-shape configuration provides substantial increase in 3D body-centric localization accuracy by using a compact 2D plane BSs placement and channel information when compared to the available methodologies reported in the literature [26–30]. The L-Shape configuration is a good trade-off between the number of BSs needed, volume occupancy of the BSs and accuracy [54].

Body-centric localization measurements were performed in a motion capture studio laboratory [3]. The wearable antennas were placed on the joints of the arms/legs and the torso region over a 3 × 3 grids to study the propagation channel type and localization results. The human subject is standing at the center of the localization area of 1.8 × 1.8 m^2 and has average built and height of 168 cm. The 3BSs are positioned near the vertices of the cuboid in an L-Shape configuration with BS1 as the reference co-ordinate at (0, 0, 0) (Fig. 6.15a). The schematic of the L-shape configuration set up is presented in Fig. 6.15b. Compact and low cost tapered slot co-planar waveguide fed UWB tapered slot antennas (TSA) were used as the body-worn sensor and the BSs. VICON motion capture system was used as reference providing standard 3D positioning data with an accuracy of <1 cm.

6.4.1 L-shape Localization

The compact L-Shape configuration (Fig. 6.15b) requires only 3 BSs to achieve accurate 3D localization [51]. First TOA data fusion method is used to obtain the estimated values for the x_m and z_m coordinate of the target. To obtain the third coordinate y_m, the following equations are used:

$$\theta = \cos^{-1}\frac{z_m}{R\_est} \tag{6.17}$$

$$S = \frac{z_m}{\tan(90 - \theta)} \tag{6.18}$$

$$y_m = \sqrt{S^2 - x_m^2} \tag{6.19}$$

where R_est is the estimated distance from the BS1 (origin) to the target P. θ is the zenith angle and S is the projected distance of point P from origin in the x–y plane. Four variations of the L-Shape BS configuration are analyzed namely: (L1, L2, L3 and L4) positioned in the *x–z plane* and *y–z plane* as shown in Fig. 6.15c BS1,2_A and BS1,2_B are common BSs positions for Fig. 6.15 [I]–[II] and Fig. 6.15 [III]–[IV] respectively. BS3 is denoted as $BS3_{c1}/BS3_{c2}$ and its location can be in the *x–z/y–z plane* depending on the orientation in which the configuration is positioned.

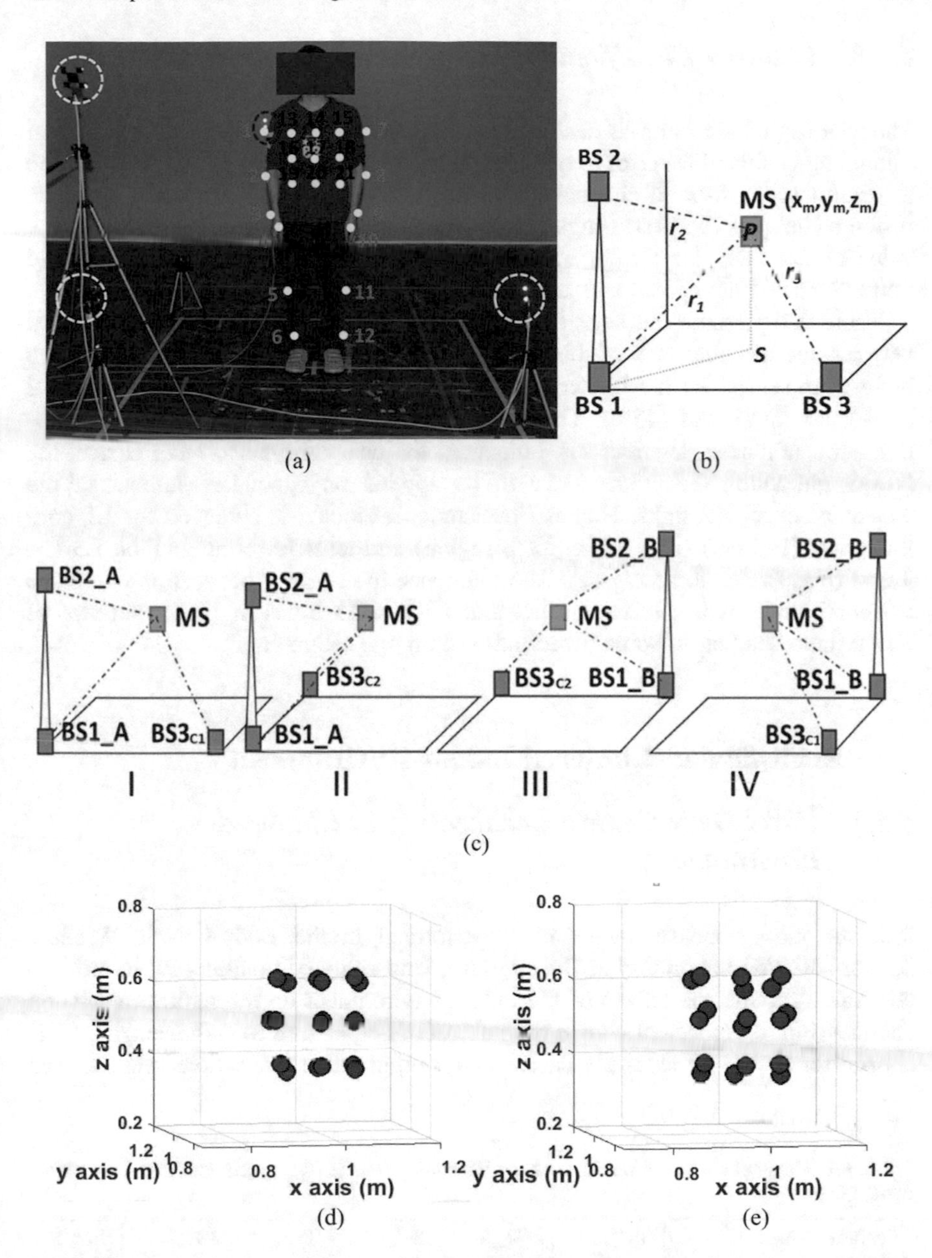

Fig. 6.15 **a** Experimental set up: x–z plane L-Shape BS configuration and human subject with wearable sensor. **b** L-Shape base station configuration schematic. **c** L-Shape base station configurations: (I) L1: x–z plane, facing forward, (II) L2: y–z plane, facing side right, (III) L3: x–z plane, facing backward, (IV) L4: y–z plane, facing side left. 3D localisation of the torso region on-body nodes using IR-UWB (red) and VICON motion capture system as reference (blue) L-Shape configuration **d** L1, **e** L3 (Reprinted with permission from IEEE [54])

6.4.2 Channel Classification and Localization Accuracy

The localization algorithm is designed to consider the NLOS and MPCs effects and mitigating positive bias error to obtain high accuracy body-centric localization. The algorithm uses off-body channel information and TOA peak detection and data fusion techniques [3]. To distinguish between total LOS and NLOS scenarios, peak value of the PDP, for further classification of the links into partial (P)LOS and PNLOS, rms delay spread and kurtosis is considered.

Table 6.1 presents the classification of the type of body centric links formed between the different BSs of the four L-Shape configurations and location of the body-worn sensor. L1 has higher number of LOS links (70%) in comparison to L2 (57%), L3 (30%), L4 (42%). The plane (*x–z* or *y–z*) in which the L-Shape configuration is placed, significantly influences the channel type formed. Hence, the NLOS mitigation techniques have to be applied on a smaller number of the wearable sensor-BS links. Highest localization accuracy is achieved for L1 configuration (1–3 cm) (Fig. 6.15a, torso region) and least for configuration L3 (2–4 cm) (Fig. 6.15b, torso region). The difference in accuracy between the L-Shape BS configurations is due to the fact that L3 has 2.5 times higher occurrence of NLOS links leading to some deterioration in range estimation.

6.5 Realistic and Cluttered Indoor Environment

6.5.1 UWB Body Centric Localization in Cluttered Environments

Realistic indoor measurements were performed in the Body-Centric Wireless Sensor (BCWS) Laboratory at Queen Mary, University of London [29] in order to take into account, the effects of the indoor environment on the radio propagation channel (Fig. 6.16a). Real human test subject of height 1.68 m and average built is chosen for localizing antennas on the body. Eight antenna locations were chosen

Table 6.1 Channel classification for L-Shape BS configurations (Reprinted with permission from IEEE [54])

Wearable sensor location	BS1_A	BS2_A	$BS3_{c1}$	$BS3_{c2}$	BS1_B	BS2_B
Right Limbs S1-S6	LOS	LOS	NLOS PNLOS	LOS	NLOS PNLOS	NLOS PNLOS
Left Limbs S7-S12	NLOS PNLOS	NLOS PNLOS	LOS	NLOS	LOS	LOS
Torso S13-S19	LOS PLOS	LOS PLOS	LOS	NLOS	NLOS	NLOS

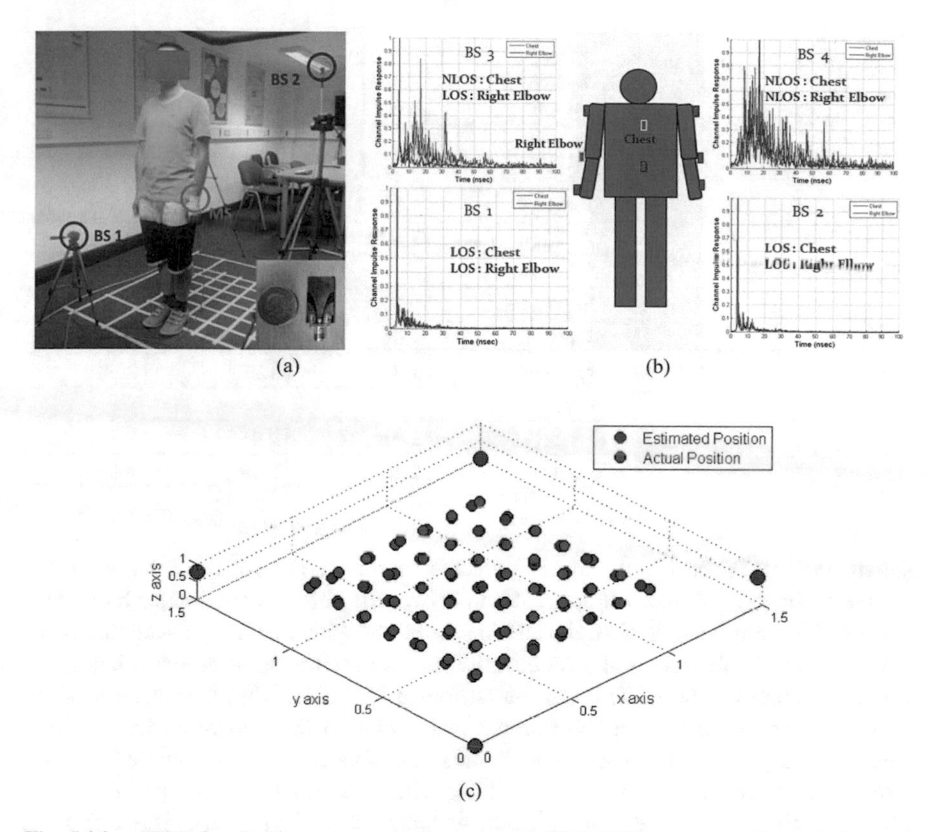

Fig. 6.16 **a** UWB localization measurement set up with TSA UWB antenna in inset, body worn sensor on the left wrist and BSs for localization of the body-worn antennas. **b** Different kind of channel impulse response observed: Line-of-sight direct path, line-of-sight with high multipath and detectable non-line-of-sight path. **c** Comparison of actual (blue), estimated (pink) positions of the wearable antennas: Right elbow (Reprinted with permission from IEEE [29])

with six at the joints of the arm (wrist, elbow, shoulder) and two on the torso (chest, waist). The distance between the human body surface and the antenna is around 5 mm. The body-worn antennas localisation was performed over the frequency band of 3–10 GHz using UWB compact and cost-effective tapered slot antennas (TSAs) [36] which act as the body-worn antenna and base stations. Different channel conditions such as line-of-sight (LOS), partial non-line-of-sight (PNLOS) and non-line-of-sight (NLOS) have been observed from the measured data for each antenna location on the body (Fig. 6.16b) [3, 8].

UWB body centric localization technique as reported in Sect. 6.2 is also applied in this work using received signal amplitude, rms delay spread and kurtosis as channel classification parameters along with time of arrival-based localization algorithms. Table 6.2 lists the accuracy achieved by applying TOA least square localization algorithm [3] in 3D for the eight antenna locations chosen on the upper body when compared with actual coordinates that have been calculated

Table 6.2 Average localisation accuracy for the body-worn antennas placed on the upper half of the human body (Reprinted with permission from IEEE [29])

Average localization error (cm)			
Antenna position	X axis (cm)	Y axis (cm)	Z axis (cm)
Left arm			
Shoulder left	2.16	1.26	2.34
Elbow left	1.14	1.13	1.93
Wrist left	1.52	1.21	2.45
Right arm			
Shoulder right	2.27	1.23	2.51
Elbow right	1.09	1.14	1.76
Wrist right	1.48	1.29	2.68
Torso			
Chest	2.31	2.25	3.21
Waist	2.10	2.18	3.06

geometrically. Highest accuracy results have been observed for the antenna placed on the elbows (right and left arm). Accuracy results below a centimeter have been observed in more than 50% of the antenna locations with an average accuracy of 1–2 cm. Figure 6.16c shows the localization results for the right elbow when compared to actual positions. This can be attributed to the fact that the position of the elbow falls in the center of the volume of the cuboid in z-direction, hence along with x- and y- axis best accuracy results are obtained in the z-direction also. Low DOP values are obtained for all the antenna locations in the range of values (1–2) which is considered as excellent accuracy values [49, 50]. The GDOP is computed for different positions of the antenna locations for each BS configuration in an area of $1.5 \times 1.5\ \text{m}^2$. The average GDOP, HDOP and VDOP values for Cuboid-Shape configuration used is 2.15, 1.22 and 1.64, respectively.

6.5.2 UWB Body Centric Localization Using Hybrid Antenna Configuration

Measurements are performed in a cluttered office indoor environment in the 4–8 GHz of bandwidth of the UWB range [55]. As shown in Fig. 6.17a, a human subject of average height and medium build was chosen and positioned at the center of the localization area. The human subject is made to rotate from 0 to 360° with an interval of 15° in-order to obtain a spatial analysis of the channel characteristics. The three base station antennas were positioned at the corner of an equilateral triangle having side length of 3 m as show in Fig. 6.17b. Various communication links involving each pair of (Tx-Rx) antennas and different human subject orientations (0°–360°) were considered. The antenna at BS1 acts as the reference co-ordinate to find the position of the human subject present in the localization area.

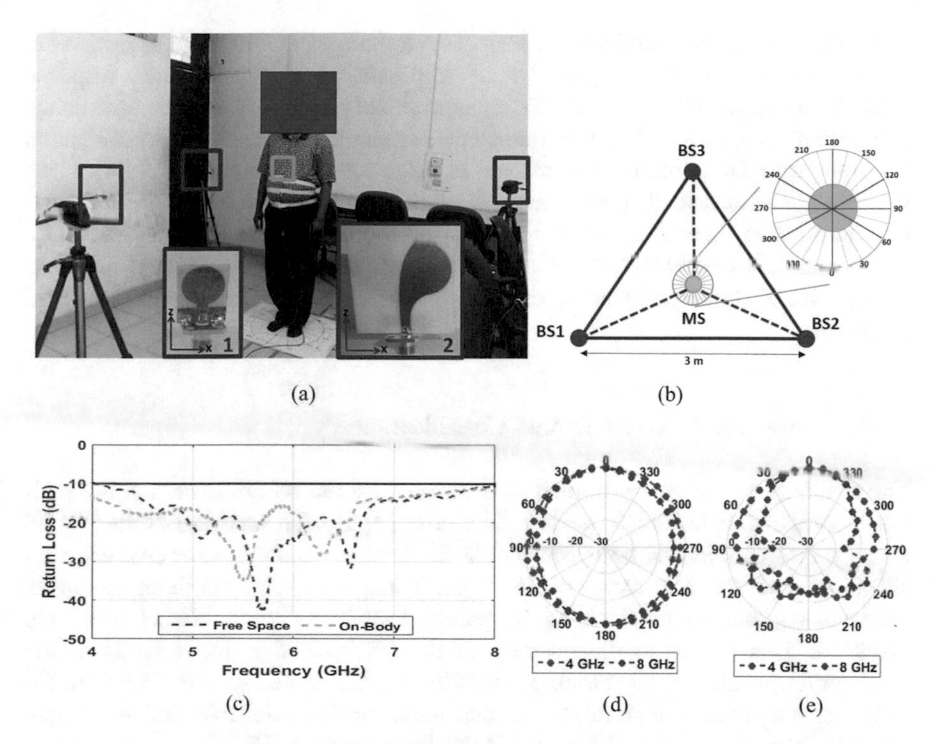

Fig. 6.17 **a** Experimental set up for UWB body centric localization. The compact circular CPW-fed UWB monopole antenna (inset 1) is used as the wearable antenna and the UWB antipodal Vivaldi antenna (inset 2) as the three BSs. **b** The schematic of the triangular shape BSs configuration. **c** The return loss of the UWB Circular monopole wearable antenna in free space and on-body. Radiation pattern for **d** the Circular monopole antenna, x–y plane, **e** Vivaldi antenna, y–z plane (Reprinted with permission from IEEE [55])

Substrate with $\varepsilon_r = 3.2$ and thickness of 1.524 mm was used for the antenna realization with dimensions of 14×18 mm^2. Figure 6.17 (inset 1 illustrates the designed and fabricated very compact UWB coplanar waveguide (CPW) fed monopole antenna. The return loss over the band of operation is well below −10 dB and overall, the antenna has good performance for free space and on-body scenarios as shown in Fig. 6.17c. The antenna has omni-directional radiation characteristics in the x–y plane which is depicted in Fig. 6.17d. The designed and fabricated UWB antipodal Vivaldi antenna (Fig. 6.17a (inset 2)) has dimensions of (40×36 mm^2) and its radiation pattern is directive in nature as seen in Fig. 6.17e. The antenna performs well in the frequency range with −10 dB return loss in free space and indoor environments.

A two-step positioning approach is considered in which certain parameters (such as signal strength, angle of arrival, or time of flight) are first extracted from the signals traveling between the target and the given base stations which is further used to estimate the position.

Firstly, for the range estimation, frequency domain measurements are carried out using a vector network analyzer (VNA), to obtain the channel impulse response (CIR) by using the UWB signals. Then, appropriate parameters such as time delay, or amplitude, are acquired by the channel parameter extraction. NLOS identification and mitigation is carried out to enhance range estimation accuracy [43, 44]. Once the channel is classified, peak detection algorithm is applied for LOS links and threshold-based time of arrival estimate algorithms are applied to partial NLOS/NLOS channels [3]. Secondly, the computed range estimates for each MS-BS links are used for TDOA data fusion to estimate the location of the target node by signal processing.

6.5.2.1 Fine and Coarse Channel Classification

The accuracy of the range estimation is affected by the NLOS links and the multipath components leading to positive bias error [3]. Spatial variation of the PL and σ_τ for the three BSs has been considered as channel parameters for channel classification. The variation for each BS of the Triangular-Shape configuration gives different regions over which LOS, partial LOS/NLOS and NLOS links are observed. This is due to the position of the BSs placed at the vertices of the equilateral triangle and the variable orientation of the human subject over 0°–360° angles at the centroid region. Hence, depending on the orientation of the human subject, different angles will lead to NLOS links for each BS.

The cumulative distribution function (CDF) for the PL magnitude is depicted in Fig. 6.18a for the three base stations classified into LOS and NLOS scenarios. It can be observed that, higher magnitude of the PL represents obstruction caused by the human subject leading to attenuation of the signal. PL values in the range of 50–55 dB represent LOS scenarios and values in the range of 60–65 dB represent NLOS scenarios. As observed, the number of NLOS links are 40% and LOS links are 60% for each MS-BS links studied.

For fine classification of the channel, rms delay spread is computed with very low magnitude for LOS links (0.5–4 nsec), mid-values in the range of (5–10 nsec) for partial NLOS links and highest values for NLOS links (11–25 nsec). A graph which represents the trend of variation of the σ_τ for the three BSs-MS links for various positions of the human subject is presented in Fig. 6.18b. It can be observed that the region of total NLOS is different for each BS link. This is since, the BS1-BS3 are located at different positions. Hence, depending on the orientation of the human subject, different angles will lead to NLOS links for each BS. For example, total NLOS for BS1/BS2/BS3 will occur when the subject is at 120°/240°/0° position respectively. Direct LOS links will be observed at locations 300°/60°/180° for BS1/BS2/BS3, respectively.

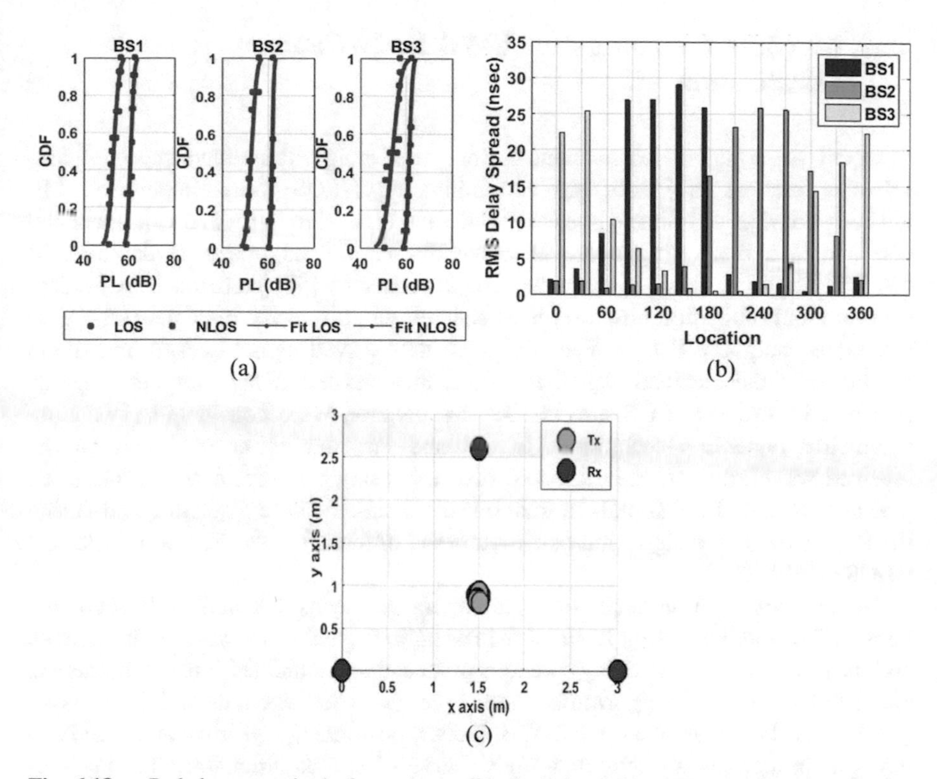

Fig. 6.18 **a** Path loss magnitude for various off-body links: BS1, BS2 and BS3. **b** Rms delay spread for different human subject locations 0° to 360°: BS1, BS2 and BS3 [55]. **c** UWB body-centric localization accuracy for triangular-shape base station configuration (Reprinted with permission from IEEE [55])

6.5.2.2 Localization Results

High accuracy localization results are obtained with an average of 3 cm using channel information and TDOA data fusion algorithm [56, 57]. The scenarios in which two BSs out of three are in LOS with the wearable antenna, higher accuracy is obtained. In these scenarios, BS1 and BS2 are in direct line-of-sight propagation and BS3 is in NLOS region. When two out of three BSs are in NLOS scenario lower localization accuracy is obtained. Figure 6.18c depicts the localization accuracy obtained for the various locations of the human subject with respect to the reference position.

6.6 Machine Learning and UWB Body-Centric Localization

In recent years supervised machine learning (ML) algorithms have received considerable interest in identifying and mitigating NLOS links. Supervised ML methods consists of training phase in which the algorithm is trained using labelled data and then the testing phase in which the learning algorithm is deployed to discriminate the unlabelled test data. The approach in [58] performs NLOS identification and mitigation based on least-squares support-vector machine (LS-SVM) technique using UWB tags. Features, such as received signal energy, maximum amplitude of the received signal, rise time, mean excess delay, rms delay spread (σ_τ) and kurtosis for LOS and NLOS classification were considered. Two non-parametric regression techniques to estimate the ranging error, based on the received waveform and the estimated distance, using tools from ML. Miao et al. [59] utilized only LOS signals to train one class classification algorithm and during the testing phase the algorithm can capture the differences between both LOS and NLOS conditions.

NLOS identification technique employing K-Means Clustering (KMC) and kernel KMC methods using features such as kurtosis, the mean excess delay spread, and the root mean-square delay spread is proposed and validated through numerical simulations [60]. ML algorithms have been proposed such as a Naive Bayes (NB) algorithm developed for UWB indoor positioning systems and validated through simulations with the area under curve of the classification characteristics around 87% [61]. Deep learning methods represented by Convolutional neural network (CNN) and Long Short-Term Memory (LSTM) was employed in the UWB NLOS/LOS signal classification in [62] with accuracy ranging from 57 to 82% for one to four convolution layers, respectively. ML classifiers utilizing Multi-Layer Perceptron (MLP) and Boosted Decision Trees (BDT) to improve NLOS detection in dense multipath industrial environments was reported in [63].

NLOS error mitigation is proposed in [64] using method by k-Nearest Neighbours (k-NN) regression to reduce complexity with an accuracy of 87%. NLOS mitigation method based on Sparse Pseudo-input Gaussian Process (SPGP) is proposed in [65] which mitigates the bias of LOS/NLOS links and validated through Monte-Carlo simulations. This approach directly mitigates the bias of both LOS and NLOS conditions. An equality constrained Taylor series robust least squares (ECTSRLS) technique is proposed to suppress residual NLOS range errors by introducing robustness to Taylor series least squares method [66]. The correct rates of NLOS and LOS identification were 93.9% and 92% respectively which is evaluated through static and mobile localization experiments.

Keeping in view the current technologies and state-of-the art techniques, the main gap lies in proposing a low complexity algorithm suitable for wearable sensor/antenna localization in an indoor environment. This is possible by having good understanding of the body-centric propagation channel, using a smaller number of features for training the ML algorithm, optimum training dataset, simple and effective NLOS mitigation techniques to achieve high localization accuracy. Major work reported in open literature deals with NLOS identification using only UWB tags in indoor environment with obstacles such as concrete/brick wall, door, window etc. Work related to application of ML for wearable UWB communication and localization is very limited.

A simple and effective body-centric localization algorithm has been proposed in [67] using UWB wearable technology. The algorithm has been validated through measurement campaigns with human subject volunteer in an indoor environment The algorithm considers statistical channel parameter analysis, machine learning algorithms and time of arrival (TOA) based range estimation/data fusion techniques. Two channel parameters namely path loss magnitude and rms delay spread are proposed as classification features to be applied to the ML algorithm to accurately classify the off-body channel links into LOS, PNLOS and NLOS scenarios. Multi-class support vector machine (MC-SVM) classifier along with Synthetic Minority Oversampling Technique (SMOTE) algorithm to consider class imbalance is applied with the channel classification accuracy of 98.63%. Threshold based range estimation algorithms are applied to mitigate NLOS scenarios caused mainly due to the presence of the human subject. Results report human localization accuracy in 0.5–3 cm range using TDOA data fusion technique for target estimation.

6.6.1 Measurement Set Up

Body-centric channel localization measurements were performed in an indoor office environment. The frequency range chosen for measurements was 4–10 GHz. The human subject chosen is of average height (165 cm) and built (60 Kgs) with a body mass index (BMI) of 22. The wearable antenna was placed on the torso region of the human subject leading to scenarios of high obstruction between the mobile station (MS)-base-station (BS) link.

Substrate with permittivity 3.2 and thickness 1.524 mm was chosen to design the compact, low cost octagonal-shape UWB antenna. Substrate dimensions of the proposed antenna are 25 × 18 × 1.524 mm^3 and has good performance in free space and on-body scenario across the UWB spectrum with return loss below −10 dB. The antenna has relatively omnidirectional radiation pattern in free space and directive radiation pattern when placed on the human subject leading to an increase in the front to back ratio. The proposed antenna acts as the BS and MS. The antennas are connected to the two-port vector network analyser (VNA) using low-loss cables for measuring the S_{21} parameters. The sweep is performed over 3201 equally spaced frequency points for the frequency range of 4–10 GHz at the center frequency of 7 GHz.

6.6.2 Algorithm and Localization Results

Learning from the large amount of UWB signal data through ML models enables an evolving scheme to enhance localization accuracy. Statistical channel analysis of the experimental results reported in [67] supports the application of ML algorithms for channel classification. To improve the performance of PNLOS, NLOS identification and mitigation, multi class Support Vector Machine (MC-SVM) is chosen due to its high performance and robustness [68–70].

Imbalanced classification involves developing predictive models on classification datasets that have a severe class imbalance. The challenge of working with imbalanced datasets is that most ML techniques will ignore, and in turn have poor performance on, the minority class. To deal with imbalance problem, oversampling techniques add samples for the minority class to change the distribution balance of original data. A representative work is the Synthetic Minority Oversampling Technique (SMOTE) [71, 72]. Instead of replicating minority samples, it generates new samples based on feature space similarities. The occurrence of partially obstructed links is much lower than the occurrence of LOS and total NLOS links through channel analysis carried out in [67]. This will lead to reduction in the performance of the classification algorithm. Hence, SMOTE is applied to generate synthetic data in the training phase of the classification process to obtain more balanced class data. This will enhance the robustness of the model which will be further applied to the test data. The proposed body centric localization algorithm comprises of the analysis of the channel information between various MS-BS off-body links, ML algorithms and TOA data fusion techniques to achieve the accurate localization of the human subject.

Figure 6.19 presents the proposed algorithm with various steps for accurate human body target localization. Initial steps include computation of the CIR from the measured S_{21} parameters and further obtaining the received signal by convolution between the CIR and the transmitted pulse. As the measured channels have distinct characteristics depending on the LOS, PNLOS or NLOS scenario, this information is considered for the accurate classification of various off-body links. There are several parameters based on statistical information and large/small scale fading that can be selected as potential features for distinguishing between the links. In order to reduce the complexity and computational time, only two channel parameters are selected as features for the classification of LOS, PNLOS and NLOS links by applying MC-SVM ML algorithm. Path loss magnitude describes the fluctuations of the received signal with respect to the distance and the σ_τ indicates the amount of spread of the signal due to presence of multipath. Based on the analysis carried out, the parameters give an indication of the attenuation or the spread of the signal and have distant dependent characteristics.

The measured data consisting of more than 1000 measurements is divided into training phase (66.6%) and test phase (33.4%) to apply the ML algorithm. In the training phase, the SVM classifier is trained to obtain optimum classification model to classify the three channel types. From the measured data, it can be observed that

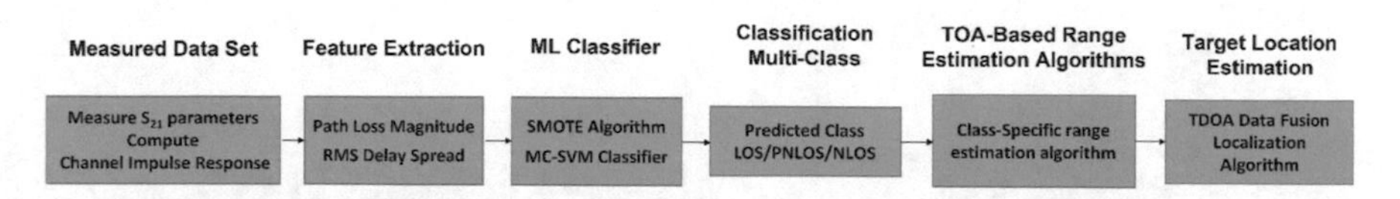

Fig. 6.19 Body-centric localization algorithm considering channel information, multi-class SVM classifier and time of arrival range estimation and data fusion techniques (Reproduced from [67])

the occurrence of PNLOS links is significantly less in comparison to the LOS or NLOS, which leads to class imbalance. SMOTE algorithm is applied to the training data to deal with the issue, leading to more accurate and robust classification of all the channel link types. Once, the classifier is trained into a model and optimized, the test data is applied to evaluate the performance of the MC-SVM classifier.

After identification of the PNLOS and NLOS links from the LOS scenarios, PNLOS/NLOS mitigation techniques are applied to the classified channels. Class-specific TOA estimation algorithms are applied which are broadly divided into peak detection and threshold based algorithms. For LOS links, the strongest peak of the CIR is chosen as an estimate of the TOA/range and gives accurate results for scenarios with low multipath components and interference levels. For partial obstructed and total obstructed links caused mainly due to the human subject in the current set up, the direct signal between Tx and Rx is significantly attenuated and has high multipath. The strongest path in such scenarios does not give the correct range estimate, leading to large ranging errors. To mitigate NLOS errors and accurately estimate the range, threshold-based algorithms are used [3, 42, 43]. For partial NLOS scenarios, search back technique algorithm is chosen and for total NLOS links threshold-based leading-edge algorithm is chosen [44].

After acquiring accurate range estimates for each MS–BS link, TDOA-based mobile location algorithm is applied to determine the unknown position of the target, [73, 74]. Let d_1, d_2, d_3 represent the range measurements obtained from three TOA measurements where (x_i, y_i) is the known position of the ith BS, and (x, y) is the position of the target MS. In the case of TDOA-based positioning, each TDOA measurement determines a hyperbola for the position of the target node. For three reference nodes, two range differences (obtained from TDOA measurements) can be expressed as follows:

$$d_{i1} \triangleq d_i - d_1$$

$$= \sqrt{(x - x_i)^2 + (y - y_i)^2} - \sqrt{(x - x_1)^2 + (y - y_1)^2} \tag{6.20}$$

for $i = 2, 3$ which define the two hyperbolas. The expressions are further computed using least square solutions to obtain the location of the target.

The performance of the MC-SVM classifier can be evaluated through the confusion matrix and related performance parameters. The confusion matrix is used to

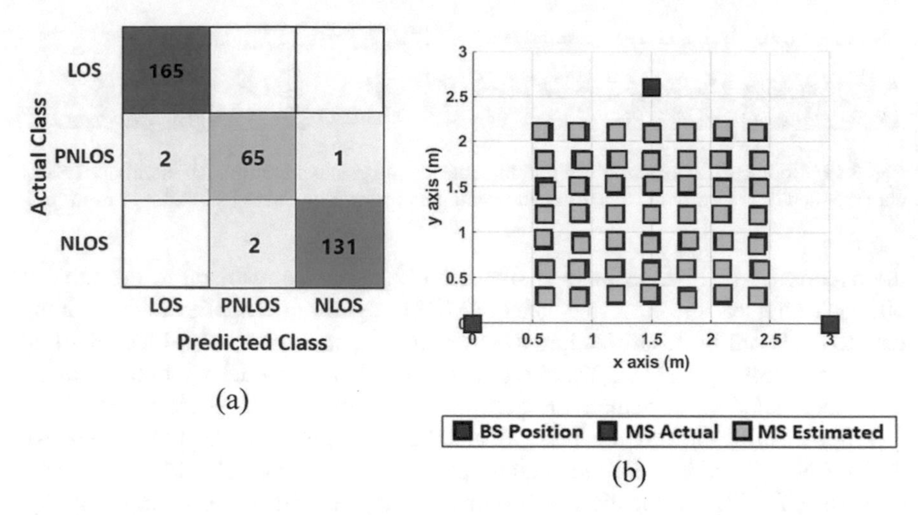

Fig. 6.20 **a** Example of Confusion Matrix of multi-class channel type classification using MC-SVM model. Class 1: LOS, Class 2: PNLOS, Class 3: NLOS **b** Localization results (Reproduced from [67])

represent how effectively different classes have been predicted. From the confusion matrix in Fig. 6.20b, it is observed that all the classes have been classified accurately. Class 1 (LOS) has highest accuracy as it has distinct characteristics when compared with Class 2 (PNLOS) or Class 3 (NLOS) links. Class 2 and Class 3 links were classified with 95% and 98% accuracy respectively due to some overlap of some of the feature values for PNLOS and NLOS scenarios. By applying the SMOTE algorithm the accuracy of estimation of Class 2 is significantly increased enhancing the classifier performance.

These results can then be used to evaluate the model based on several ratios between them, calculated as where TP is a true positive, TN is a true negative, FP is a false positive, and FN is a false negative [75]. Table 6.3 lists the performance of the classifiers with the parameters detailed for each of the classes under consideration. Namely, the Accuracy, Specificity, Sensitivity, Precision together with the F1-score, were calculated to evaluate the model. Along with the high accuracy, MC-SVM classifier also gives high precision and recall, which suggests that the model is highly reliable and robust. Overall accuracy of classification is around 98.63%.

Table 6.3 Classifier performance metrics (Reproduced from [67])

	Accuracy	Sensitivity	Specificity	Precision	F1_score
Class 1	0.995	1.000	0.990	0.988	0.994
Class 2	0.986	0.956	0.993	0.970	0.963
Class 3	0.992	0.985	0.996	0.992	0.989

After accurate classification of the channel types, NLOS mitigation techniques are applied through which the range estimates of such links becomes comparable to that of the LOS links. Once, all the range estimates have been obtained, TDOA algorithm is applied which results in very high accuracy localization results of the human target. Figure 6.20b shows the results obtained for the estimated and actual position of the subject in the area of interest. In both the scenarios very high accuracy results in the range of 0.5–3 cm is obtained showing the superiority of the proposed algorithm.

6.7 Conclusion

Accurate 3D limb movement tracking and whole-body localization has been achieved using compact, and cost-effective body-worn antennas in an indoor environment. High accuracy is obtained in the centimeter range (1–3 cm) suitable for human motion tracking, patient monitoring, physical rehabilitation, and motion-capture applications. Average localization accuracy as small as 0.5–2.5 cm has been obtained which is comparable to common commercial optical systems. Simple and robust localization schemes have been proposed for motion tracking and localization applications considering NLOS mitigation and identification techniques. In-depth analysis of various wireless channel parameters such as path loss magnitude, number of multipath components, received signal amplitude, rms delay spread, and kurtosis have been carried out. Investigations on the effect of random placement of base-stations on 3D body-centric localization in the ultra-wideband frequency range is also presented which shows the significance of the optimum placement of the base stations to enhance the positioning accuracy. Experimental and analytical studies are reported for body centric localization in cluttered indoor environments for various base station configurations and the channel-based time of arrival localization algorithms. The results obtained have an average accuracy of the order of 1–3 cm, demonstrating time of arrival based high precision UWB localization. Machine learning approach for enhancing localization accuracy has also been presented in this chapter. The work signifies the influence of the indoor environment, selection of the antenna type and the human subject on the body-centric propagation channel links, positions of the base stations which has direct effect on the localization accuracy.

References

1. Roehr S, Gulden P, Shmakov D, Bilous I (2015) Wireless local positioning—existing solutions and novel system concepts. In: IEEE MTT-S international conference on microwaves for intelligent mobility (ICMIM), Heidelberg, pp 1–4

2. Yassin et al (2017) Recent advances in indoor localization: a survey on theoretical approaches and applications. IEEE Commun Surv Tutor 19(2):1327–1346
3. Bharadwaj R, Swaisaenyakorn S, Parini CG, Batchelor JC, Alomainy A (2017) Impulse radio-ultra wideband communications for localization and tracking of human body and limbs movement for healthcare applications. IEEE Trans Antennas Propag 65(12):7298–7309
4. Zhang D, Xia F, Yang Z, Yao L, Zhao W (2010) Localization technologies for indoor human tracking. In: Proceedings of the IEEE international conference on future information technology (FutureTech'10), Pusan, Korea, May 2010
5. Oguntala G, Abd-Alhameed R, Jones S, Noras J, Patwary M, Rodriguez J (2018) Indoor location identification technologies for real-time IoT-based applications: an inclusive survey. Comput Sci Rev 30:55–79
6. Farid Z, Nordin R, Ismail M (2013) Recent advances in wireless indoor localization techniques and system. J Comput Netw Commun
7. Yan S, Soh PJ, Vandenbosch GAE (2018) Wearable ultrawideband technology—a review of ultrawideband antennas, propagation channels, and applications in wireless body area networks. IEEE Access 6:42177–42185
8. Bharadwaj R, Swaisaenyakorn S, Parini CG, Batchelor JC, Alomainy A (2014) Localization of wearable ultrawideband antennas for motion capture applications. IEEE Antennas Wirel Propag Lett 13:507–510
9. Otim T, Bahillo A, Díez LE, Lopez-Iturri P, Falcone F (2020) Towards sub-meter level UWB indoor localization using body wearable sensors. IEEE Access 8:178886–178899
10. Ahmed SNA, Zeng Y (2017) UWB positioning accuracy and enhancements. In: TENCON 2017—2017 IEEE Region 10 Conference, Penang, pp 634–638
11. Ridolfi M, Van de Velde S, Steendam H, De Poorter E (2018) Analysis of the scalability of UWB indoor localization solutions for high user densities. Sensors 18:1875
12. Monica S, Ferrari G (2014) Accurate indoor localization with UWB wireless sensor networks. In: 2014 IEEE 23rd international WETICE conference, Parma, pp 287–289
13. Zhang H, Zhang Z, Gao N, Xiao Y, Meng Z, Li Z (2020) Cost-effective wearable indoor localization and motion analysis via the integration of UWB and IMU. Sensors 20:344
14. Monica S, Bergenti F (2019) Hybrid indoor localization using WiFi and UWB technologies. Electronics 8:334
15. Taponecco L, D'Amico AA, Mengali U (2011) Joint TOA and AOA estimation for UWB localization applications. IEEE Trans Wirel Commun 10(7):2207–2217
16. Precise Indoor Positioning is finally here, Thanks to Ultra-wideband (UWB) RTLS technology, https://locatify.com/blog/in-practice-precise-indoor-location-detection-with-uwb-ultra-wideband/
17. He J, Geng Y, Pahlavan K (2014) Toward accurate human tracking: modeling Time-of-Arrival for wireless wearable sensors in multipath environment. IEEE Sens J 14 (11):3996–4006
18. DIMENSION4™ UWB Real-Time Location System (RTLS), https://ubisense.com/dimension4/
19. KIO RTLS—A UWB-based Indoor Positioning System, https://www.elikoee/products/kio-rtls/
20. BeSpoon, https://bespoon.xyz/
21. DWM1000 Module, https://www.decawavecom/product/dwm1000-module/
22. Jiménez Ruiz AR, Seco Granja F (2017) Comparing Ubisense, BeSpoon, and DecaWave UWB location systems: indoor performance analysis. IEEE Trans Instrum Meas 66(8):2106–2117
23. Xsens Motion Grid, http://www.xsenscom/en/general/motiongrid
24. Shaban HA, El-Nasr MA, Buehrer RM (2010) Toward a highly accurate ambulatory system for clinical gait analysis via UWB radios. IEEE Trans Inf Technol Biomed 14(2):284–291
25. Djaja-Josko V, Kolakowski J (2015) UWB positioning system for elderly persons monitoring. In: 23rd Telecommunications Forum Telfor (TELFOR), Belgrade, pp 169–172

26. Hyun J, Oh T, Lim H, Myung H (2019) UWB-based Indoor Localization Using Ray-tracing Algorithm. In: 2019 16th international conference on Ubiquitous Robots (UR), Jeju, Korea (South), pp 98–101
27. Minne K, Macoir N, Rossey J, Van den Brande Q, Lemey S, Hoebeke J, De Poorter E (2019) Experimental evaluation of UWB indoor positioning for indoor track cycling. Sensors 19:2041
28. Zheng Y, Zang Y, Pahlavan K (2016) UWB localization modeling for electronic gaming. In: IEEE international conference on consumer electronics (ICCE), Las Vegas, NV, pp 170–173
29. Bharadwaj R, Alomainy A, Parini CG (2015) Experimental investigation of 3D human body localisation using wearable ultra wideband antennas. IEEE Trans Antennas Propag 63 (11):5035–5044
30. Hamie J, Denis B, Maman M (2014) On-body localization experiments using real IR-UWB devices, In: IEEE international conference on ultra-wideBand (ICUWB), Paris, pp 362–367
31. Yoon PK, Zihajehzadeh S, Kang B, Park EJ (2017) Robust biomechanical model-based 3-D indoor localization and tracking method using UWB and IMU. IEEE Sens J 17(4):1084–1096
32. Xu Y, Shmaliy YS, Li Y, Chen X (2017) UWB-based indoor human localization with time-delayed data using EFIR filtering. IEEE Access 5:16676–16683
33. Hanssens et al (2018) An indoor variance-based localization technique utilizing the UWB estimation of geometrical propagation parameters. IEEE Trans Antennas Propag 66(5):2522–2533
34. Li M et al (2018) Localization of an UWB mobile station with sparse antenna array and reference station. In: 2018 IEEE international conference on computational electromagnetics (ICCEM), Chengdu, pp 1–3
35. Mahfouz MR, Zhang C, Merkl BC, Kuhn MJ, Fathy AE (2008) Investigation of high-accuracy indoor 3-D positioning using UWB technology. IEEE Trans Microw Theory Tech 56(6):1316–1330
36. Alomainy A, Sani A, Rahman A, Santas JG, Hao Y (2009) Transient characteristics of wearable antennas and radio propagation channels for ultrawideband body-centric wireless communications. IEEE Trans Antennas Propag 57(4):875–884
37. Swaisaenyakorn S, Kelly SW, Batchelor JC (2014) A study of factors affecting wrist channel characteristics for walking postures using motion capture. IEEE Trans Antennas Propag 62 (4):2231–2237
38. Bharadwaj R, Parini C, Alomainy A (2016) Analytical and experimental investigations on ultrawideband pulse width and shape effect on the accuracy of 3-D localization. IEEE Antennas Wirel Propag Lett 15:1116–1119
39. Guvenc I, Chong CC (2009) A survey on TOA based wireless localization and NLOS mitigation techniques. IEEE Commun Surv Tutor 11(3):107,124
40. Rappaport TS (1996) Wireless communications principles and practice. Prentice Hall, Inc, New Jersey
41. Ghassemzadeh S, Tarokh V (2002) A statistical path loss model for in-home UWB channels. In: IEEE conference on ultra wideband systems and technologies, May 2002, pp 59–64
42. Marano S, Gifford WM, Wymeersch H, Win MZ (2010) NLOS identification and mitigation for localization based on UWB experimental data. IEEE J Sel Areas Commun 28(7):1026–1035
43. Silva B, Hancke GP (2016) IR-UWB-based Non-Line-of-Sight identification in harsh environments: principles and challenges. IEEE Trans Industr Inf 12(3):1188–1195
44. Guvenc I, Sahinoglu Z (2005) Threshold-based TOA estimation for impulse radio UWB systems. In: IEEE international conference on ultra-wideband, ICU 2005, 5–8 Sept 2015, pp 420–425
45. Kuhn MJ, Turnmire J, Mahfouz MR, Fathy AE (2010) Adaptive leading-edge detection in UWB indoor localization. IEEE Radio Wirel Symp (RWS) 10–14:268–271
46. Sayed AH, Tarighat A, Khajehnouri N (2005) Network-based wireless location: challenges faced in developing techniques for accurate wireless location information. IEEE Signal Process Mag 22(4):24–40

47. Cheung KW, So HC, Ma WK, Chan YT (2004) Least squares algorithms for time-of-arrival-based mobile location. IEEE Trans Signal Process 52(4):1121–1130
48. Sahinoglu Z, Gezici S, Guvenc I (2008) Ultra-wideband positioning systems: theoretical limits, ranging algorithms, and protocols. Cambridge University Press
49. Li A, Dempster AG, Wang J (2011) 3D DOPs for positioning applications using range measurements. Wirel Sens Netw 3(10):334–340
50. Langley RB (1999) Dilution of precision. GPS World 10(5):52–59
51. Bharadwaj R, Parini CG, Alomainy A (2014) Ultrawideband-based 3-D localization using compact base-station configurations. IEEE Antennas Wirel Propag Lett 13:221–224
52. Fort DC, De Doncker P, Wambacq P, Van Biesen L (2006) Ultra wideband body area propagation: from statistics to implementation. IEEE Trans Microwave Theory Tech 54 (4):1820–1826
53. Bharadwaj R, Swaisaenyakorn S, Batchelor JC, Koul SK, Alomainy A (2018) Base-station random placement effect on the accuracy of ultrawideband body-centric localization applications. IEEE Antennas Wirel Propag Lett 17(7):1319–1323
54. Bharadwaj R, Swaisaenyakorn S, Parini CG, Batchelor JC, Koul SK, Alomainy A (2020) UWB channel characterization for compact L-shape configurations for body-centric positioning applications. IEEE Antennas Wirel Propag Lett 19(1):29–33
55. Bharadwaj R, Koul SK (2019) UWB channel analysis using hybrid antenna configuration for BAN localization applications. In: IEEE Indian conference on antennas and propogation (InCAP), Ahmedabad, India, pp 1–4
56. Prorok, Tomé P, Martinoli A (2011) Accommodation of NLOS for ultra-wideband TDOA localization in single- and multi-robot systems. In: 2011 international conference on indoor positioning and indoor navigation, pp 1–9
57. Monica S, Ferrari G (2013) Optimized anchors placement: an analytical approach in UWB-based TDOA localization. In: 29th international wireless communications and mobile computing conference (IWCMC), pp 982–987
58. Wymeersch H, Marano S, Gifford WM, Win MZ (2012) A machine learning approach to ranging error mitigation for UWB localization. IEEE Trans Commun 60(6):1719–1728
59. Miao ZM, Zhao LW, Yuan WW, Jin FL (2016) Application of one-class classification in NLOS identification of UWB positioning. In: Proceedings of International Conference on Information System Artificial Intelligence (ISAI), Hong Kong, pp 318–322
60. Zeng H, Xie R, Xu R, Dai W, Tian S (2019) A novel approach to NLOS identification for UWB positioning based on kernel learning. In: IEEE 19th international conference on communication technology (ICCT), Xi'an, China, pp 451–455
61. Che F, Ahmed A, Ahmed QZ, Zaidi SAR, Shakir MZ (2020) Machine learning based approach for indoor localization using Ultra-Wide Bandwidth (UWB) system for Industrial Internet of Things (IIoT). In: 2020 international conference on UK-China Emerging Technologies (UCET), Glasgow, United Kingdom, pp 1–4
62. Jiang A, Shen J, Chen S, Chen Y, Liu D, Bo Y (2020) UWB NLOS/LOS classification using deep learning method. IEEE Commun Lett 24(10):2226–2230
63. Krishnan S, Xenia Mendoza Santos R, Ranier Yap E, Thu Zin M (2018) Improving UWB based indoor positioning in industrial environments through machine learning. In: 15th International Conference on Control, Automation, Robotics and Vision (ICARCV), Singapore, pp 1484–1488
64. Zhang Q, Zhao D, Zuo S, Zhang T, Ma D (2015) A low complexity NLOS error mitigation method in UWB localization. In: IEEE/CIC International Conference on Communication in China (ICCC), Shenzhen, pp 1–5
65. Yang X (2018) NLOS mitigation for UWB localization based on sparse pseudo-input gaussian process. IEEE Sens J 18(10):4311–4316
66. Yu K, Wen K, Li Y, Zhang S, Zhang K (2019) A novel NLOS mitigation algorithm for UWB localization in harsh indoor environments. IEEE Trans Veh Technol 68(1):686–699

67. Bharadwaj R, Alomainy A, Koul SK, (2021) Experimental investigation of body-centric indoor localization using compact wearable antennas and machine learning algorithms. IEEE Trans Antennas Propag
68. Hsu CW, Chang CC, Lin CJ (2003) A practical guide to support vector classification, Tech Rep Department of Computer Science National Taiwan University, pp 1–16
69. Vapnik VN (1995) The nature of statistical learning theory. Springer, Berlin, Germany
70. Platt JC, Cristianini N, Shawe-Taylor J (2000) Large margin DAGs for multiclass classification. In: Advances in neural information processing systems. The MIT Press, Cambridge, MA, USA, pp 547–553
71. He H, Garcia EA (2018) Learning from imbalanced data. IEEE Trans Knowl Data Eng 21 (9):1263–1284
72. Chawla NV, Bowyer KW, Hall LO, Kegelmeyer WP (2002) SMOTE: Synthetic Minority Over-Sampling Technique. J Artif Intell Res 16:321–357
73. Zandian R, Witkowski U (2018) NLOS detection and mitigation in differential localization topologies based on uwb devices. In: International conference on indoor positioning and indoor navigation (IPIN), Nantes, pp 1–8
74. Laoudias C, Moreira A, Kim S, Lee S, Wirola L, Fischione C (2018) A survey of enabling technologies for network localization, tracking, and navigation. IEEE Commun Surv Tuts 20 (4):3607–3644
75. Sammut C, Webb GI (2011) Encyclopedia of machine learning. Springer, New York, NY, USA

Chapter 7
Wearable Technology for Human Activity Monitoring and Recognition

7.1 Introduction

Wearable technology has brought about rapid advancement in the field of human movement estimation and assessment with potential applications emerging in the field of rehabilitation, sports, physical activity monitoring, robotics, human-robot-interaction, digital entertainment to biomechanics analysis for clinical and sports applications [1–3]. Compact and wearable wireless communication devices suitable for body-centric communication do not constraint the user's motion and enhance performance of the way users interact with each other and the surrounding environment. Wearable devices have the advantage of being portable, lightweight and can be integrated with other devices [3] which form an important part of the wireless body area networks (WBANs). The physical activities are classified based on the frequency, strength and flexibility required to carry out the activities on daily basis which need regular progress assessment, monitoring and classification [4, 5]. Figure 7.1 illustrates various physical activities undertaken related to general daily activities, fitness exercises, yoga, sports, and weight training.

An upcoming trend in healthcare domain is having more home-centric treatment/remote monitoring instead of hospital centric medical checks keeping in view of the patient's comfort, reducing hospital costs and medical staff burden [6]. This involves compact body-worn devices, data processing centres, and increased connectivity beyond clinic between doctor-patient and further expanding to patient-to-patient/relatives-patient communication [7, 8]. Figure 7.2 illustrates the overall schematic of human activity recognition and remote monitoring process. Various day-to-day activities and specific exercises can be monitored/classified/recognised such as standing, walking, falling, lying, running, stair climbing, sitting, and bending. Apart from recognizing or monitoring an activity, vital-sign information can also be obtained in real-time through such sensors. Such information is shared over personal assistant devices, smart gadgets and transmitted over the internet to family, relatives, doctors, hospitals, and data server. In the case of an

S. K. Koul and R. Bharadwaj, *Wearable Antennas and Body Centric Communication*, Lecture Notes in Electrical Engineering 787,
https://doi.org/10.1007/978-981-16-3973-9_7

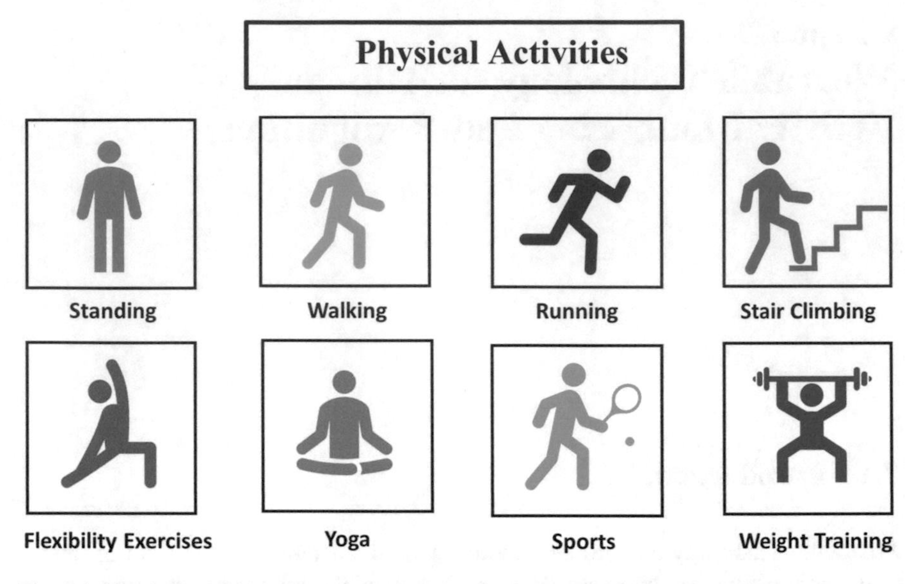

Fig. 7.1 Physical activities consisting of static, dynamic activities. Daily Activities: Standing, Walking, Running, Stair climbing, Physical Exercises: Flexibility exercises, Yoga, Sports and fitness related, and Weight training

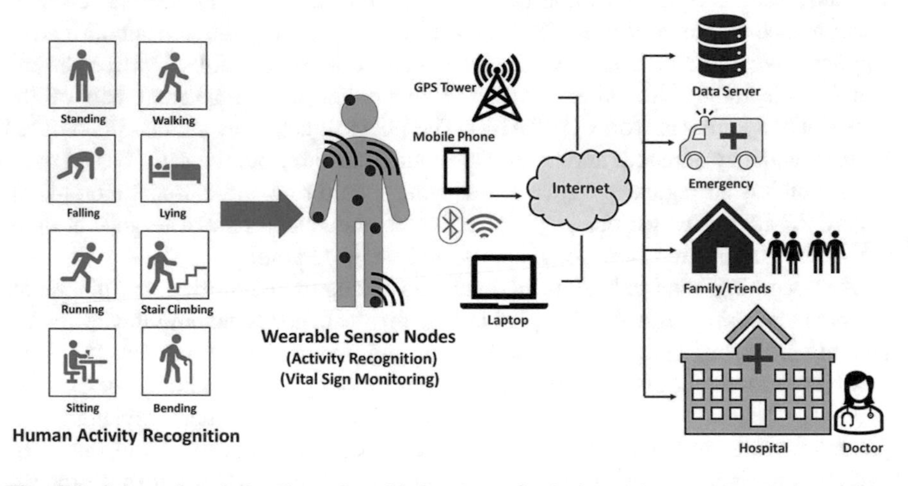

Fig. 7.2 Schematic of functions for elderly care, physical activity tracking and real-time monitoring of vital signs

emergency, an alarm will be triggered, and appropriate action can be taken to prevent any mishap or provide appropriate treatment to the subject under monitoring. This is feasible through several wearable sensors that are placed on the human subject communicating through on-body and off-body channels in the WBAN network. Wearable devices provide the opportunity to gather information about their users with unprecedented details. The availability of a large amount of personal data, including movement patterns and physiological measurements, paved the way to the development of novel applications in well-being and telemedicine.

7.1.1 State-of-the-Art-Technologies

Various technologies such as optical, magnetic, ultrasound-based motion tracking systems, inertial and ultra-wideband (UWB) are used for human motion tracking, posture detection, gait analysis, activities of daily living (ADL) for classification and monitoring purposes [9–13]. Optical systems which are camera-based such as Vicon [14] and BTS [15] are highly accurate and are considered as benchmark for various tracking and motion capture applications. The main disadvantage of these systems is that they need specific laboratory set up, complex settings, and highly skilled operators. In addition, they have the line-of-sight (LOS) restriction, where if the marker is not detected by at least two cameras its position is not recorded [10, 11].

Inertial sensors are commonly used for estimating the joint angle of the limbs during movements carried out by the human subject [16, 17] and pose-estimation during physical activities [18] using data fusion, Kalman filter algorithms and multiple sensors placed on specific joints. Inertial sensors have also been used for posture detection. Accelerometery-based gait analysis systems and emerging applications have been reported in open literature using inertial sensors which give information about the movement related disorders of the human subject and the pattern in which the human subject walks [19, 20].

A posterior-adapted class-based weighted decision fusion algorithm which combine multiple accelerometer data for physical activity recognition (8 daily activities) is proposed in [21] with an accuracy of 88–91%. The joint angles position for detection of three upper limb movements using two magnetic, angular rate, and gravity (MARG) sensors [22] could achieve an accuracy of 88–93/80–83% in controlled/semi-naturalistic environment. A wearable capacitive printed accelerometer [23] provides distinctive features for classifying three limb movement postures during rubber band exercises.

Low cost IMU and MARG sensors give accurate results but are often prone to misalignment/offset errors and suffer from calibration errors. These technologies generally use complex data fusion/machine learning algorithms, advance signal processing, large training sets and features (10–50) extracted from each sensor attached to the human subject to assess the activity.

A multi-sensor approach is carried out in [24] to improve accuracy of estimating knee flexion kinematics for rehabilitation purposes. A modified complementary filter is used for hand position monitoring using a multi-sensor system with results comparable to an optical tracking system [25]. Multiple sensors are mounted on the human subject and are focused on activity recognition and joint angle measurements with accuracy in the range of 80–95%. Human activity classification with transmission and reflection coefficients of on-body antennas through deep convolutional neural networks (DCNN) is reported in [26].

A textile-based smart undershirt consisting of 11 body worn sensors is presented in [27] to detect upper body activities with an accuracy of 94%. Work related to full-body motion sensing is reported in [28] using a sensing suit made of stretchable fabric and 20 soft strain sensors. Flexible magnetic induction–sensors [29, 30] are used for recording different inductance signals to detect physical movements. Pressure, strain, and flexible-textile sensors also suffer from reduction in performance accuracy due to repeated use.

Non-contact human activity classification [31–33] and physical activity monitoring such as gait analysis [34] based on UWB radar system combines range, frequency, and time information to obtain spectrograms results with accuracy between 75 and 95% [31–33]. Techniques include deep convolutional neural networks (DCNN)/principal component analysis (PCA)/k-means clustering algorithm used for feature extraction and classification. UWB radar-based technology focuses on activity classification and recognition with an accuracy in the range of 75–95% using complex DCNN and machine learning (ML) algorithms. It also confines the user's position/movement to a restricted short-range environment.

Most of the sensor-based technologies use large number of feature sets, complex algorithms, lot of experimental data for training along with advance machine/deep learning algorithms to obtain high accuracy monitoring and classification results. Ultra-wideband technology brings out huge research and development interests for potential healthcare monitoring applications due to its fine time resolution, high data rate, low power consumption, resolving multipath components, penetration through obstacles and ease of integration with other devices [35–37]. Wearable UWB technology is very suitable for body-centric localization and tracking [38–40], joint angle estimation [41], activity monitoring [42, 43], and gait analysis [44] applications due to the high time resolution of the UWB signal, low power, compact wearable antennas, accurate detection of variation in the movements through channel related parameters, high sensitivity, and precision. UWB has also proved to be suitable for activity recognition and classification purposes which have been reported for ADLs [45] and gait identification of individuals [46].

The use of ultra-wideband (UWB) radios in wearable devices has been investigated in detail, as it brings about accurate indoor localisation and limb movement tracking, flexion angle measurements of limb joints using wearable antennas and various aspects of monitoring gait activity. This chapter presents work related to monitoring and assessment of physical activities using wearable UWB technology. Detailed investigation of monitoring progress of physical exercises for upper and lower limb movements has been carried out in this chapter. Various works related to gait analysis have been described which use compact UWB trans-receivers to monitor the walking patterns of individuals and assess various parameters such as step length, base-of-support (BOS) etc. Methodology for joint angles estimation during extension and flexion of limbs is also presented. The works presented consider channel information, time-of-arrival/ranging data inter-distance information between the wearable devices/antennas using very compact UWB antennas placed on the human subject. Algorithms and methods used for classification and monitoring of daily physical activities and gait movement using UWB wearable technology is also presented in this chapter.

7.2 Assessment of the Physical Activities

Monitoring of physical activities is challenging and must take into account various parameters such as body-effects, location of the wearable device/antenna on the body, the type of physical activity performed. Activity assessment represents monitoring of the activity which gives an estimate of the progress in the physical exercise performance in relation to the displacement between the wearable nodes. It represents how well a subject can perform the limb movement activities over the whole range of motion that can be accomplished by a human subject. The work presented in [42] has proposed UWB body-centric channel parameters for assessing the performance of the physical activities suitable for rehabilitation and physiotherapy purposes in the healthcare domain.

Due to the large signal bandwidth and ultra-fine pulses, UWB technology is suitable to monitor minor changes in distance variation, which can be observed by evaluating various channel parameters [42, 43]. The measured magnitude and phase of the S_{21} is converted to the time domain by applying an inverse Fast Fourier transform (IFFT) to determine the channel impulse response (CIR). Further convolution is carried out with the UWB Gaussian pulse to obtain the received pulse [36]. Channel parameters, namely, path loss magnitude (PL) and rms delay spread (σ_τ) have been proposed as good indicators to monitor the physical activity of the human subject and access the progress/trend while performing the physical exercises. The path loss is obtained as mean path gain over the measured frequency band, [47]

$$PL(d(p)) = -20.\log_{10}\left\{\frac{1}{K}\frac{1}{N_f}\sum_{j=1}^{K}\sum_{n=1}^{N_f}\left|H_j^p(n)\right|\right\} \tag{7.1}$$

where $PL(d(p))$ is the path loss at the position of p, at which the distance d between the Tx and Rx is a function of the position p. N_f is the number of frequency samples of the VNA and K is the number of sweeps. $H_j^p(n)$ is the measured frequency response of the channel for the position p, jth snapshot, and nth frequency sample. PL varies with distance due to the signal attenuation with increase in distance, leading to higher PL magnitude.

The rms delay spread [48] is defined as:

$$\sigma_{\tau} = \sqrt{\frac{\sum_k (\tau_k - \tau_m)^2 . \ |h(\tau_k; d)|^2}{\sum_k |h(\tau_k; d)|^2}} \tag{7.2}$$

where τ_k are the multipath delays relative to the first arriving multipath component, $h(\tau_k; d)$ is the channel impulse response (CIR), τ_m is the mean excess delay. The rms delay spread is higher for NLOS links due to presence of higher multipath components in comparison to LOS links, and increases with distances between the Tx-Rx due to increase in spread of the signal due to larger distance.

7.2.1 Measurement Set Up

Various human activity measurements are performed in an indoor environment in the 4–8 GHz range as shown in Fig. 7.3. The wearable antenna used is a compact and miniaturized tapered slot antenna (TSA) (Fig. 7.3, inset) which has good performance in free space and on-body in the desired frequency band (Fig. 7.4a–c). The size of the antenna is 12×18 mm^2 and is inspired by [49].

Six human subjects (3 females (SUB 1–3) and 3 males (SUB 4–6) are considered for performing various exercises and physical activities. Female and male average height is 160 ± 2.5 cm and 173 ± 6.7 cm respectively. Female and male average weight is 55 ± 3.1 kg and 67 ± 5.7 kg with the body mass index (BMI) in the range of 20–23. The position of the wearable antennas (transmitter (Tx) and receiver (Rx)) considered is the wrist for the upper limbs and ankle for the lower limb activities. For limb bending exercises shoulder/thigh region is chosen as the Tx location and wrist/ankle as the Rx. These wearable locations give maximum displacement during physical activity, through which the variation in the channel parameters over distance will be more prominent. Three orientations of the wearable antennas are considered as shown in Fig. 7.5 i.e. outer (OUT), inner (IN), front (FR) of the wrist/ankle (WR/AK) giving rise to different types of on-body channel links which are denoted as: OUT_WR/OUT_AK, IN_WR/IN_AK, FR_WR/FR/AK. The antennas are connected to the 2-port vector network analyser (VNA) using low loss flexible cables to provide freedom in limb movement activities. The sweep is performed over 1601 frequency points for 4–8 GHz range. Data is recorded at different positions for each activity (five repetitions) and the angles are estimated through digital protractor with shoulder/thigh or elbow/knee joint as reference.

The exercises studied in this work are related to rehabilitation, physiotherapy, and general fitness exercises [50]. The type of movements chosen for activity monitoring focus on the upper and lower limbs range-of-motion exercises performed by the human subject [51, 52]. The exercises performed are categorized into both limb movements (A1-A4)/(L1-L5), single limb movement (A5-A7)/(L6-L9)

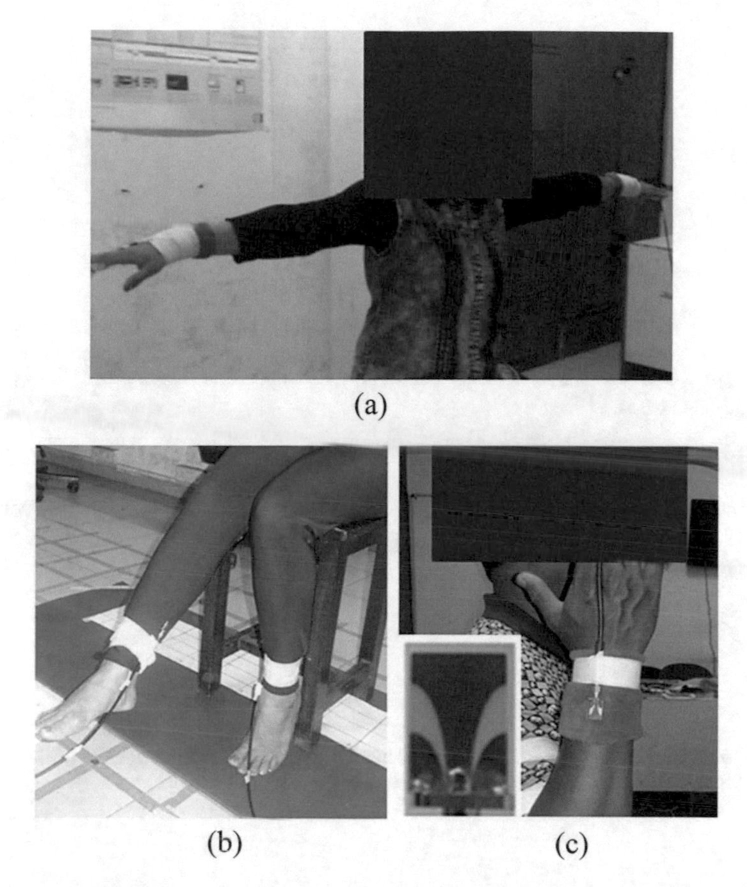

Fig. 7.3 Activity monitoring in an indoor environment: **a** Upper limbs, both arms movement; **b** Lower limb, single leg movement; **c** Upper limb bending movement, inset mini-TSA antenna (Reprinted with permission from IEEE [42])

and limb bending (A8-A11)/(L10) scenario. The measurements are recorded at 15°/30° interval during the exercise. While performing various physical exercises, the human subject postures vary from sitting, standing, and sleeping position.

Each set of physical activity measurement gives a channel parameter pattern for the full range of motion, through which the progress of the person's ability to perform the exercise accurately can be detected. Received signals for both upper limb movement activity A4 (IN_WR), upper limb bending A8 (OUT_WR) are depicted in Fig. 7.6a, b respectively. Lower limbs both limb activity L2 (FR_AK) and single limb movement activity L6 (FR_AK) received signal plots are presented in Fig. 7.7a, b respectively. It can be observed that each activity performed has distinctive trend and features of the received signal. The received signals plotted are with respect to the angles formed which in turn are related to varying distance.

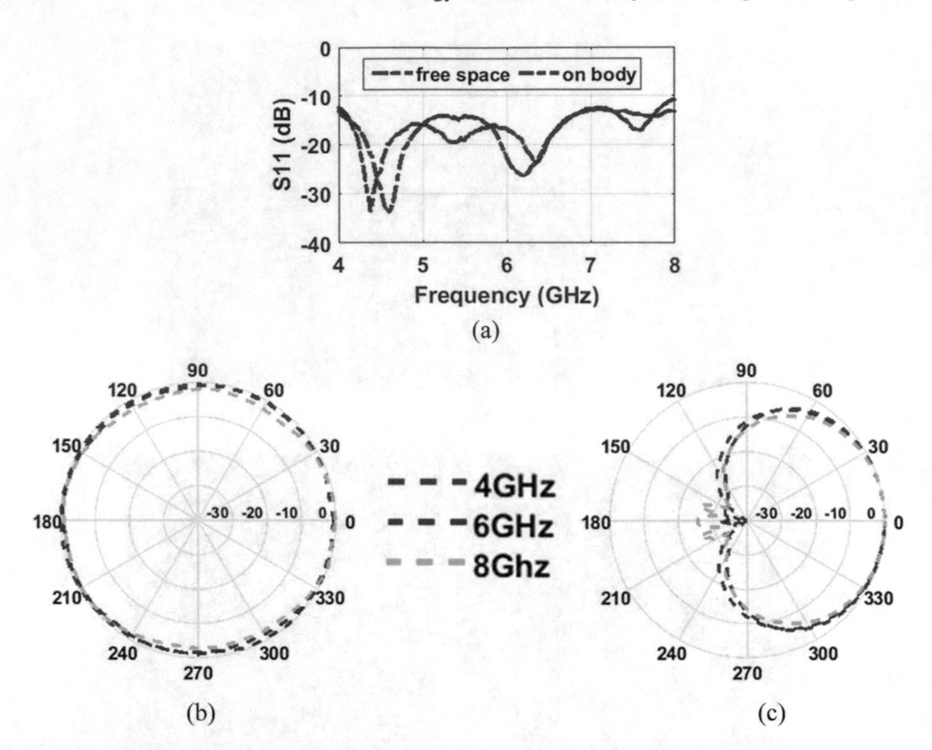

Fig. 7.4 **a** S_{11} for free space and on-body scenarios. Azimuth radiation pattern for **b** free space and **c** on-body scenario (Reprinted with permission from IEEE [42])

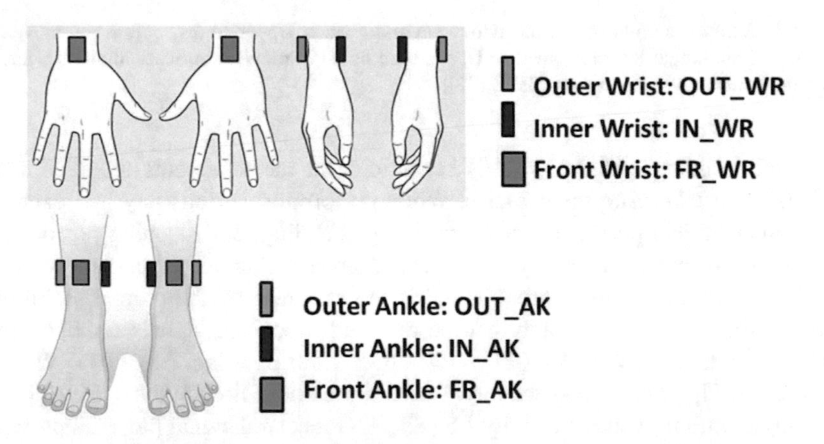

Fig. 7.5 Orientations of the wearable antenna for the wrist and ankle region of the human subject (Reprinted with permission from IEEE [42])

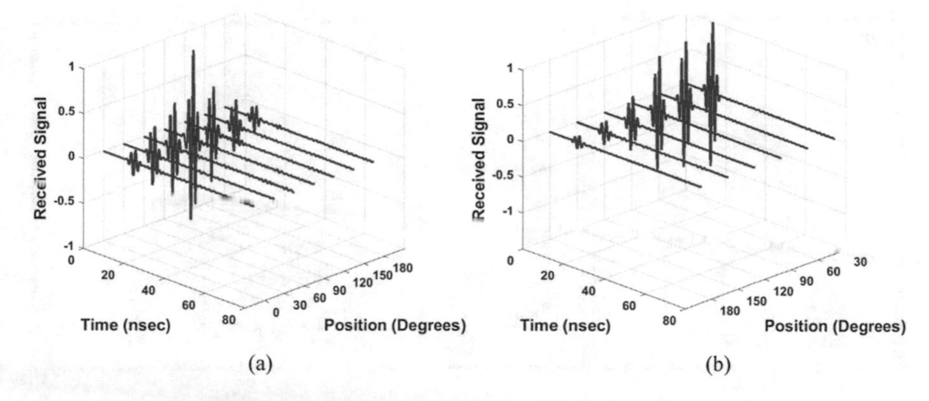

Fig. 7.6 Normalized received signal for **a** Both upper limb activity A4, IN_WR; **b** Limb upper bending activity A8, OUT_WR (Reprinted with permission from IEEE [42])

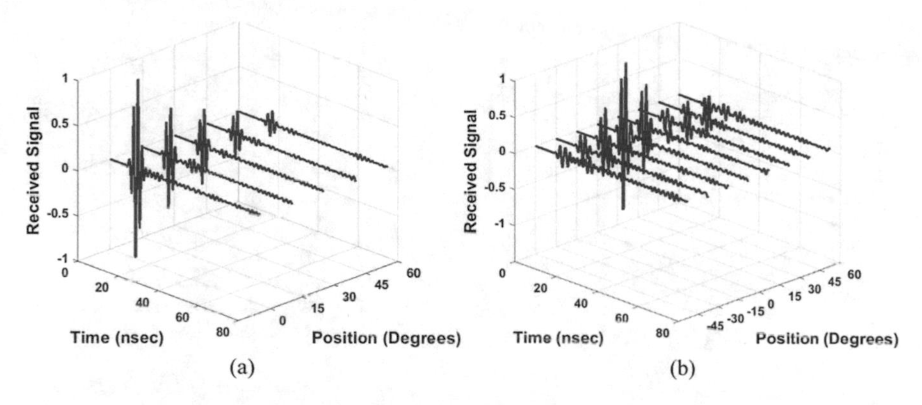

Fig. 7.7 Normalized received signal for **a** Both lower limb activity L2, FR_AK; **b** Single lower limb activity L6, FR_AK (Reprinted with permission from IEEE [42])

The characteristics of the signal are a function of the distance between the Tx and Rx wearable antenna and the type of link (LOS/NLOS) which indicates the obstruction caused by the body region which is mainly the human torso for on-body communication.

7.2.2 Activity Assessment Results

7.2.2.1 Upper Limbs Activity

Figure 7.8 presents the cases related to upper limb activities categorized into both limb movement and limb bending. Examples of the computed and statistically analysed channel parameter trends for PL and σ_τ for selected activities is shown in Fig. 7.9.

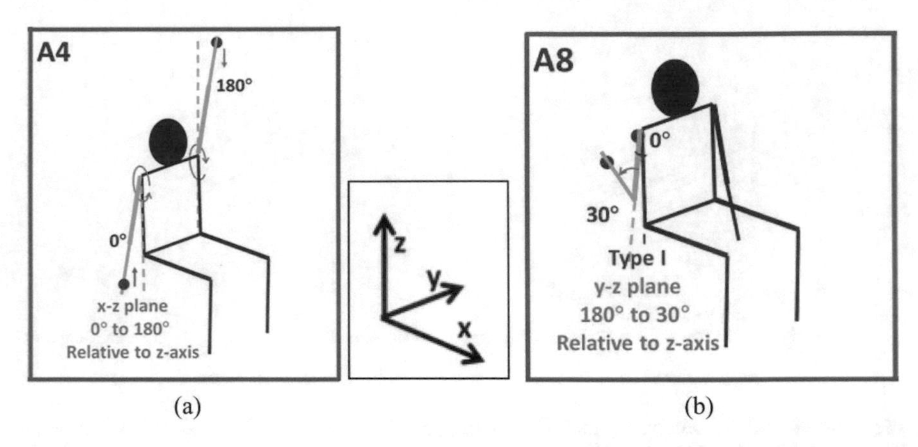

Fig. 7.8 Upper limb activity: **a** Both limb movement (A4); **b** Limb bending (A8) (Reprinted with permission from IEEE [42])

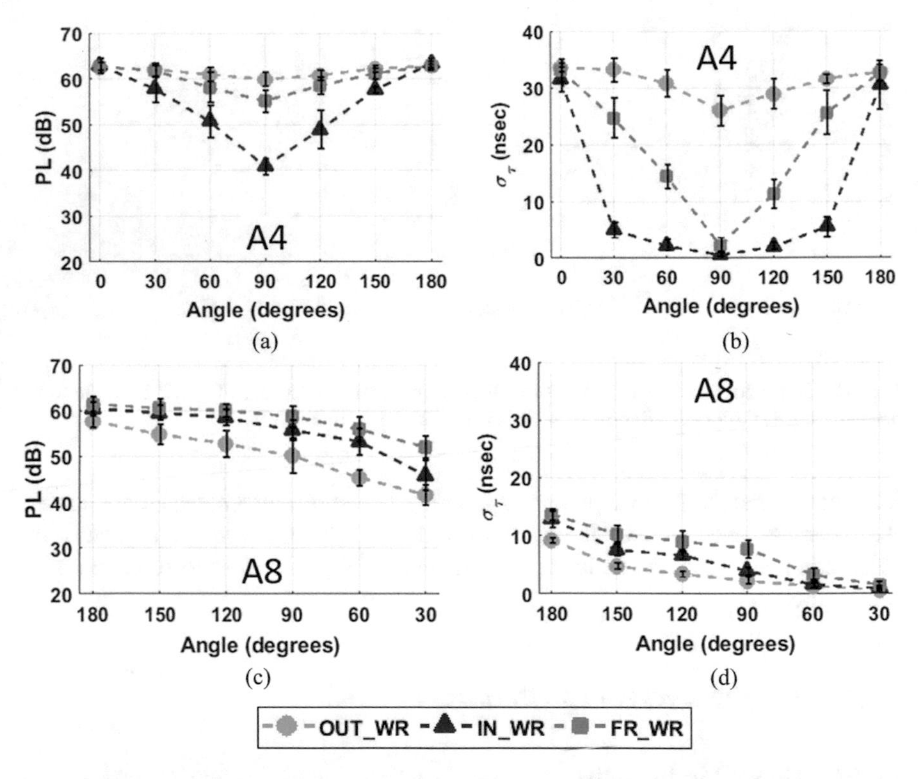

Fig. 7.9 Relation between path loss magnitude (dB) and rms delay spread (nsec) with respect to limb position (angles): Upper limb activity assessment. Six subject's average: **a**, **b** A4: Both arms activity; **c**, **d** A8: Arm bending activity ([42], Reprinted with permission from IEEE)

Activity A4

The wearable antennas are placed on the left and right wrist for both limb activities (activity A4), opposite right and left arm movement is carried out (right/left: 0°/180°–180°/0° which gives rise to a transition from NLOS → LOS → NLOS links. As seen in Fig. 7.9a, b initial (0°) and final (180°) positions lead to highest magnitude of the channel parameters and for 90° position minimum values are observed. IN_WR has least magnitude of the PL and σ_τ due to the occurrence of maximum number of LOS links. OUT_WR suffers maximum obstruction as the antenna position is placed on the outer region of the wrist, hence high PL and σ_τ is obtained.

Activity A8

The wearable antennas are placed on the wrist and shoulder of the right arm for the limb bending activity (activity A8). In this scenario, the initial angle made by the right arm is 180° and during the bending activity reaches a minimum of 30°. The reference angle is the elbow joint with the arm bending in the y–z plane. Type I and II indicates the angle made by the upper limb with the torso region which is 0° and 90° respectively. It is observed that the PL and σ_τ reduces with decrease in bending angle (180° → 30°), as the distance between the antennas placed on the wrist and shoulder reduces. Figure 7.9c, d presents the trend of the channel parameters observed for activity A8. FR_WR depicts maximum magnitude of the channel parameters due to the NLOS links formed in comparison to IN_WR/OUT_WR.

7.2.2.2 Lower Limb Activity

Figure 7.10 presents cases related to the upper limb activities categorized into both limb movement and single limb movement. Examples of the computed channel parameter trends for PL and σ_τ for selected activities is shown in Fig. 7.11.

Activity L2

Activity L2 presents both leg movements i.e., spreading of the legs in the y–z plane, which leads to an increase in distance between the legs. Hence, it will lead to an increase in PL and σ_τ magnitude as the angle between the legs increases from 0° to 60° which can be seen in Fig. 7.11a, b for activity L2. The variation is more prominent in IN_AK/FR_AK due to the LOS links between the two wearable antennas. The least variation is observed for OUT_AK due to the obstructed signal propagation between the Rx and Tx.

Activity L6

Activity L6 depicts right leg forward and backward movement, with the left leg static as shown in Fig. 7.10b with the human subject in standing position. There will be variation of the channel comprising of NLOS/partial NLOS links for OUT_AK, LOS links for IN_AK and partial NLOS → LOS → partial NLOS links for FR_AK as the leg position varies from −45° to 60° in the x–z plane.

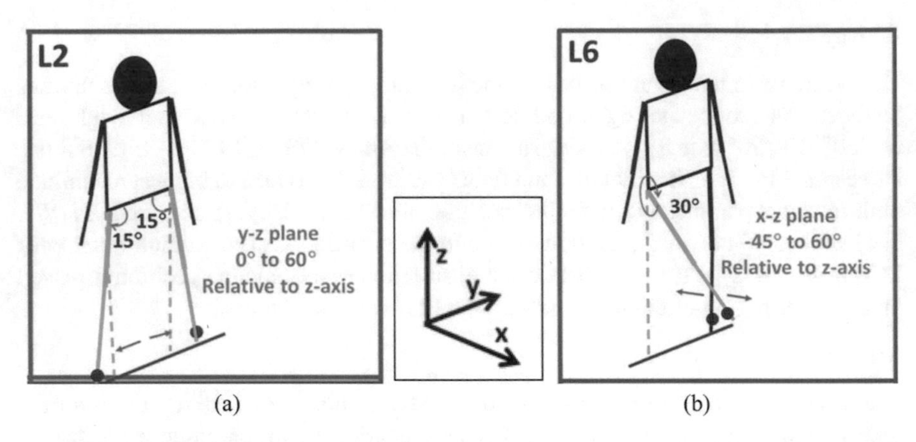

Fig. 7.10 Lower limb activity: **a** Both limb movement (L2); **b** Single limb movement (L6) (Reprinted with permission from IEEE [42])

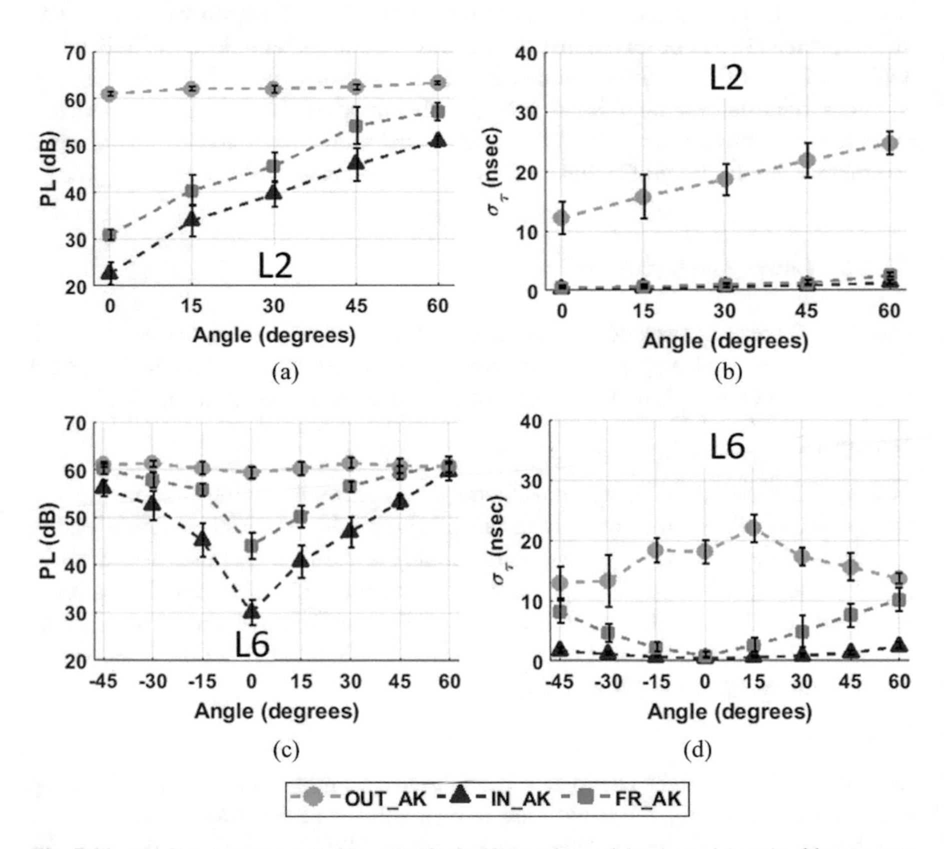

Fig. 7.11 Relation between path loss magnitude (dB) and rms delay spread (nsec) with respect to limb position (angles): Lower limb activity assessment. Six subjects' average: **a**, **b** L2: Both legs activity; **c**, **d** L6: Single leg activity (Reprinted with permission from IEEE [42])

Figure 7.11c, d presents the results for the channel parameters variation backward and forward limb movement which are well correlated with the distance variation for IN_AK and FR_AK.

7.2.3 Activity Monitoring Performance

The performance accuracy is measured in terms of the correlation between the trend observed for the wearable nodes inter-distance variation and the trend for the corresponding channel parameter measurements recorded for each physical activity. Each set of physical activity measurement gives a channel parameter pattern for the full range of motion, through which the progress of the person's ability to perform the exercise accurately can be detected.

Figure 7.12 presents the proposed procedure for estimation of the progress of limb movements carried out by the human subject. Initial steps include placement of the wearable antennas at appropriate locations of the human subject such as wrist or ankle so that maximum displacement and direct path links are observed during the activity. The subject can choose a specific activity it wants to assess from the data-base of activities fed into the system. The user can also choose an activity category wise based on the expected trend such as (increasing, decreasing, constant or increasing → decreasing, decreasing → increasing etc.) which is related to the inter-distance separation between the two wearable nodes. The S_{21} parameters are measured from the VNA through which the channel impulse response is computed for different angles/ positions of the limb during the activity. Further the channel parameters PL and σ_τ are computed and optimised for all the measured positions of the limb.

The results are compared with the normalized inter-distance node values and correlation between the two is computed. High correlation between the values shows that the assessment is correctly predicted, and the subject has accurately performed the exercise. The correlation values have been categorized with values greater than 0.95 as excellent performance and values below 0.75 as poor performance.

Received signals for both limb movement activity A4 (IN_WR), is depicted in Fig. 7.12. The received signals plotted are with respect to the angles formed which in turn are related to varying distance. The characteristics of the signal are a function of the distance between the Tx and Rx wearable antenna and the type of link (LOS/NLOS) which indicates the obstruction caused by the body region which is mainly the human torso for on-body communication. The average (PL and σ_τ) correlation coefficient values obtained is higher for IN/WR (0.96/0.97) and lowest for OUT (0.93) orientation of the limb orientation. This is because, LOS links are generally observed for IN/WR orientation of the wearable antennas leading to more distinct variation in PL and σ_τ values for different angles of the limb movement.

Correlation coefficient values in the range of 0.9–0.99 are considered as high accuracy results in estimating the performance of the limb movement activity. Values below 0.75 are considered as low correlation, and such trends obtained when compared with the trends of the inter-distance values, are considered as

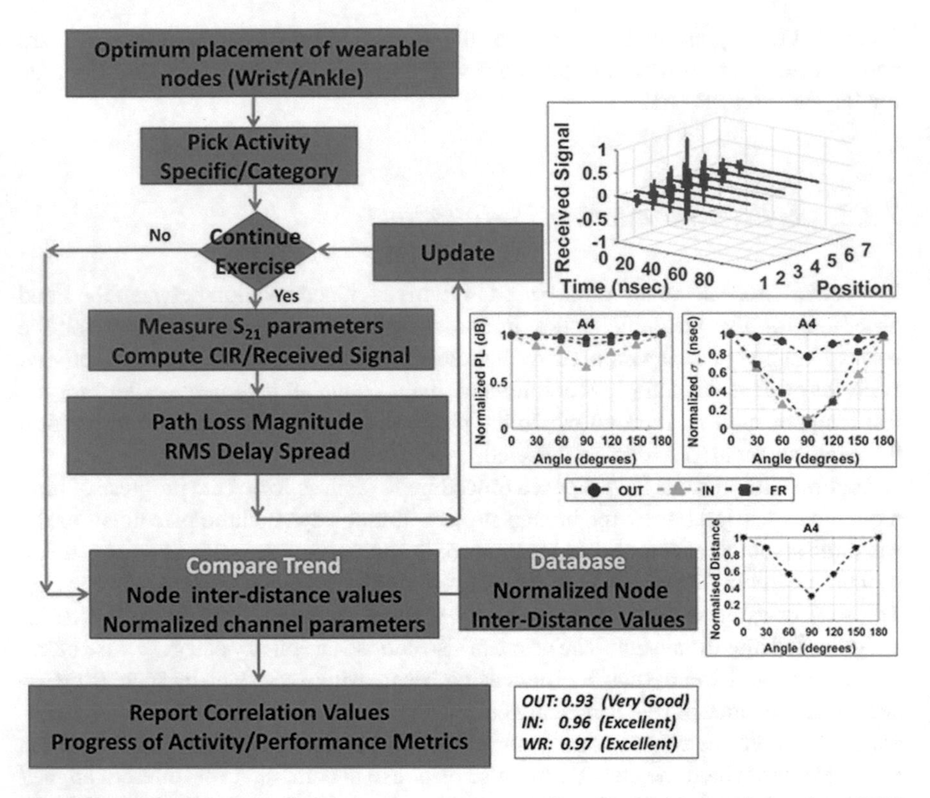

Fig. 7.12 Methodology for limb movement assessment (Reprinted with permission from IEEE [42])

inaccurate results. Best results are obtained for links with LOS or partial LOS links due to direct line of path propagation. Worst cases are observed when only total NLOS links which leads to attenuation of the signal mainly observed due to the torso region of the body. Performance accuracy has been computed for the combined channel parameters patterns trends of OUT, IN and FR orientation. For upper limb activities high monitoring performance (95–98%) is observed for both limbs (A1-A4) and limb bending (A8-A11) exercises. For lower limb activities high monitoring performance (96–99%) is observed for both limb movements (L1-L5) and single limb movement (L6-L9) activities.

7.3 Daily Physical Activity Recognition

Identification and tracking of daily physical activities (PA) by using wearable devices is of great importance, since it can provide health related information to individuals and remotely to relatives, doctors, and medical data centres. Accurate classification and assessment of PAs is required to accurately access general well-being of patients, elderly and provides with fitness related information. Daily activity recognition using wearable ultra-wideband technology is presented in [45]. Channel parameters spread are analysed for various postures occurring during the daily activity which act as key features to estimate the activity trend. Compact wearable antennas are placed on suitable locations on the human subject for each activity studied to have maximum direct path propagation between the two wearable on-body nodes.

A female human subject of height 1.6 m, weight of 60 kg with a body mass index (BMI) of 23 has been chosen for performing various physical activities. The location of the on-body nodes was chosen thoughtfully to obtain maximum line-of-sight links and to avoid occurrence of obstructed links between the nodes. The on-body antennas were positioned 5 mm away from the body at specific locations. The node positions were chosen with regards to possible applications such as hearing aid instrument, smartphone, fitness tracker etc. Measurements were performed in an indoor office environment in the 4–8 GHz of the UWB band. Miniaturised Tapered slot UWB antennas are used as wearable antennas having dimensions of 18×12 mm^2 and having -10 dB return loss over the entire frequency band of operation. Daily activities such as standing, sitting, and walking are analysed through placement of nodes at optimum locations on the body such as wrist, shoulder and ankle as shown

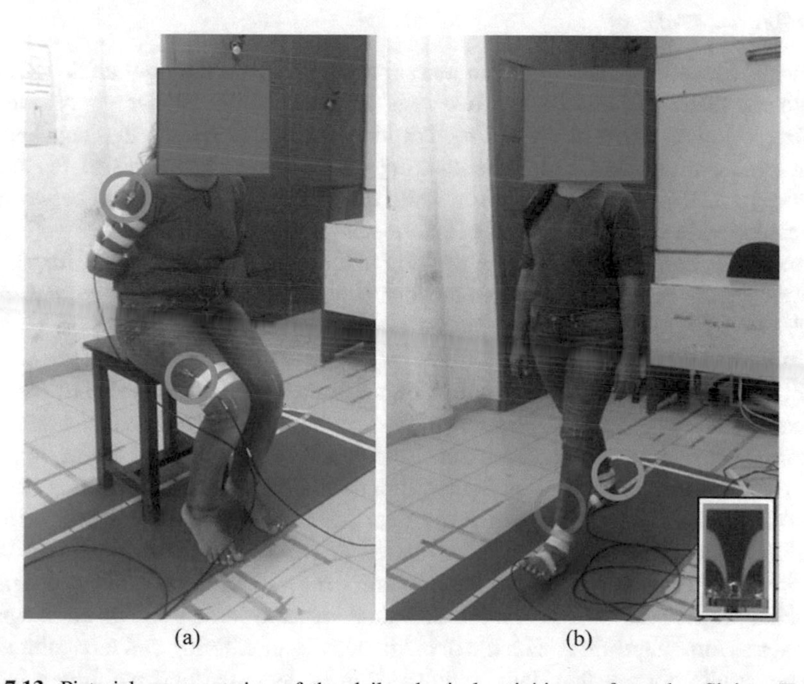

Fig. 7.13 Pictorial representation of the daily physical activities performed **a** Sitting (Tx/Rx: Right Shoulder/Thigh facing outwards); **b** Walking (Tx/Rx: Right/Left ankle facing inwards). (Inset) Compact and low-cost miniaturized UWB tapered slot antenna used as wearable antennas (Reprinted with permission from IEEE [45])

in Fig. 7.13. Measurements are recorded in pseudo-dynamic mode for various postures occurring during the activity. The activity has been performed several times to insure robustness and validation of the proposed methodology.

UWB channel parameters have been used to monitor daily activities such as sitting, standing, and walking. Optimal sensor positions have been chosen on the limbs of the human subjects to obtain accurate estimate of the trend of activity. It is observed that each physical activity has a unique pattern of the channel parameters (PL and σ_τ) analysed as observed in Chap. 2, Sects. 2.5.6 for standing → sitting → standing activity and standing → walking activity.

Standing → Sitting → Standing

The sensor location chosen is the right shoulder (R. SH) and right lower thigh region (R. TH). Maximum distance between the shoulder and lower thigh region leads to highest value of PL and σ_τ when in standing position (P1) and decreases as the subject goes from standing to sitting position (P1→P4). For sitting position P4, lowest PL and σ_τ values are observed due to least distance between the body-worn sensors while the subject is bending forward. The distance increases slightly as the subject leans backward while sitting as seen for P5. For sitting to standing activity similar values of PL and σ_τ are observed. Hence, completing a standing → sitting → standing activity set.

Standing → Walking

The sensor locations chosen are the left ankle (L. AK) and the right ankle (R. AK). Minimum distance between the two legs is observed for standing position P1, leading to lowest value of PL and σ_τ. Large distance between the two legs leads to higher value of PL and σ_τ for the walking position P2. The values of PL and σ_τ reduces for P3 and P4 in comparison to P2 due to decrease in distance between the ankles during walking. The channel parameters further increase due to increase in distance between the legs during walking with a maximum distance for P6. P7 shows decrease in distance between the legs which tends to further increase from P7 to P9. P7 to P9 shows increase in the channel parameter values PL/σ_τ with P9 returning to initial walking position P1.

Statistical analysis of various activities performed are categorised into dynamic scenario considering various postures and static which focuses only on the initial/final posture (Table 7.1). It is observed that each activity has its own distinct statistical values with dynamic activities showing higher variation in the channel parameters in comparison to static. Also, the variance is more for the sitting activity in comparison to the standing dynamic activity. This is attributed to the varied postures formed while transiting from standing to sitting position and the location of the on-body antennas. The channel patterns formed and the statistical values information can be integrated together to form algorithms which will aid in activity monitoring and recognition.

The results portray that daily activities can be assessed and monitored by the means of low-cost antennas and simple assessment methodology by using a smaller number of features. The findings focused the significance of the effects of body movement and UWB on-body channel characteristics when designing the

Table 7.1 Daily activities: PL/ and σ_τ statisitics ([45], Reprinted with permission from IEEE)

Activity	PL (dB) and σ_τ (nsec)								
	Node position	Mean		Stdv.		Max		Min	
		PL	σ_τ	PL	σ_τ	PL	σ_τ	PL	σ_τ
Dynamic									
Sitting	R. SH-R. TH (OUT)	57.5	1.3	1.9	1.2	60.3	3.6	54.2	0.2
Walking	R. AK-L. AK (IN)	48.7	0.60	8.7	0.50	55.6	1.45	28.9	0.06
Static									
Sitting	R. SH-R. TH (OUT)	55.8	0.58	1.1	0.16	57.1	0.80	54.2	0.40
Standing I	R. SH-R. TH (OUT)	60.3	2.97	0.7	0.43	61.0	3.64	59.0	2.40
Standing II	R. AK-L. AK (IN)	30.2	0.14	2.0	0.10	33.2	0.33	27.5	0.06

healthcare-monitoring devices. It is observed that the three activities analysed: walking, standing, and sitting show distinct variation in the channel features analysed making it possible to classify the activities through statistical analysis and inter-distance measurements between the wearable nodes. Low correlation is observed between the activity patterns with 0.1–0.3 correlation coefficient values. This indicates that the activities can be easily distinguished from each other using channel information. The work carried out is suitable for tracking, rehabilitation, and activity monitoring applications in the healthcare domain.

7.4 Joint Angle Estimation Using UWB Wearable Technology

A methodology for measuring and monitoring human body joint angles is proposed in [41], which uses wearable ultrawideband (UWB) transceivers mounted on specific regions of the human body. The model is based on providing a high ranging accuracy (inter-sensor distance) between a pair of transceivers placed on the adjacent segments of the joint center of rotation. The measured distance is then used to compute the joint angles based on the law of cosines which is then converted to corresponding flexion/extension angles. Figure 7.14 shows a simplified diagrammatic representation for the placement of a pair of the transceivers which form a triangle along with the joint (elbow or knee). α and β are the elbow or knee flexion/extension angle, which is defined as the relative angle between two adjacent segments of knee or elbow joint, respectively. Since the two sides of the triangle, i.e., d_1 and d_2, are constant during human movement with transceivers fixed on the body, the joint angles can be estimated using the law of cosines. The transmitter and receiver were used to measure the distance d which will be converted to angle estimation by the law of cosines. As shown in Fig. 7.14, elbow joint acts as a virtual point that consists of a triangle with transmitter and receiver. The following equation demonstrates how angle α is calculated from the triangle in Fig. 7.14. During segment movement, modelling the variation of the arm yields:

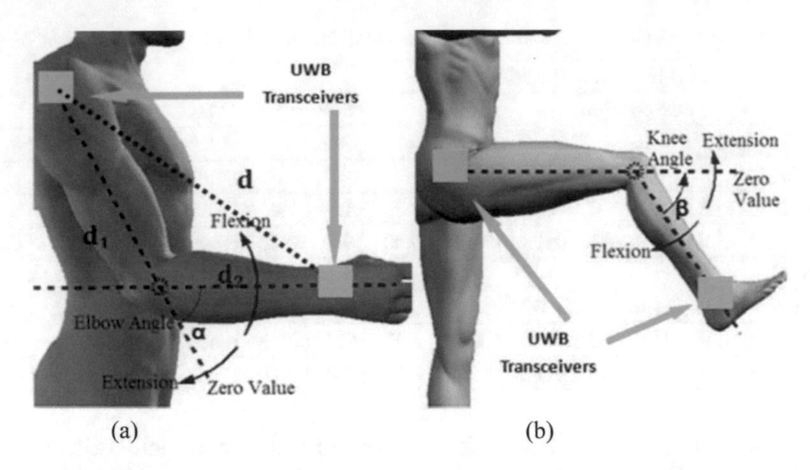

Fig. 7.14 Configuration of elbow/knee flexion/extension angle α angle β, respectively. d_1 and d_2 denote the distance between elbow, joint, and transceivers, respectively; d is the ranging data required in the proposed system (Reprinted with permission from IEEE [41])

$$\cos\alpha = -\frac{d_1^2 + d_2^2 - d^2}{2d_1 d_2} \tag{7.3}$$

where d_1 and d_2 have been measured in the initial or static phase, d is the transmitter–receiver separation distance.

To investigate the accuracy of the proposed approach, real on-body UWB measurements were obtained. A pair of UWB transceivers were attached to the thigh and shank, or forearm and upper arm, of the test subject to estimate the distance between the two points based on TOA of the received pulses. The original UWB pulse generated from Picosecond Pulse Labs 3500D impulse generator has a null-to-null pulse width of around 200 ps. The centre frequency of the received pulse from the antennas attached on arms, as shown in Fig. 7.15, is around 4.5 GHz with a −10 dB bandwidth that is approximated to 3 GHz. Skycross SMT-3TO10M UWB antennas cover 3.1 to 10 GHz bandwidth, have been used due to their small-size (25.0 $\times$ 18.5 $\times$ 1 mm^3) and low-profile characteristics, which are suitable for body sensors requirements. The received waveform is sampled by real-time oscilloscope DSO80804B with a sampling rate of 40 GHz.

To benchmark the performance of the proposed system, a comparison with a goniometer probe (PS-2137 from PASCO) attached on a human leg/arm is carried outo. The goniometer consists of two metal arm links and a potentiometer. As the angle between the leg/arm changes, the resistance of the potentiometer changes [13]. The accuracy of the equipment is $\pm 1^\circ$ when calibrated, with resolution of 0.042° at a sampling rate of 500 Hz. The subject wearing UWB antennas and goniometer probe was instructed to perform flexion and extension of his forearm and shank for 1 min, then repeated 10 times. The attachment of the goniometer and UWB antennas to the arm and leg is as shown in Fig. 7.15a, b. A Goniometer was

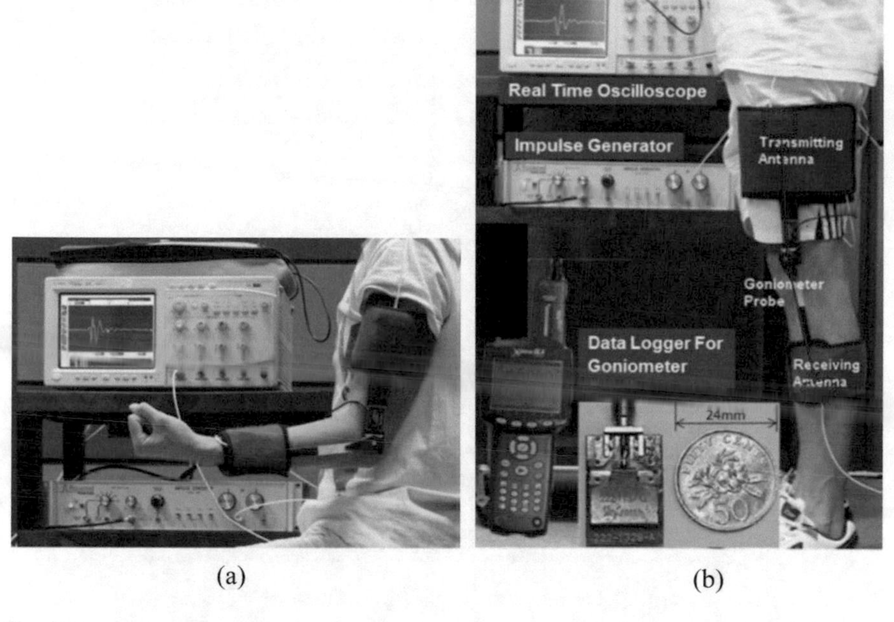

Fig. 7.15 **a** View of antennas and goniometer placement on a human arm, and **b** View of attachment of antennas and goniometer on thigh and shank (Reprinted with permission from IEEE [41])

mounted on the lateral side of the leg at the centre of knee joint. Simultaneously, the data were recorded with the oscilloscope for later analysis. The results of the subject for the arm motion and leg movement are shown in Fig. 7.16a, b. The performance of the method was compared with a flexible goniometer by simultaneously measuring joint flexion–extension angles at different angular velocities, ranging between 8° and 90°/s. The measurement errors were evaluated by the average differences between two sets of data (ranging from 0.8° for slow movement to 2.8° for fast movement), by standard deviation (ranging from 1.2° to 4.2° for various movement speeds) and by the Pearson correlation coefficient (greater than 0.99) which demonstrates the suitability of the UWB technology for accurate joint angle estimation.

7.5 Gait Activity Assessment

Gait analysis refers to the measurement, description, and assessment of the quantities that characterize human locomotion, where musculoskeletal functions are quantitatively evaluated through the measurement of joint kinematics and kinetics

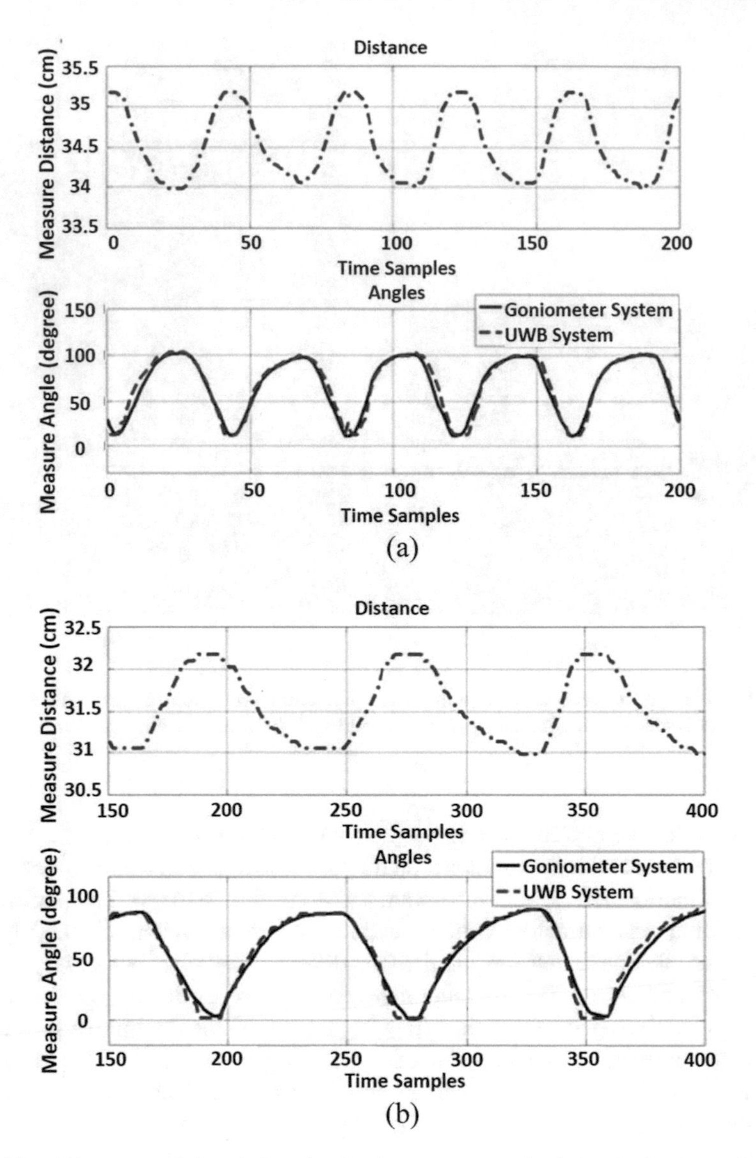

Fig. 7.16 **a** Distance variation during the shank movement and relative flexion/extension angle for goniometer and UWB system. **b** Distance variation during the forearm movement and relative flexion/extension angle for goniometer and UWB system (Reprinted with permission from IEEE [41])

[53]. Typically, the evaluation of abnormal gait requires the knowledge of normal movement biomechanics, and gait abnormalities which occur due to pain, abnormal range of motion, and leg-length discrepancy [54].

7.5.1 Gait Activity Analysis

A low-cost and low-complexity wireless ambulatory human locomotion tracking system suitable for the assessment of clinical gait analysis using wearable ultra-wideband (UWB) transceivers is presented in [44]. The work is based on on body fixed sensors using UWB technology and TOA ranging techniques with applications for long term monitoring of mobility issues in indoor or outdoor environments. The system design and transceiver performance are presented in additive-white-Gaussian noise (AWGN) and simulations were carried out based on realistic channel models for body area networks. The UWB radios measure the distances between different points on the body during movement. It includes an initial measurement phase, carried out through the measurement of the subject's height, weight, etc. Then, ranging data are acquired between different nodes while the subject is walking through the estimation of the TOA of the first path, which is then converted to a distance estimate. The measurements include the intersegmental distances that have LOS links and provide measurements for important parameters such as BOS (base-of-support) which is defined as the distance from the heel-to-heel while walking and step length. The proposed system is theoretically capable of providing a ranging accuracy error of 0.11 cm at distances equivalent to intermarker distances, at an 18 dB SNR in realistic on-body UWB channels. The experimental results have shown that the system has sufficient accuracy for clinical applications, such as rehabilitation.

7.5.2 Step Length Estimation of Human Gait

Step length estimation (SLE) of human gait has attracted the attention of many researchers because of the possibility of using this information in a variety of applications, such as human gait analysis and rehabilitation, activity monitoring or pedestrian navigation. A preliminary evaluation of the accuracy of UWB technology for step length estimation purposes is carried out in [55].

Regarding the UWB nodes, several TREK1000 development kits manufactured by Decawave were used. They are fully compliant with the IEEE 802.15.4–2011 UWB standard and make it possible to achieve ranging measurements with an accuracy of ± 10 cm using two-way ranging (TWR) TOF measurements. Five nodes were used, three of them configured as anchors and the other two as tags. However, due to the operation of the software provided by the manufacturer, the nodes configured as anchors A1 and A2 are placed on the feet and the nodes configured as T0, T1 and A0 are placed in fixed positions. In this way, in addition to the ranges from the feet to the fixed nodes, the inter-feet ranges can also be obtained. Figure 7.17a presents how the nodes were mounted on the feet using Velcro straps and wires. They were mounted as rigidly as possible to avoid vibrations of the UWB node.

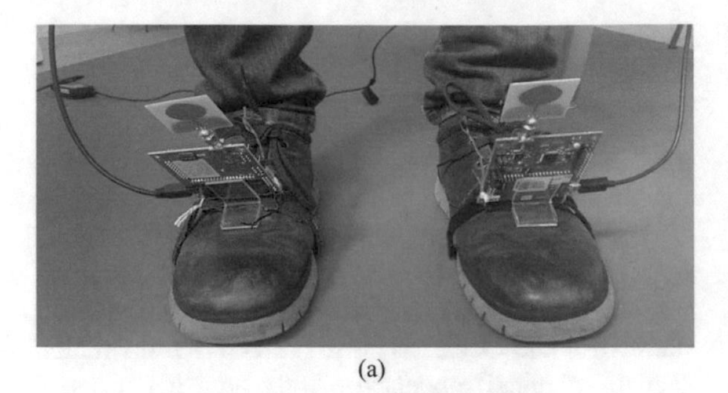
(a)

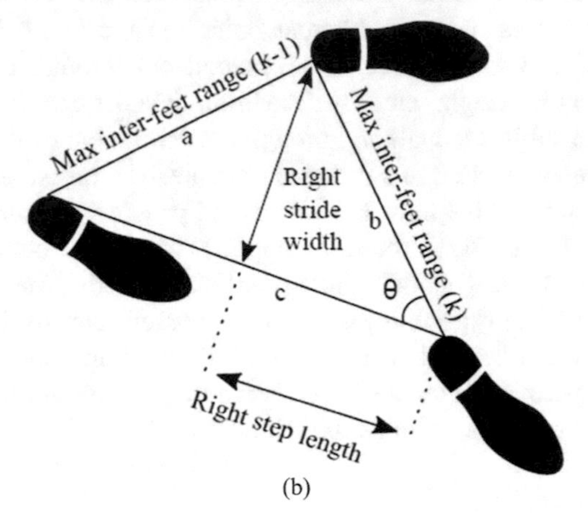

(b)

Fig. 7.17 **a** Decawave UWB nodes mounted on the feet. **b** Triangle defined by the consecutive footsteps. The inter-feet range presented as a periodic behaviour with consecutive local maxima (Reprinted with permission from IEEE [55])

Step Length Definition

To calculate the step length, the definition from Huxham et al. [56] is used, which exploits the triangle defined by the positions of the feet when resting on the ground, as shown in Fig. 7.17b. By applying the cosine rule, these step length expressions can be obtained:

$$\textit{Right step length } (SL) = b\,\cos\theta = \frac{b^2 + c^2 - a^2}{2c} \tag{7.4}$$

$$\textit{Right stride width } (SW) = \sqrt{b^2 - SL^2} \tag{7.5}$$

The advantage of this definition is that it is not necessary to walk in a straight line, which is to say that the width of the steps can vary between successive steps. As noted above, it is not necessary to know the positions of the feet in a global reference system but only their relative positions, since only the dimensions of the triangle defined by the footsteps are necessary. After analysing the performances of the different SLE methods, the first observation can be highlighted: since the step is a relative measure between the foot positions, the use of inter-feet ranges has proven to be essential when it comes to reducing the estimation error.

Four different methods are implemented and tested: one based on the variations of the ranges to a fixed point, another based on the inter-feet ranges and two more methods based on the relative positions between footsteps. The method based on the inter-feet ranges obtained an average step length error of 9.9 cm (±7.1 and a median relative error of 15%, while one of methods based on the feet positions got an average error of 11.9 cm (±16.3 cm), but only a median error of 5.9 cm if line-of-sight (LOS) scenarios are only considered. These preliminary results confirm that errors of a few centimetres can be achieved.

7.5.3 Foot Clearance Analysis During Walking

Foot clearance above ground is a key factor for better understanding of the complicated relationship between falls and gait. This work proposes a wearable system using UWB transceivers to monitor the vertical heel/toe clearance during walking.

The proposed foot clearance measurement approach uses a pair of wearable UWB transceivers because of its high ranging and positioning capacity [57]. The foot clearance is defined as the vertical distance between a foot and the ground during walking. The distance during walking is widely used in many fields extending from gait analysis to rehabilitation. First, a pair of very small and light weight antennas is placed on a point approximating to the heel/toe of the foot, acting as a transmitter and receiver. Then, the reflected signal from ground is captured and propagation delay is detected using noise suppressed Modified-Phase-Only-Correlator (MPOC). The performance of the UWB-based system was compared with an ultrasound system for stationary movements.

To demonstrate the performance of the proposed system, on-body UWB measurements are conducted in indoor environment. Two UWB transmitting and receiving antennas, manufactured by the Skycross corporation, are attached on the heels of test subject. The reflected signal is sampled by Agilent Real Time Oscilloscope DSO80804B with sampling rate of 40 GHz/s and postprocessed by MATLAB. Normal gait of healthy person consists of two major phases: swing phase and stance phase. Specific features of foot clearance during gait 6, such as Minimum Toe Clearance (MinTC), Maximum Toe Clearance (MaxTC) and Maximum Heel Clearance (MaxHC) are reported which are defined as the minimal/maximal vertical distance between the toe/heel and the ground during the swing phase. Swing phase during gait is the transition from toe/heel-off to toe/heel-strike point, and stance phase is from toe/heel-strike to toe/heel-off. The experimental results show that an overall mean difference between these two systems is about 0.634 mm with correlation coefficient value of 0.9604. The UWB-based system is then used to measure foot clearance during walking which shows promising results for gait events detection.

7.5.4 Gait Activity Identification

Everybody has a different way of walking, and for this reason, gait has been studied in the last few years as an important biometric information source. This study explores a novel approach, based on ultra-wideband (UWB) technology, for user identification via gait analysis [46]. In the proposed method, the user is supposed to wear two or more devices embedding a UWB transceiver. During gait, the distances between the devices are estimated via UWB and then analysed by means of a machine learning classifier, which provides automatic identification. Experiments were carried out on 12 volunteers, who walked while wearing four UWB boards (placed on the head, wrist, ankle, and in a trouser pocket). The off-line evaluation considered a set of different possible configurations in terms of number and position of the wearable devices. Despite a relatively low sampling frequency of 10 Hz, the results are promising: average identification accuracy is as high as $\sim$96% with four devices, and above 90% with three devices (wrist, trouser pocket, and ankle). This novel approach may enhance the accuracy of inertial-based systems for continuous user identification.

Interdistance information between wearable devices, collected during gait periods, represent valuable information for user identification. Thus, information originating from UWB-enabled transceivers can be combined with information coming from accelerometers and gyroscopes to achieve even better identification results. A prototype of the proposed method was implemented using four Decawave EVB1000 boards. The EVB1000 board is based on a DW1000 IEEE 802.15.4–2011 UWB compliant wireless transceiver and an STM32F105 ARM Cortex M3 microcontroller. Let XY be the interdistance between the devices X and Y. Interdistance XY is computed as the average of the measurements provided by the devices at its two ends (namely X and Y). A system based on four devices is considered, and thus characterised by six interdistance cases. With a period T, each device estimates the distance from all the other devices using two-way ranging.

Interdistance samples are organised in non-overlapping blocks. A walking detection technique is supposed to be used to retain only the blocks containing gait data. This is a reasonable assumption, as previous work showed that it is possible to achieve high accuracy in walking detection with body-worn UWB sensors. In the proposed identification method, blocks associated with gait periods are processed to extract a vector of relevant features (feature extraction). More precisely, the following features are calculated for each interdistance: mean, standard deviation, skewness, root mean square, min–max (the difference between the maximum and minimum value in a block), kurtosis, mean crossing rate (the number of times the mean value is crossed in a block), inter-quartile range, mean absolute deviation (a robust measure of statistical dispersion), average absolute acceleration variation (AAV), min, max. Five classification methods were considered: subspace k-nearest neighbour (subspace k-NN), weighted k-nearest neighbour (weighted k-NN), bagged tree, ensemble subspace discriminant (ESD), and support vector machine (SVM). The accuracy obtained for various classifiers is considered with precision rates above 90% for all users and recall rates above 90% for 11 users out of 12.

7.6 Conclusion

UWB technology has been successfully applied for monitoring and classification of human movements, joint angle estimation, gait movement analysis and limb movement activity using compact wearable antennas. Various methodologies and algorithms for human physical activity monitoring using wearable UWB communication technology have been described. The selection of the wearable antennas/sensor and location of the wearable node is an important aspect to obtain accurate monitoring of the limb movement activity. A less complex solution provided by using only two channel parameter PL and σ_τ have been used as features to estimate the limb movement progress and effectively assess the performance of the activity. The methodology proposed based on UWB channel parameters is low cost, simple, requires less computational time and uses fewer complex algorithms and less number of features to assess the physical activities. It also provides good performance in generalization ability and can successfully monitor the limb activity. Methodology for estimation of joint angles during limb movements is also discussed in this chapter. Various methods for gait analysis and to evaluate specific gait-related performance parameters have also been evaluated in this chapter. Novel methods for activity recognition and classification have been described based on data analysis, statistical and machine learning techniques. Based on such data, physical activities and limb movement can be tracked by an individual or remotely to relatives, doctors, healthcare departments and data centres. Furthermore, the investigation provides guidelines and procedure on evaluation of the performance of specific and daily physical activities which can be incorporated in the form of algorithms in wearable devices/sensors.

References

1. Mukhopadhyay SC (2015) Wearable sensors for human activity monitoring: a review. IEEE Sens J 15(3):1321–1330
2. Nag A, Mukhopadhyay SC, Kosel J (2017) Wearable flexible sensors: a review. IEEE Sens J 17(3):3949–3960
3. Movassaghi S, Abolhasan M, Lipman J, Smith D, Jamalipour A, Wireless body area networks: a survey. IEEE Commun Surv Tutor 16(3):1658–1686
4. Dimensions of Physical Activity, http://getmovingvcepe2015weebly.com/dimensions-of-physical-activity.html
5. Types of Physical Activity, https://www.nwhu.on.ca/ourservices/HealthyLiving/Pages/Types-of-Physical-Activity.aspx
6. Jiang S, Xue Y, Cao Y, Iyengar S, Bajcsy R, Kuryloski P, Wicker S, Jafari R (2008) CareNet: an integrated wireless sensor networking environment for remote healthcare. In: Third International Conference on Body Area Networks, Tempe, Arizona, USA 2008
7. Wang Z, Yang Z, Dong T (2017) A review of wearable technologies for elderly care that can accurately track indoor position recognize physical activities and monitor vital signs in real time. Sensors 17(2):341

8. Patel S, Park H, Bonato P, Chan L, Rodgers M (2012) A review of wearable sensors and systems with application in rehabilitation. J Neuro Eng Rehabil 9(1)
9. Zihajehzadeh S, Yoon PK, Kang B, Park EJ (2015) UWB-aided inertial motion capture for lower body 3-D dynamic activity and trajectory tracking. IEEE Trans Instrum Meas 64 (12):3577–3587
10. Bharadwaj R, Swaisaenyakorn S, Parini CG, Batchelor J, Alomainy A (2014) Localization of wearable ultrawideband antennas for motion capture applications. IEEE Antennas Wirel Propag Lett 13:507–510
11. Zhang D, Xia F, Yang Z, Yao L, Zhao W (2010) Localization technologies for indoor human tracking. In: Proceedings of IEEE International Conference on Future Information Technology (FutureTech), Busan, South Korea, May 2010, pp1–6
12. Widagdo PAC, Lee H-H, Kuo C-H (2017) Limb motion tracking with inertial measurement units. In: IEEE international conference on systems man and cybernetics (SMC) 2017, pp 582–587
13. Kyrarini M, Wang X, Gräser A (2015) Comparison of vision-based and sensor-based systems for joint angle gait analysis. In: IEEE International Symposium on Medical Measurements and Applications (MeMeA) 2015, pp 375–379
14. Vicon Motion Capture Systems [Online] Available: http://www.vicon.com
15. BTS Bioengineering [Online] Available: http://www.btsbioengineering.com
16. Weygers I, Kok M, Konings M, Hallez H, De Vroey H, Claeys K (2020) Inertial sensor-based lower limb joint kinematics: a methodological systematic review. Sensors (Basel) 20(3):673
17. El-Gohary M, McNames J (2015) Human joint angle estimation with inertial sensors and validation with a robot arm. IEEE Trans Biomed Eng 62(7):1759–1767
18. Lisini Baldi T, Farina F, Garulli A, Giannitrapani A, Prattichizzo D (2020) Upper body pose estimation using wearable inertial sensors and multiplicative kalman filter. IEEE Sens J 20 (1):492–500
19. Chen D et al (2020) Bring gait lab to everyday life: gait analysis in terms of activities of daily living. IEEE Internet Things J 7(2):1298–1312
20. Jarchi D, Pope J, Lee TKM, Tamjidi L, Mirzaei A, Sanei S (2018) A review on accelerometry-based gait analysis and emerging clinical applications. IEEE Rev Biomed Eng 11:177–194
21. Chowdhury K, Tjondronegoro D, Chandran V, Trost SG (2018) Physical activity recognition using posterior-adapted class-based fusion of multiaccelerometer data. IEEE J Biomed Health Inf 22(3):678–685
22. Mazomenos EB et al (2016) Detecting elementary arm movements by tracking upper limb joint angles with MARG sensors. IEEE J Biomed Health Inf 20(4):1088–1099
23. Park JS, Robinovitch S, Kim WS (2016) A wireless wristband accelerometer for monitoring of rubber band exercises. IEEE Sens J 16(5):1143–1150
24. Tannous H, Istrate D, Benlarbi-Delai A, Sarrazin J, Gamet D, Ho Ba Tho MC, Dao TT (2016) A new multi-sensor fusion scheme to improve the accuracy of knee flexion kinematics for functional rehabilitation movements. Sensors 16:1914
25. Abbasi-Kesbi R, Nikfarjam A (2018) A miniature sensor system for precise hand position monitoring. IEEE Sens J 18(6):2577–2584
26. Kim Y, Li Y (2017) Human activity classification with transmission and reflection coefficients of on-body antennas through deep convolutional neural networks. IEEE Trans Antennas Propag 65(5):2764–2768
27. Mokhlespour Esfahani MI, Nussbaum MA (2018) A smart undershirt for tracking upper body motions: task classification and angle estimation. IEEE Sens J 18(18):7650–7658
28. Kim D, Kwon J, Han S, Park Y, Jo S (2019) Deep full-body motion network for a soft wearable motion sensing suit. IEEE/ASME Trans Mechatron 24(1):56–66
29. Liming et al (2019) Whole system design of a wearable magnetic induction sensor for physical rehabilitation. Wiley Adv Intel Syst 26:1–11
30. Golestani N, Moghaddam M (2020) Human activity recognition using magnetic induction-based motion signals and deep recurrent neural networks. Nat Commun 11:1551

31. Chen W et al (2019) Non-contact human activity classification using DCNN based on UWB radar. In: 2019 IEEE MTT-S International Microwave Biomedical Conference (IMBioC), Nanjing, China, 2019, pp 1–4
32. Ding C et al (2018) Non-contact human motion recognition based on UWB radar. IEEE J Emerg Sel Topics Circuits Syst 8(2):306–315
33. Qi F et al (2019) Position-information-indexed classifier for improved through-wall detection and classification of human activities using UWB bio-radar. IEEE Antennas Wirel Propag Lett 18(3):437–441
34. Seifert AK, Amin MG, Zoubir AM (2019) Toward unobtrusive in-home gait analysis based on radar micro-doppler signatures. IEEE Trans Biomed Eng 66(9):2629–2640
35. Bharadwaj R, Parini CG, Koul SK, Alomainy A (2019) Effect of limb movements on compact UWB wearable antenna radiation performance for healthcare monitoring. Prog Electromag Res C 91:15–26
36. Bharadwaj R, Koul SK (2019) Experimental analysis of ultra-wideband body-to-body communication channel characterization in an indoor environment. IEEE Trans Antennas Propag 67(3):1779–1789
37. Yan S, Soh PJ, Vandenbosch GAE (2018) Wearable ultrawideband technology—a review of ultrawideband antennas, propagation channels, and applications in wireless body area networks. IEEE Access 6:42177–42185
38. Bharadwaj R, Swaisaenyakorn S, Parini CG, Batchelor JC, Alomainy A (2017) Impulse radio ultra-wideband communications for localization and tracking of human body and limbs movement for healthcare applications. IEEE Trans Antennas Propag 65(12):7298–7309
39. Bharadwaj R, Parini CG, Alomainy A (2015) Experimental investigation of 3D human body localisation using wearable ultra wideband antennas. IEEE Trans Antennas Propag 63 (11):5035–5044
40. Bharadwaj R, Swaisaenyakorn S, Parini CG, Batchelor JC, Koul SK and Alomainy A (2019) UWB channel characterization for compact L-shape configurations for body-centric positioning applications. IEEE Antennas Wirel Propag Lett 1–5
41. Qi Y, Soh CB, Gunawan E, Low K, Maskooki A (2014) A novel approach to joint flexion/ extension angles measurement based on wearable UWB radios. IEEE J Biomed Health Inf 18 (1):300–308
42. Bharadwaj R, Koul SK (2021) Assessment of limb movement activities using wearable ultra-wideband technology. IEEEs Trans Antennas Propag 69(4):2316–2325
43. Bharadwaj R, Koul SK (2019) Monitoring of limb movement activities during physical exercises using UWB channel parameters. In: International Biomedical Microwave Conference (IMBIOC 2019), Nanjing, China, 6–8 May, 2019, pp 1–3
44. Shaban HA, El-Nasr MA, Buehrer RM (2010) Toward a highly accurate ambulatory system for clinical gait analysis via UWB radios. IEEE Trans Inform Technol Biomed 14(2):284–291
45. Bharadwaj R, Koul SK (2020) Wearable ultra wideband technology for daily activity recognition. In: 2019 International Biomedical Microwave Conference (IMBIOC 2020), Toulouse, France, 14–17 Dec, 2020, pp 1–3
46. Vecchio A, Cola G (2019) Method based on UWB for user identification during gait periods. Healthc Technol Lett 6(5):121–125
47. Fort A, Desset C, De Doncker P, Wambacq P, Van Biesen L (2006) An ultra-wideband body area propagation channel Model-from statistics to implementation. IEEE Trans Microw Theory Tech 54(4):1820–1826
48. Rappaport TS (1996) Wireless communications: principles and practice Upper Saddle River. Prentice-Hall, NJ, USA
49. Alomainy A, Sani A, Rahman A, Santas JG, Hao Y (2009) Transient characteristics of wearable antennas and radio propagation channels for ultrawideband body-centric wireless communications. IEEE Trans Antennas Propag 57(4):875–884
50. Reider B (2015) Physical medicine and rehabilitation: principles and practice, Volume 1, Orthopaedic rehabilitation of the athlete: getting back in the game treatment of chronic pain conditions: a comprehensive handbook, Saunders, 1 edn

51. Active Range of Motion Exercises, https://www.drugscom/cg/active-range-of-motion-exercises.html
52. Joint Actions, https://brookbushinstitute.com/article/joint-actions/
53. Cappozzo A, Croce U, Leardini A, Chiari L (2005) Human movement analysis using stereophotogrammetry part 1: theoretical background. Gait Posture 21(2):186–196
54. Gross J, Fetto J, Rosen E (eds) (2002) Musculoskeletal examination, 2nd edn. Blackwell Science Inc., Malden, MA
55. Díez L E, Bahillo A, Otim T, Otegui J (2018) Step length estimation using UWB technology: a preliminary evaluation. In: 2018 international conference on indoor positioning and indoor navigation (IPIN), Nantes, 2018
56. Huxham F, Gong J, Baker R, Morris M, Iansek R (2006) Defining spatial parameters for non-linear walking. Gait Posture 23(2):159–163
57. Qi Y, Soh CB, Gunawan E, Low K, Maskooki A (2013) Using wearable UWB radios to measure foot clearance during walking. In: 35th annual international conference of the IEEE engineering in medicine and biology society (EMBC), Osaka, 2013, pp 5199–5202

Chapter 8
UWB and 60 GHz Radar Technology for Vital Sign Monitoring, Activity Classification and Detection

8.1 Introduction

Recent developments in miniaturised devices, smart sensors, Internet-of-Things (IoT), sophisticated signal processing algorithms and machine learning techniques has increasingly contributed to the unobtrusive sensing and activity monitoring applications. Radar technology is an upcoming, useful, and safe technology in the field of wireless body area networks (WBANs) and takes into account user privacy issues in comparison to camera-based approaches [1]. Non-invasive assessment and monitoring are gaining importance in various domains such as healthcare, elderly care, emergency operations and smart living [1–3]. It is an important aspect in scenarios such as electrocardiogram (ECG) monitoring for infants, monitoring vital parameters of burn victims, detection of people in emergency and rescue situations to avoid the complicated wired connections. The UWB and mmWave radar technologies has seen recent advancement in terms of compact and high-performance antenna design, sophisticated system on chip developments and advance signal processing algorithms which has made it suitable for several monitoring and sensing applications [4, 5]. The radar exploits the fine time resolution of these technologies which aid in collection and extraction of the vital information in the presence of clutter and noise.

UWB technology is quite suitable for biomedical applications because it radiates and consumes little power, coexists well with other instruments, and is robust to interference and multipath [5–8]. Due to the unique characteristics such as higher penetration capabilities, extremely precise ranging, low power requirement, low cost, simple hardware, and robustness to multipath interferences, impulse radio ultra-wideband (IR-UWB) technology is appropriate for non-invasive medical applications. Such systems are based on the transmission and reception of sub-nano second pulses without carriers or modulated short pulses with carriers and can be conveniently processed in time domain [9–11]. IR-UWB sensors can detect the macro as well as micro movement inside the human body due to its fine range resolution such as breathing and heart rate.

S. K. Koul and R. Bharadwaj, *Wearable Antennas and Body Centric Communication*, Lecture Notes in Electrical Engineering 787,
https://doi.org/10.1007/978-981-16-3973-9_8

Recently, the 60-GHz millimetre-wave (mmWave) frequency band has been extensively studied for short-range Gigabit communication applications [12]. The use of an EM wave with higher frequency may have better phase-modulation sensitivity of the backscattered EM wave signal during the radar measurements. This increased sensitivity may allow for better breathing and heartbeat signals to be extracted from the mmWave life detection systems [12]. In addition, at the higher mmWave frequency range, a smaller chip size, compact high gain reconfigurable phased array antennas with directional beam forming, high precision are some of the other features which has made it an attractive candidate for monitoring and sensing applications.

Radar based noncontact detection, identification, classification, and recognition of human activities and behaviour methodologies have attracted several research interests. This chapter presents recent advances in IR-UWB and mmWave radar system design for healthcare, such as vital signs measurements, through-wall vitals measurement and detection, daily activity monitoring, fall detection, sleep monitoring, gait analysis and gesture recognition. The feasibility of estimating vital signs specifically breathing rate and heartbeat frequency from the spectrum of recorded waveforms, using robust algorithms and sophisticated system design has been reported. The aim of this chapter is to present various state-of-the-art techniques and algorithms for non-invasive detection and monitoring applications which enhance the sensing and detection capabilities. Various state-of-the-art techniques and algorithms for non-invasive detection and monitoring applications such as vital-sign monitoring, classification and activity recognition, localization and tracking of human subjects have been discussed. This chapter provides insight into the recent developments in UWB and mmWave radar technology for physiological parameters monitoring and human activity recognition.

8.2 Vital Sign Monitoring

Respiration and heartbeat rates are critical physiological parameters for human being, especially in medical applications, public safety, and rescue missions. The block diagram of a basic radar system for monitoring applications is presented in Fig. 8.1. The principle of detection of vital sign monitoring is based on the tiny vibration induced on the chest wall due to heart or lung displacements. The phase of the reflected signal from the chest contains chest wall displacement information and is used for the extraction of vital-signs information. Valuable physiological information such as the heartbeat and respiration rates can be obtained from such sensors [13–15]. Various advancements have been discussed in this section in terms of algorithms, methodologies, and system design.

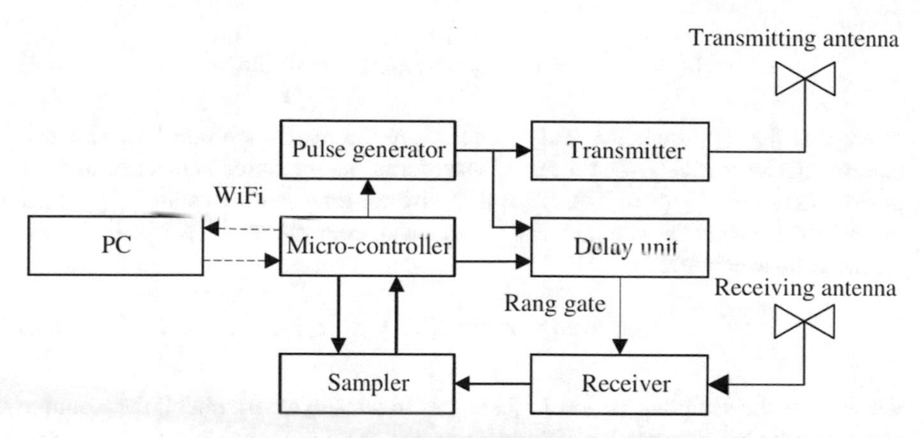

Fig. 8.1 The block diagram of the IR-UWB radar system (Reproduced courtesy of The Electromagnetics Academy [14])

8.2.1 Mathematical Model

A mathematical model is developed so that the spectrum of the detected signal can be obtained and understood. When the transmitted pulse hits the human target, part of it is reflected due to the high reflectivity of the body [13, 14]. The time-of-flight (TOF) or arrival (TOA) of this pulse is denoted by τ_0, and depends on the antenna distance, d_0. Due to respiration and heart motion, the chest cavity expands and contracts periodically, so the distance travelled, $d(t)$, varies periodically around the nominal distance d_0. For vital signs monitoring, the body movement caused by both respiration and heartbeat must be detected:

$$d(t) = d_0 + m(t) = d_0 + m_b \sin(2\pi f_b t) + m_h \sin(2\pi f_h t) \tag{8.1}$$

where m_b and m_h are the respiration and heartbeat displacement amplitudes, f_b and f_h are the respiration and heartbeat frequencies. The received signal can be represented as the sum of the responses of the channel, and the variation due to the respiration and heartbeat:

$$r(t, \tau) = \sum_i A_i p(\tau - \tau_i) + Ap(\tau - \tau_d(t)) \tag{8.2}$$

where $p(t)$ is the normalized received pulse, A_i is the amplitude of each multipath component, τ_i its delay, and A is the amplitude of the pulse reflected on the body. From (8.2) it is evident that respiration and heart movements modulate the received signal. The time delay τ_d associated with the vital sign is modelled as the sum of the time-of-flight τ_0 plus two sinusoidal delays associated to respiration and heartbeat displacements:

$$\tau_d(t) = \frac{2d(t)}{c} = \tau_0 + \tau_b \sin(2\pi f_b t) + \tau_h \sin(2\pi f_h t) \tag{8.3}$$

where c is the light velocity, and τ_b and τ_h are the respiration and heartbeat displacements, respectively. The received waveforms are measured at discrete instants in slow time $t = nT_s$ $(n = 1, 2. \ldots.., N)$. N discrete-time sequences are stored after the received signal is sampled and these values are stored in a matrix $\mathbf{R}$, the elements of which are:

$$R[n, m] = r(\tau = nT_f,\ t = mT_s) \tag{8.4}$$

where T_f is the sampling period in fast-time. In a static environment, the resulting clutter can be considered as a DC-component in the slow-time direction. The only movement is caused by the person's respiration and heart activity, from (8.2) background clutter does not depend on slow-time t. Thus, the background clutter can be removed by filtering the signal. This can be done by subtracting the average of all received waveforms from the original signal (a new matrix $\mathbf{X}$ can be obtained by subtracting the average of all the rows in $\mathbf{R}$ from each row).

$$x(t, \tau) = r(t, \tau) - \lim_{T \to \infty} \frac{1}{T} \int_0^T r(t, \tau) dt = Ap(\tau - \tau_d(t)) - r_0(\tau) \tag{8.5}$$

The DC component $r_0(\tau)$ is blocked by subtracting the average of all samples in fast-time (the result is saved in a new matrix $\mathbf{Y}$ that is obtained by subtracting the average of all columns in $\mathbf{X}$ from each column). The signal of interest is:

$$y(t, \tau) = x(t, \tau) - x_0(\tau) = Ap(\tau - \tau_d(t)) \tag{8.6}$$

In the case of body movement, the distance d_0 depends on the time. To compensate random body movements, after the static background has been removed, the first waveform (or column) is used as reference. Then, the cross correlation between the reference waveform and the next waveforms will be carried out. The delay position of the peak of cross correlation is used to shift each waveform to align the waveforms and compensate the delay due to the body movement. It can be found that the Fourier transformed slow-time $y(t, \tau)$ is:

$$Y(f, \tau_0) = A \sum_{k=-\infty}^{+\infty} \sum_{l=-\infty}^{+\infty} C_{kl} \delta(f - kf_b - lf_h) \tag{8.7}$$

$$C_{kl} = \int_{-\infty}^{+\infty} P(v) J_k(\beta_b v) J_l(\beta_h v) dv \tag{8.8}$$

where $\beta_b = 4\pi m_b/c$ and $\beta_h = 4\pi m_h/c$ and τ_0 correspond to the delay for the maximum amplitude of the received pulse. The spectrum in (8.7) is a discrete function, consisting of a train of delta functions centred at the frequencies of the harmonics of f_b, f_h and their intermodulation products. The amplitude of each intermodulation product for a frequency of $f = kf_b + lf_h$ is controlled by the coefficient C_{kl}. The fundamental component of breathing is C_{10}, whereas the fundamental component of the heartbeat signal is C_{01}. The harmonic components of breathing can be obtained for the index $l = 0$.

8.2.2 Algorithms and Techniques for Vital Sign Monitoring

8.2.2.1 Algorithms, System Design and Signal Processing Methodologies

The research on vital information extraction meets a lot of challenges due to the narrow bandwidth and low signal-to-noise ratio (SNR) of a vital sign in real environment. A non-contact vital sign detection method based on multiple higher order cumulant (HOC) is presented in [16]. According to the characteristic of vital sign for IR-UWB radar, the quasi-periodic reflected echo in slow-time is analysed. The method is theoretically deduced from fourth-order cumulant and has proved to provide accurate results as seen from simulations and experiments when compared with the reference Fast Fourier transform. The new vital detection method with the properties of simplicity, high output SNR, and suppression of higher harmonics is appropriate for detecting respiration or heartbeat periodic signals. By using the proposed technique, the range position and frequency information of life can be extracted accurately and automatically.

This work reports a system-on-chip (SoC) UWB pulse radar for respiratory rate monitoring implemented in 90 nm complementary metal oxide semiconductor (CMOS) technology [17] and its experimental tests in operating scenarios. A system-on chip UWB pulse radar based on correlation receiver has been implemented in 90 nm bulk CMOS technology, characterized experimentally, and applied to respiratory rate monitoring. The experimental tests on adult (both genders) and baby have been carried out successfully in different scenarios, demonstrating that the UWB pulse radar sensor can detect the respiration activity associated with sub-centimetre chest movements.

In this work an IR-UWB system is presented, which can be used as a radar sensor for highly precise object tracking and breath-rate sensing [18]. The hardware consists of an impulse generator integrated circuit (IC) in the transmitter and a correlator IC with an integrating baseband circuit as correlation receiver. To localize objects precisely in front of the sensor, three impulse tracking methods are compared: Tracking of the maximum impulse peak, tracking of the impulse slope, and a slope-to-slope tracking of the object's reflection and the signal of the static direct coupling between the transmit and receive antenna; the slope-to-slope tracking showing the best performance. The breathing signals of male humans and a seven-week-old infant are presented, hence, validating the usage of IR-UWB radar technology for breath-rate estimation.

Harmonic Path (HAPA) algorithm for estimation of heart rate (HR) and respiration rate (RR) with IR–UWB radar is introduced in [19]. Most of the existing methods try to identify the fundamental component to estimate the HR and/or RR. However, often the fundamental is distorted or cancelled by interference, such as RR harmonics interference on the HR fundamental, leading to significant error for HR estimation. The algorithm utilizes not only the fundamental component but also its harmonics to improve the estimation accuracy of the vital signs which has been demonstrated experimentally. The harmonics are referred to as the heart components, and are equidistant and separated by a frequency equal to the HR fundamental. Secondly each heart component is at a multiple of that inter-peak distance. HAPA will first detect in the received spectrum a path, defined as a set of three or more consecutive approximately equally spaced spectral peaks, such that their frequencies are approximately an integer multiple of the average inter-peak distance in terms of frequency; that distance of the most powerful path is the rate estimate. An accurate HR estimate is obtained even if the heart fundamental peak is completely missing.

The Spectrum-averaged Harmonic Path (SHAPA) algorithm for estimation of heart rate (HR) and respiration rate (RR) with low cost IR-UWB radar is introduced in [20]. The SHAPA algorithm (1) takes advantage of the HR harmonics, where there is less interference, and (2) exploits the information in previous spectra to achieve more reliable and robust estimation of the fundamental frequency in the spectrum under consideration. The new algorithm thus retains all the properties of HAPA, including robustness against RR harmonic interference, and provides an accurate estimate of HR if the fundamental or one of the harmonics is missing.

A CMOS integrated IR-UWB based radar transceiver applied in remote respiration- and heart-rate monitoring is presented in [21]. The combination of high bandwidth and direct RF sampling offers fine spatial resolution and accuracy along with flexibility in the application-level signal processing. The high-speed RF sampling is enabled by a time-interleaved Analog-to-Digital Converter (ADC) relying on the inherent coherency of radar Tx/Rx. The topology is based on Continuous Time Binary Valued (CTBV) signal processing moving design constraints from the amplitude domain to the time domain adapting to the very short time-constants offered by modern CMOS technology. The radar sensor is capable of measuring very small movements in demanding scenarios by combining high Tx bandwidth with a very sensitive Rx front-end and high-speed samplers. The Pulse-Doppler (PD) signal processing technique was utilized to further enhance the ability to isolate small movements in the presence of larger static or slow-moving objects. Experimental results including a scenario with respiration sensing of 13 individuals during sleep show very good alignment with the gold-standard reference where 95% of the compared samples were within 5% deviation from the data reported by the PSG equipment.

Ultra-wideband (UWB) pulse Doppler radars have be used for accurate non-contact vital signs monitoring in work reported in [22]. The complex signal demodulation (CSD) and arctangent demodulation (AD) techniques used for accurately detecting the phase variations of the reflected signals of continuous wave radars to UWB pulse radars is proposed for vital signs monitoring. These detection techniques reduce the impact of the interfering harmonic signals, thus improving the SNR of the detected vital sign signals. To further enhance the accuracy of the HR estimation, a recently developed state-space method (SSM) has been successfully combined with CSD and AD techniques and over 10 dB improvements in SNR is demonstrated. To validate the phase-based methods, two experiments on human subjects and a third one using an actuator have been carried out. In these experiments, horn antennas were used, which were 1.5 m above the ground. These two antennas were located 50 cm away from each other for adequate decoupling, and the human target was 0.8 m away (Fig. 8.2).

In the first experiment, subject 1 held his breath and sat still in front of the radar system; in the second experiment, subject 2 breathed normally and kept stationary in front of the UWB radar. The data collected by a reference commercial sensor serve as a reference for the heartbeat signal. The six algorithms, i.e., direct FFT, CSD, AD, SSM, SSM-CSD, and SSM-AD, are applied to the radar data collected in the experiments and their accuracies as well as their SNRs are evaluated as shown in Fig. 8.3a–f.

8.2.2.2 Vital Sign Monitoring in Complex Environments

Identifying vital signs and locating buried survivors is possible using UWB radar technology. It is difficult to identify a human's vital signs (breathing and heartbeat) in complex environments due to the low signal-to-noise ratio of the vital sign in radar signals. Advanced signal-processing approaches are used to identify and to extract human vital signs in complex environments. Curvelet transform is used to remove the source-receiver direct coupling wave and background clutters [23].

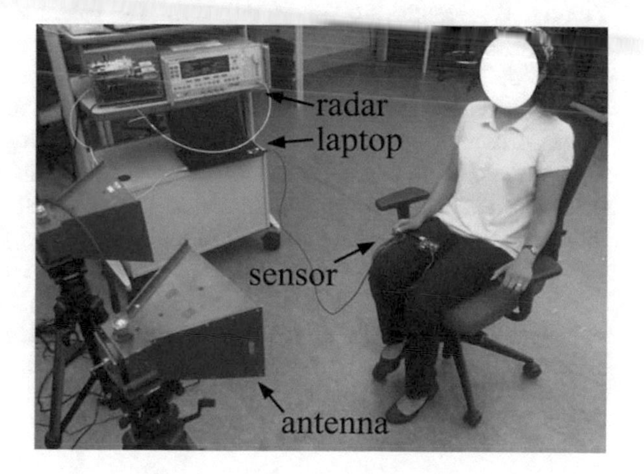

Fig. 8.2 Experimental setup with human subject during vital-sign monitoring (Reprinted with permission from IEEE [22])

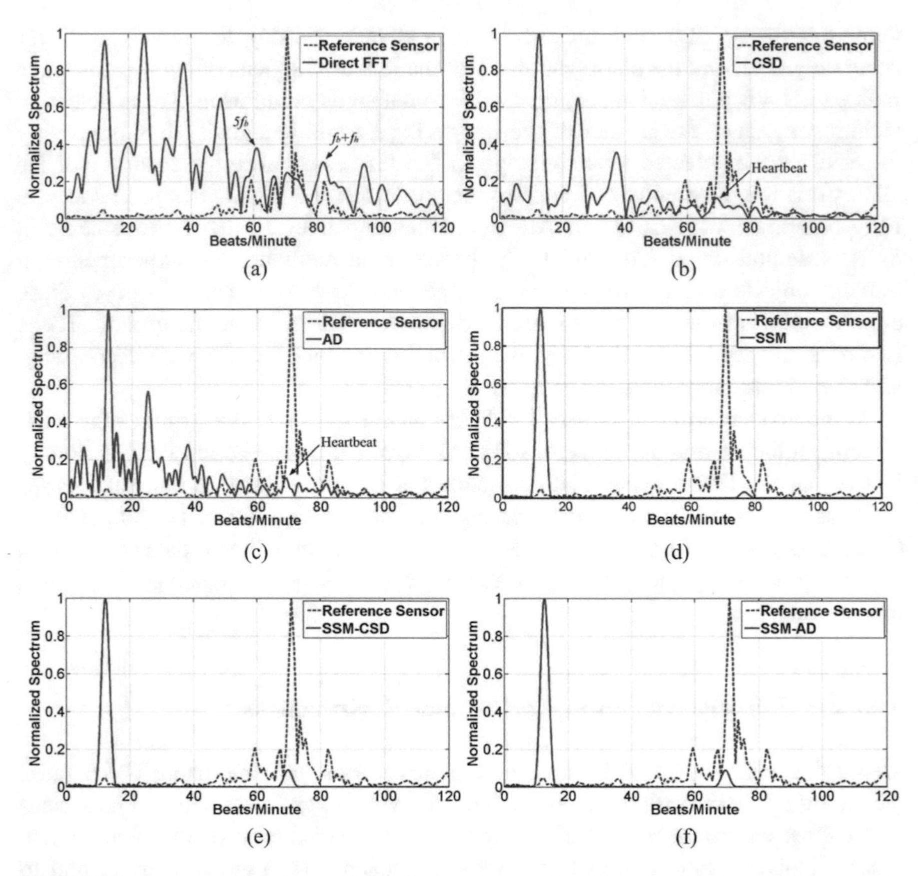

Fig. 8.3 Spectrum graphs when subject 2 breathed normally, detected HR using **a** reference FFT, **b** CSD, **c** AD, **d** SSM, **e** SSM-CSD, and **f** SSM-AD (Reprinted with permission from IEEE [22])

Next, singular value decomposition (SVD) is used to de-noise in the life signals. Finally, the results are presented based on Fast Fourier Transform (FFT) and Hilbert-Huang transform to separate and to extract human vital sign frequencies, as well as the micro-Doppler shift characteristics. The proposed processing approach is first tested by a set of synthetic data generated by finite difference time domain (FDTD) simulation for UWB radar detection of two trapped victims under debris at an earthquake site of collapsed buildings. Besides, the characteristic frequency of the two humans' breathing and heartbeat is also extracted. Then, it is validated by laboratory experiments data showing the potential of UWB radar technology for efficient vital sign detection and location in search and rescue for trapped victims in complex environment. The above signal-processing methods are applied in the experimental data. Although the effect of the background disturbance and the human's life signal is very weak, the algorithm can also extract the human's breathing frequency from the FFT result effectively.

Various signal processing algorithms and the IR-UWB radar system is proposed to detect through wall human respiration [24]. Such analysis and techniques can be useful for battlefield and victim search applications. Results are presented for through wall respiration detection through reinforced concrete wall with detection range of 0.7–2.5 m.

Techniques based on IR-UWB radar that can be helpful in developing methods for preventing car crashes by monitoring of vital signs related to drowsiness driving and the detection of mobile phone usage by the driver. Fast time locations have better information about the vital signal of the driver and the signals are divided at those locations into segments and vital signals are constructed based on the correlation concept. After construction of the vital signal, an FFT algorithm was applied, and the respiration and heart rate were found. Another objective of this work was to detect the use of mobile phones while driving. The proposed algorithm distinguishes and detects the driver's cell phone use from various other actions or changes inside the car using the dual mode background subtraction method [25].

In the experiments, commercially available single-chip IR-UWB radar transceiver NVA6201 made by NOVELDA (Novelda AS, Kviteseid, Norway) is used. The radar has a center frequency of 6.8 GHz, a bandwidth of 2.3 GHz, and a transmission output power of −53 dBm/Hz. The pulse repetition frequency (PRF) is 100 MHz, and the slow time sampling frequency (measurement rate) is 110 samples/s. Figure 8.4a shows the experimental set up where the region is divided in the mobile detection region and the vital sign detection region. Figure 8.4b shows the transceiver module with antennas attached. In Fig. 8.4c, the frequency domain signal is shown. The highest peak i.e., the peak at 21 cycles/minutes represents the breathing frequency whereas the second highest peak in the heart frequency range i.e., at 62 cycles per minute represent the heart rate of the human.

8.2.2.3 Advance Signal Processing Trends for Vital-Sign Monitoring

Detection of the subject chest-wall movement and non-contact estimate of respiration and heartbeat rates of the subject is proposed in [26]. IR-UWB radar is chosen to achieve low power consumption, flexible detection range and desirable accuracy. A noise reduction method based on improved ensemble empirical mode decomposition (EEMD) and a vital sign separation method based on the continuous-wavelet transform (CWT) are applied together on experimental data to improve the signal-to-noise ratio (SNR) to obtain accurate vital signs monitoring.

IR-UWB radar is used for monitoring respiration and the human heart rate in [27]. The vital signs were estimated for the signal reflected from the chest, as well as from the back side of the body in different experiments. A Kalman filter is applied to reduce the measurement noise from the vital signal and an algorithm applied to separate the heart rate signal from the breathing harmonics. An auto-correlation based technique is chosen for detecting random body movements (RBM) during the measurement process. Vital signal obtained during different

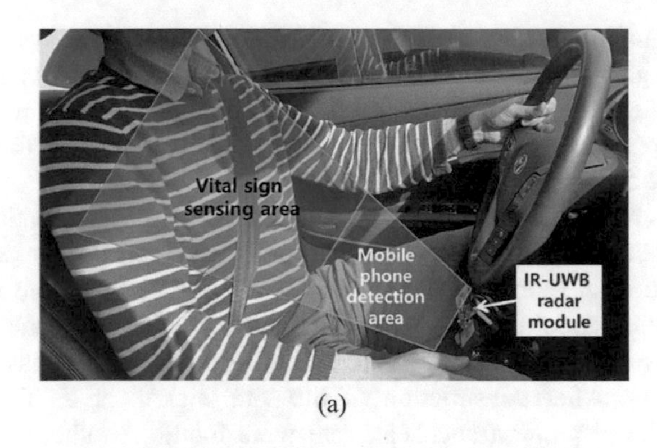

(a)

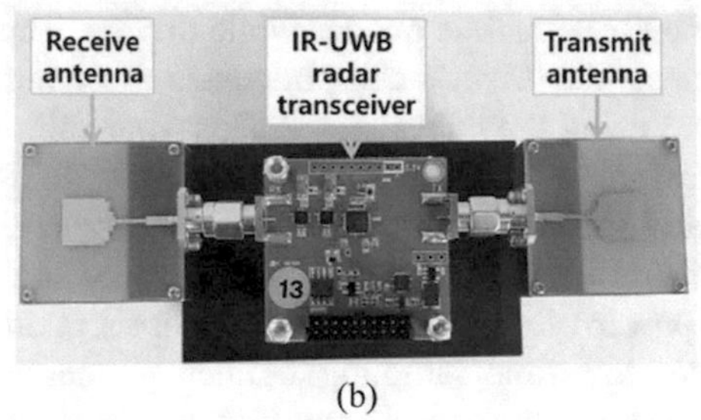

(b)

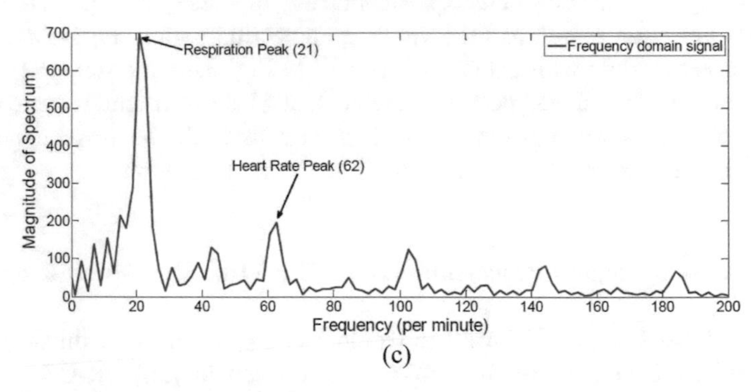

(c)

Fig. 8.4 **a** Experimental setup inside the car, **b** IR-UWB radar module, **c** spectrum obtained by applying FFT algorithm to the signal obtained by proposed algorithm (Reprinted with permission from Sensors [25])

motion states of the body: (a) vital signal with different body states; (b) when body is stationary; (c) while speaking; (d) moving head slightly; (e) moving whole body slightly.

IR-UWB radar is used to continuously monitor a resting subject's respiration and heart rates and gather information about the health status [28]. An algorithm comprising of noise reduction, respiration rate extraction, and heart rate extraction is used to process the recorded waveforms to extract vital signals information. The algorithm also addresses the effects of harmonics and intermodulation between the breathing and heartbeat signals without requiring the implementation of filters. Numerical results are presented to validate the proposed algorithm's accuracy.

This work analyses and discusses the capability of human being detection using IR-UWB radar with an improved detection algorithm [29]. The multiple automatic gain control (AGC) technique is employed to enhance the amplitudes of human respiratory signals. Two filters with seven values averaged are used to further improve the signal-to-noise ratio (SNR) of the human respiratory signals. The maximum slope and standard deviation are used for analysing the characteristics of the received pulses, which can provide distance estimates for human being detection. Based on the distance estimate, the interested region containing human respiratory signals can be determined, which can be used to improve the SNR and the accuracy of the frequency estimate of the human respiratory movement.

The method proposed in [30] uses the feature time index with the first valley peak of the energy function of intrinsic mode functions (IMF) (FVPIEF) calculated by pseudo bi-dimension ensemble empirical mode decomposition method and extracts the vital signals by the ensemble empirical mode decomposition (EEMD). Both simulation and experiment results evidently show that the proposed FVPIEF based two-layer EEMD method is effective for separating the small heartbeat signal from the large breath signal and significantly improves the evaluation of heart and breathing rates in both hold-breathing and breathing conditions. The ensemble empirical mode decomposition (EEMD) is a nonlinear, non-stationary and noise-assisted data analysis method for solving the mixing mode problem in empirical mode decomposition (EMD).

In this work instead of picking the highest peak as in the conventional methods, a novel first valley-peak of IMF energy function (FVPIEF) method is proposed for feature time index (FTI) identification. Using FVPIEF based EEMD with Pseudo-Bi-Dimensional Ensemble Empirical Mode Decomposition (PBDEEMD) method can divide the received echo signals off not only the clutter and the breathing signal, but also the tiny heartbeat signal effectively. The proposed scheme has the following steps: Remove the system static clutter from original data of UWB receiver by using the PBDEEMD. The first level PBDIMF matrix is regarded as the system static clutter and the specific second level PBDIMF is selected appropriate physiological signals. Calculate the second level PBDIMF matrix to obtain the energy function of fast time axis, then use the FVPIEF method to determine the time region of interest (TROI) and extract the feature time index (FTI) by the peaks and valleys of the energy function. Decompose the slow time signals of the second level PBDIMF at FTI into the IMF, using EEMD. Those IMFs are divided into the high frequency clutters, the low frequency breathing and heartbeat signals. Heartbeat detection in breathing state is detected and its reliability is verified with normal vital sign's value. The corresponding peak rates of breathing and heartbeat are 0.25 Hz (15 breath/min) and 1.15 Hz (69 beat/min), respectively.

A miniature inductor less impulse-radio ultra-wideband transmitter-receiver and radar for wireless short-range communication and vital-sign sensing is proposed in [31]. The setup for measuring the human respiration by the radar mode is shown in Fig. 8.5a. The distance between the antenna and body is about 40 cm. The pulse rate is 1 Mbps while the chest movement rate is under 10 Hz. The chest is approximately still for many pulses and the movement can be reconstructed by time of flight (TOF) estimation. Directional antennas with 10 dB gain are used to increase the transmission gain. The digital output of the receiver is sampled and recorded by a Tektronix DPO 71254 oscilloscope. The frame sample rate is 3 Hz and the time resolution is 4 ps. The displacement of the chest versus time during normal respiration is shown in Fig. 8.5b. Fourier transformation reveals the respiratory rate as shown in Fig. 8.5c). The respiration frequency is 0.25 Hz (15 rpm) for normal respiration. To verify the accuracy of the measurement results, a commercial hospital-standard Philips IntelliVue X2 module is used to measure the respiration rate simultaneously as the reference.

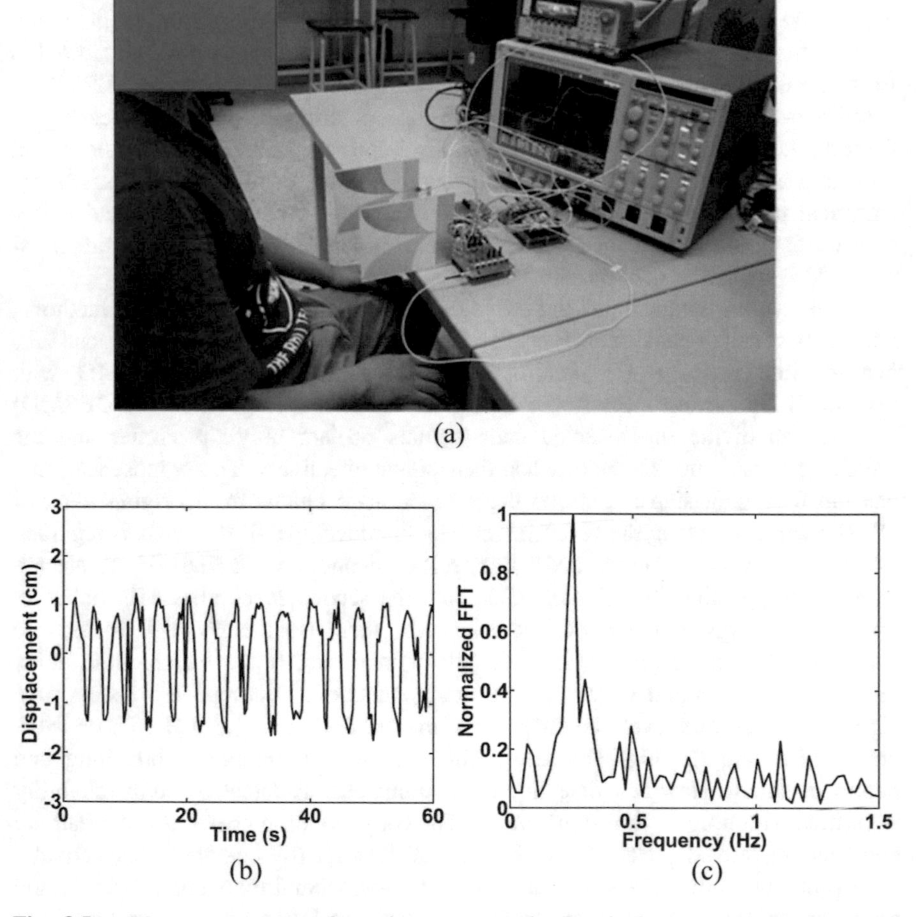

Fig. 8.5 **a** Measurement setup for radar sensor mode. Respiration measurement **b** normal respiration in time domain, **c** normal respiration in frequency domain (Reprinted with permission from IEEE [31])

The work reported in [32] presents a method based on autocorrelation to measure the RR and HR using IR-UWB radar. The correlation coefficient waveform contains the vital sign signals, overcoming the effect of noise and clutter. Experiments are carried out using a PulsOn410 UWB radar. The waveform with high signal-to-noise clutter ratio (SNCR) can be derived directly from the autocorrelution computation, without performing any pre-processing of the original radar data. Applying the conventional FFT method, respiration rate can be easily obtained. In general, heartbeat signal is difficult to detect due to its small amplitude. Variational Mode Decomposition (VMD) algorithm is chosen that can separate center frequencies of RR and HR. By dividing the radar data matrix into several blocks in fast time direction and moving one block out every time, the autocorrelation coefficient of the remainder matrix in slow time direction can be produced. Subsequently the subject location can be estimated from the correlation results.

UWB through-wall radar for vital signals detection with an improved sensing algorithm for random-noise de-noising and clutter elimination is reported in [33]. One filter is used to improve the signal-to-noise ratio (SNR) of these vital sign signals. Using the wavelet packet decomposition, the standard deviation based spectral kurtosis is employed to analyse the signal characteristics which provides the targets distance estimate. The system efficiency is improved by reduction in data size defined w.r.t region of interest (ROI). The respiration frequency is estimated using a multiple time window selection algorithm. Experimental results are presented which illustrate the effectiveness of this method.

In this work, the harmonic multiple loop detection (HMLD) algorithm for heart rate (HR) and respiration rate (RR) estimation along with location estimation with impulse IR-UWB radar is introduced [34]. The harmonic multiple discriminant principle is used to detect whether the peak point in the fundamental frequency band is the true vital sign signal. The cyclic spectrum updating process can remove the error peak point according to the results obtained from the part one. In this algorithm, threshold setting is not needed, and the estimation error caused by the erroneous threshold can be reduced. The algorithm uses the fundamental and second harmonic frequencies and does not require any threshold setting leading to robust performance and less complicated radar hardware.

8.2.2.4 Monitoring Vital-Signs of Multiple Subjects

A phase-based algorithm based on a logarithmic method, applicable to UWB radars and suitable to real-time monitoring, is proposed in [35] to detect the phase variations of reflected pulses caused by the tiny cardiac motions. Compared with conventional FFT vital signs detection method, the algorithm demonstrates advantage in respiration harmonics suppression and avoidance of intermodulation between respiration and heartbeat signals. Furthermore, it is experimentally shown that UWB Doppler radar is capable of multiple heartbeats detection and subject identification/localization.

A 2 × 2 IR-UWB distributed MIMO radar prototype system is developed to simultaneously localize multiple human objects and monitor their vital signs including respiration amplitude, RR and HR. An imaging algorithm based on back projection (B-P) is developed for the system to achieve 2-D localization of multiple objects [36]. An algorithm is devised for the IR-UWB radar to accurately estimate the chest skin range displacement caused by respiration and heartbeat. This algorithm can be efficiently implemented via simple FFT operations. The obtained chest skin range displacement signal (DS) is preferred over the conventional SS since DS contains the desirable respiration amplitude information, but SS does not. An algorithm is developed for RR and HR estimation using the robust, high resolution, and low sidelobe iterative adaptive approach (IAA) algorithm. The IR-UWB distributed MIMO radar prototype system is developed to deal with the random body movement problem which is composed of two identical radars placed in the front and back of the human torso. A chest skin range displacement estimation algorithm is introduced to cancel out the random body movement.

8.3 Activity Recognition and Classification

Human activity recognition plays an important role in a wide range of real-world applications, such as smart home, health care, indoor localization and fitness tracking. This section describes various methodologies in combination with robust machine learning (ML) algorithms and sophisticated radar system setups to obtain the desired type of results related to activity monitoring, recognition, classification, and detection applications.

8.3.1 Activity Recognition

A multiple-layer classification method is introduced to divide human motions into in-situ motions and non-in-situ motions for a separate process [37]. In the pre-screening layer, information in the time-range domain is used to distinguish in situ motions and non-in situ motions. Due to different kinds of motions, the weighted range–time–frequency transform (WRTFT) method is proposed to obtain corresponding spectrograms. Physical empirical features and principal component analysis (PCA)-based features are extracted for machine learning to enhance the performance of the classifier. Classification accuracy of 94.4% and 95.3% is obtained from experimental data for in situ motions and non-in situ motions, respectively. The key unit of the UWB radar system is a NVA6100 CMOS impulse radar chip. In general, it contains three modules: antenna module, radar module and input/output (I/O) module. The antenna module contains a set of two Vivaldi antennas of size 150 × 133 mm^2 and a beamwidth of 20° in E-plane and 50° in H-plane, one for transmitter (Tx) and one for receiver (Rx). The Bagged Trees achieves the highest classification accuracy rate of 95.3%, and the other two ensemble learning classifiers, the Boosted Trees and Subspace k-Nearest Neighbours (k-NN), both achieve good accuracy rates of 92.7% and 93.5%, respectively.

The possibility of using UWB radios for device-free human activity recognition is demonstrated in [38]. By simple machine learning classification algorithms high accuracy results are obtained (95.6%) for line-of-sight scenarios. The data collected using EVB1000 is used to train models using different classification algorithms such as Naive Baise Neural Nets, k-Nearest Neigbours, Random forests. The proposed solution uses high resolution (1 ns) channel impulse response (CIR) which is equivalent to time domain CFR. It can extract features and classification boundaries by directly processing time-frequency spectrograms. The receiver extracts CIR of the received packets as the user performs different activities and features are chosen based on unique property of the UWB propagation which are used to classify the activities such as sitting, standing and lying.

Deep convolutional neural networks (DCNN) are applied in non-contact human activity classification based on UWB radar system [39]. A weighted time–range–frequency transform (WRTFT) method was used to get the spectrograms combining time, range, and frequency information from human activity signals. Then DCNN is utilized to extract features and classification boundaries from spectrograms. DCNN method can achieve a 92.8% classification accuracy for classifying six typical human activities and shows good robustness facing individual diversity. A multi-classification algorithm for human motion recognition based on IR-UWB radar is presented in [40]. First, the k-Nearest Neighbor (k-NN) algorithm is used to classify the radial features of pre-processed signal to determine the subject's radial displacement direction. Then, the power spectrum feature extraction algorithm and Doppler shifts feature extraction algorithm are proposed to extract and visualize the characteristics from the different categories classified by the first part. Finally, the feature spectrograms obtained by the second part are sent into Convolutional Neural Networks (CNNs) for training and testing to realize the recognition of human motions (Fig. 8.6). The UWB radar used in experiments is Novelda's XeThru X4M300 sensor. The radar operates at 7.29 GHz and the pulse repetition frequency (PRF), sampling rate, frames per second (FPS), detection zone are 15.1875 MHz, 23.328 GS/s, 255, 0.4–9.4 m, respectively. To verify the performance of proposed algorithm, dataset was created from 15 persons including 12 kinds of motions. A five-fold cross validation was conducted to calculate the recognition accuracy. The average accuracy of judging the radial displacement directions of subjects was up to 99% and the average accuracy of estimating the motions of subjects reached 98%.

8.3.2 *Through Wall Radar Activity Recognition*

Noncontact penetrating detection and classification of human activities based on micro-Doppler signatures (MDs) using UWB bio-radars is presented in [41].

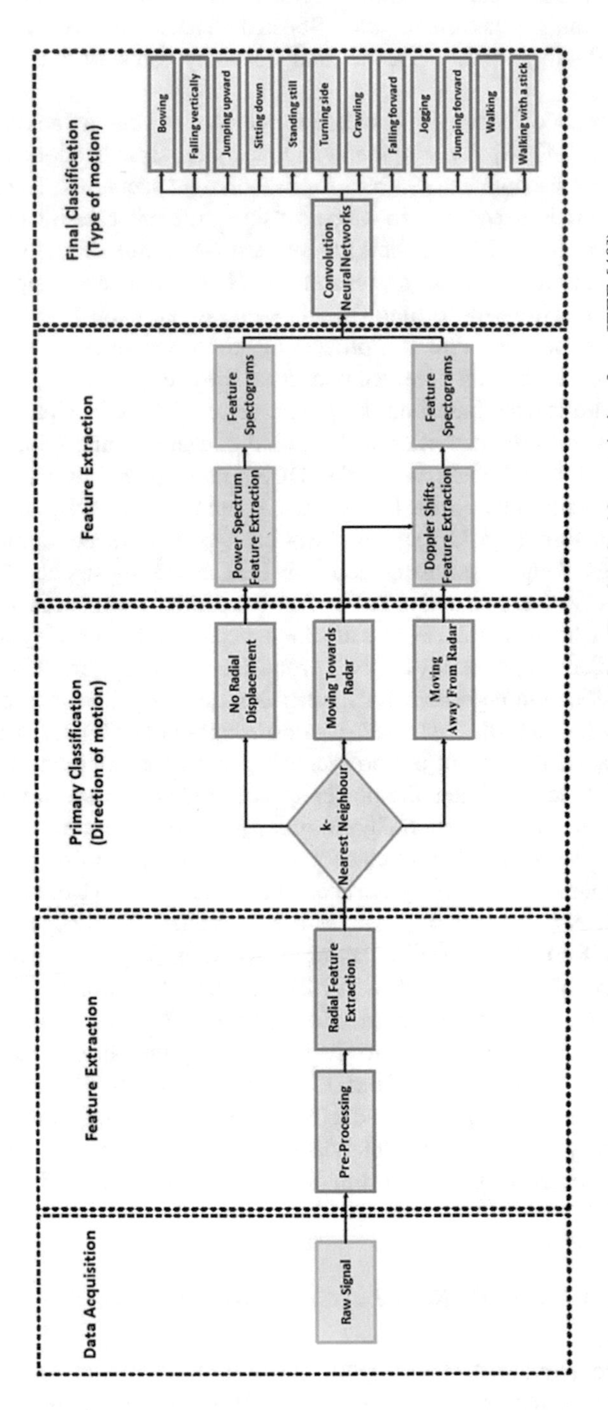

Fig. 8.6 Flowchart of multi-classification algorithm for human motion recognition (Reprinted with permission from IEEE [40])

The MD features of different-magnitude activities at different positions can result in classification errors due to MD attenuation and confusions. This method fully exploits the position information acquired by UWB bio-radar to create a position-labelled modularized database of MD features. A series of pre-processing operations are applied on the stepped-frequency continuous wave (SFCW)-UWB bio-radar echoes to eliminate noise and clutter and to remove the background. A comprehensive time-frequency analysis method combined with automatic extraction method to generate the MDs and feature parameters is chosen. Further Position-Information-Indexed Classifier (PIICs) is applied to enhance classification accuracy. Through wall detection and classification experimental results related to five activities within a range of 6 m are reported for different classifiers. All PIIC-based classifiers achieved better classifying performance than corresponding conventional classifiers with an average accuracy increase of 8.16%.

A novel quantitative method to evaluate channel micro-Doppler capacity of multiple-input and multiple-output (MIMO) system is proposed in [42]. The MIMO UWB radar human activity signal is modelled, and its corresponding time–frequency (T–F) characteristics are analysed to justify the rationality of using the new relative signal-to-noise ratio (RSNR) metric. The method is evaluated using experimental data and the capability of distinguishing the micro-Doppler (μD) capacity differences among channels is demonstrated. It can be successfully used to select the superior channels and eliminate any inferior channels or provide confidence coefficients for the collected multiple channel data of human activities. This method aids in reduction of the inferior channels' influence on further MIMO-based classification or imaging of the human activities. Micro-Doppler signatures (μDs) are seen as pronounced sidebands around the Doppler baseband in the motion spectrum, which are due to the additional modulation on the carrier signal by the movement of separate parts of the human body and can effectively describe the activity characteristics and offer distinct and distinguishable features for detailed classification (Fig. 8.7).

Fig. 8.7 Through-wall detection MIMO radar (Reprinted with permission from IEEE [42])

8.3.3 Gesture Recognition

This work investigates the feasibility of using four different classifiers for recognizing 14 different hand-gestures from a UWB impulse radar. It presented a novel framework for mapping the output of the UWB impulse radar into a sequence of range-Doppler frames that are fed to convolution neural networks (CNN)-based classifiers. 3D CNN has the same principles as the 2D version, it is composed of a series of basic structures repeated multiple times. The basic structure is primarily composed of: (a) a convolutional layer intended for feature extraction, (b) activation function for a non-linear transformation of the inputs, and (c) a pooling layer to reduce the dimension and noise of the input [43]. Four different classification architectures to predict the gesture class, namely; (1) fully connected neural network (FCNN), (2) k-Nearest Neighbours (k-NN), (3) support vector machine (SVM), (4) long short term memory (LSTM) network are compared. The classification results of the proposed architectures show a high level of accuracy above 96% and a very low confusion probability even between similar gestures. This indicates the considerable promise for utilizing UWB radar in practical hand-gesture applications.

Hand/finger gesture recognition while driving a vehicle through impulse radio ultra-wideband (IR-UWB) radar is discussed in [44]. The gestures can be used to control different electronic devices inside a vehicle. Three independent features, i.e., variance of the probability density function (PDF) of the magnitude histogram, time of arrival (TOA) variation and the frequency of the reflected signal, are chosen to classify the gestures. A data fitting method is included to differentiate between gesture signals and unintended hand or body motions. A K-means clustering technique is applied for the classification of the gestures. Moreover, the distance information is used as an additional input parameter to the clustering algorithm, such that the recognition technique will not be vulnerable to distance change. Commercially available single-chip impulse radar transceiver NVA6201 made by NOVELDA (Novelda AS, Kviteseid, Norway) was mounted on the rare mirror of the car. A human gesture recognition algorithm using IR-UWB radar was proposed in [45]. The features are extracted by using the principal component analysis (PCA) method, then neural network (NN) method is used for training and classifying gestures using the extracted features.

8.3.4 Gait Analysis

A UWB impulse Doppler radar has been developed to detect the micro-Doppler signatures of human motions [46]. Promising experimental results have been achieved from both high-resolution range profiles and Doppler spectrogram for monitoring limb movements and also detection of the vital parameters. The wideband and directional Vivaldi subarray used in the pulsed Doppler radar, is fabricated on a 31-mil thick Rogers RT5880 substrate, which has a dielectric constant of 2.2 and loss tangent of 0.0009.

Pulse Doppler radars using UWB technology are becoming very popular for their wide range of capabilities and applications [47]. UWB pulse Doppler radars provide range–time–frequency representation that provides both highly accurate range and micro-Doppler information. Detailed human gait analysis, tracking more than one subject at a time even behind walls, and respiration rate detection even for moving subjects are essential for remotely monitoring patients. The 3D matrices that include range, time, and frequency information are analysed in this work. The algorithm utilizes cross correlation to cancel associated body movement or any clock drift and enhances respiration rate detection.

8.3.5 Sleep Monitoring

Sleep monitoring is an important aspect to determine the quality of sleep and the severity of the disorders to monitor body spatial movement, especially different postures during the night [48]. This study investigates the performance of an off-the-shelf single UWB radar antenna for sleep postural transition (SPT) recognition. The proposed Multi-View Learning, entitled SleepPoseNet or SPN, with time series data augmentation aims to classify four standard SPTs. SPN exhibits an ability to capture both time and frequency features, including the movement and direction of sleeping positions. The data were collected using the Xethru X4M03 development kit, a state-of-the-art UWB radar device by Novelda [44] which operates within the range of 5.9–10.3 GHz. A method for UWB feature extraction called Weighted Range–Time–Frequency Transform (WRTFT), which included range information to a Short Time Fourier Transform (STFT). After the transformation is complete, the output image contains information from range, time, and frequency features of the human motions. The data is recorded from 38 volunteers displayed SPN with a mean accuracy of 73.7 $\pm$ 0.8%.

A wake-sleep classification algorithm based on IR-UWB radar and compared with the actigraphy based on the polysomnograpy (PSG) is presented in [49]. From the samples, the sensitivity, specificity, and agreement of the proposed algorithm and actigraphy were calculated and compared. The algorithm based on IR-UWB radar showed better specificity (35%) and agreement compared to the actigraphy results. The magnitude of the body movement is considered to distinguish sleep/wake that occurs within the IR-UWB radar detection range in real time. The collected raw signal was a mixture of noise, clutter, and vital signals from human. The algorithm is based on background subtraction to reduce clutter and noise. Detection of body movement through variation in the radar signal depending on magnitude of the actual movement and sleep/wake classification based on Time-above threshold method was proposed.

8.3.6 Daily Activity Monitoring

The assessment of many movement disorders which is heavily relied on clinical observation is proposed using four UWB radar sensors to locate the subject and estimate the amount of body movement in the test room. The XK300-MVI (Xandar Kardian, Toronto, ON, Canada) radar was used to verify the proposed algorithm. The algorithm is based on basic radar model and to further increase the detection accuracy of the targets signal, the distance between the subject and the radar is calculated from the background subtraction signal using the Constant False Alarm Rate (CFAR) algorithm [50]. The actigraphy data was measured, for comparison with the proposed IR-UWB radar base. Actigraphy sensors wGT3X-BT (ActiGraph, Florida, US) were worn on the right wrist and right ankle. The position of the subject, obtained from multiple radar sensors, can provide more information than the actigraphy sensor in diagnosing a subject's specific habits, abrupt behavior, or hyperactivity.

Investigations are carried out to show the significance of using non-wearable UWB sensors for developing non-intrusive, unobtrusive, and privacy-preserving monitoring of elderly ADLs [51]. The prototype was implemented using a non-contact UWB, and its performance was compared to conventional state-of-the-art sensing technologies including the ultrasonic and passive infrared. The classification performance was evaluated through statistical metrics and indicators revealing valuable insights into the sensing technology's ability to monitor elderly ADL. The result showed excellent performance for both systems in accuracy, sensitivity, specificity, and precision.

8.3.7 Fall Detection

Random forest (R-F) is proposed as fall detection activity classifier and threefold cross validation is applied to evaluate the performance of this classifier [52]. UWB sensors are used to generate a high-resolution pattern of the resident movement to detect falls among other types of movements. A bi-static setup is proposed for the UWB sensor, which includes both transmitter and receiver, installed over the ceiling. Novelda RS640 UWB radar is used to monitor falls in the experiments. The platform consists of the UWB trans-receiver sensor mounted over the ceiling to detect activities in its detection zone. A train of pulses will be propagated within the UWB detection zone and scattered from the obstacles including the target. The scattered signal received by the receiver antenna is further processed (relevant features are extracted and further classification algorithms are applied) to detect activities in the detection zone. The preliminary test results show that UWB technology is a promising approach to detect falls among common activities such as walking and lying down efficiently.

8.3.8 *Detection and Localization*

An effective method for monitoring people inside a vehicle by using IR-UWB is proposed in [53] which is based on the variation in the waveform of the received signal due to the arrangement of people inside the vehicle. The received IR-UWB radar signals consist of reflected and scattered signals from various parts of people or fixed objects, whose distribution varies significantly depending on the arrangement of people. Hence, the position and number of people inside the vehicle can be estimated by identifying the distribution of the received radar waveforms.

The NVA-R661 IR-UWB radar module manufactured by Novelda (Xethru) is used in this study. Based on the received signal characteristics, several features such as mean, variance, coefficient of variance, skew, kurtosis, max of signal argmax of the signal are analysed and used for classification. To determine the importance of each feature and reduce the number of features, a neighbourhood component analysis (NCA) algorithm is applied. This process helps in removing insignificant features and reduce computational costs. An ensemble learning methodology is applied with a decision tree as a base classifier to classify the various arrangement of people inside the car. The proposed method is compared with other classification algorithms, such as a decision tree, boosting with a decision tree, and an SVM. The classification results can estimate the position and number of people sitting inside a vehicle with an accuracy higher than 90%.

8.4 60 GHz Vital Sign Monitoring

An experimental study of a 60 GHz millimetre-wave life detection system (MLDS) for noncontact human vital-signal monitoring is presented in [54]. This detection system is constructed by using V-band millimetre-wave waveguide components. The block diagram of a noncontact life detection system is illustrated in Fig. 8.8a. An RF signal source produces a continuous- wave (CW) carrier, and it is fed into a directional coupler. One output of the directional coupler is amplified by a power amplifier (PA) and fed through a circulator to the antenna. The other output of the directional coupler provides a local oscillator (LO) signal for the receiver. It performs clutter cancellation for the transmitting power leakage from the circulator and background reflection to enhance the detecting sensitivity of weak vital signals. The clutter canceller consists of an adjustable attenuator and phase shifter with an input signal from the RF signal source branched through directional couplers.

The reflected signal power from the human body is measured by using the experimental setup as shown in Fig. 8.8b. The noncontact vital signal measurements have been conducted on a human subject in four different physical orientations from distances of 1 and 2 m. The constructed 60 GHz MLDS is used to measure the vital signals of the human body in four orientations. The frequency-domain results show that the measured time domain signal has a

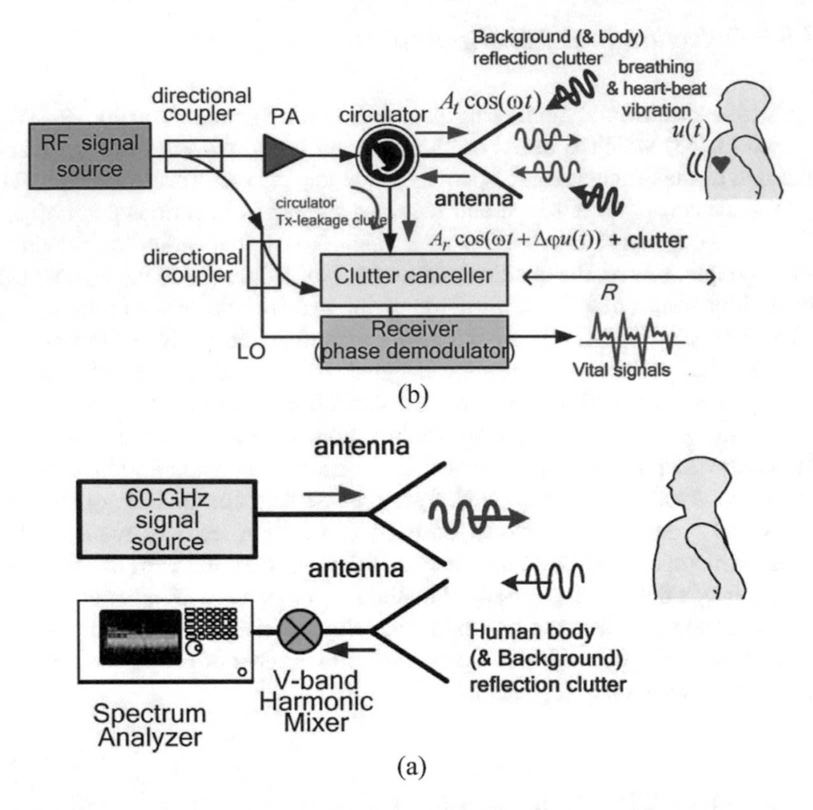

Fig. 8.8 **a** Illustration of a noncontact life detection system for human vital signal monitoring. **b** Illustration of the measurement of the human-body reflected clutter power (at 60 GHz) by using a second V-band horn antenna connected to a spectrum analyser (through a V-band harmonic mixer) (Reprinted with permission from IEEE [54])

breathing signal at about 0.3 Hz and there is a dominant peak at about 1.4 Hz, which is the heartbeat rate. It is also observed that there are second and third heartbeat harmonic peaks at about 2.8 and 4.2 Hz.

A circularly polarized sequential-rotation 2×2 patch antenna array integrated with a 60-GHz Doppler radar system is implemented on the low-temperature co-fired ceramic (LTCC) substrate. Compared to linearly polarized single-patch antenna, the 10 dB bandwidth of the antenna array on LTCC increases from 1.3 (2.4%) to 7 GHz (12.8%) at 55 GHz, and narrow 50 Ω feedline is achieved to realize flip-chip transition. The circular polarization also provides better isolation against other sensors in multiple radars applications [55]. A flip-chip-packaged and fully integrated Doppler micro-radar in 90-nm CMOS for noncontact vital-sign and vibration detection was designed. The compact 60-GHz core (0.73 mm^2) provides a 36-dB peak down-conversion gain and transmits a radar signal around 0 dBm at 55 GHz. By using single-patch antennas and without a high-power amplifier, the system demonstrates the first-pass success of human vital-sign detection at 0.3 m. The small mechanical vibration with a displacement of 0.2 mm can be detected up to 2 m away.

For vibration displacement comparable to or larger than wavelength, such as the respiration, a time-domain recovery algorithm is used to ensure robust detection. A fabricated micro-radar in 90 nm CMOS technology with Tx and Rx, and metal traces on the laminate designed for flip-chip integration is proposed in [12]. The integrated system including the chip, antennas, and bypass capacitors on the PCB, and the wires are connected to DC power supplies. The Tx and Rx patch antennas were fabricated on a 5-mil-thick RT/duroid 5870 laminate providing a gain of 6 dB each, and the angle between two 50 Ω feedlines were designed to be 90° for better Tx/Rx isolation to reduce DC offset at baseband. The integrated system is used to detect heartbeat and mechanical vibration. The integrated system including the chip, antennas, and bypass capacitors on the PCB, and the wires are connected to DC power supplies.

A 60-GHz CMOS direct-conversion Doppler radar RF sensor with a clutter canceller for single-antenna noncontact human vital-signs detection is proposed in [56] (Fig. 8.9). A high isolation quasi-circulator (QC) is designed to reduce the Tx power leakage to the Rx. The clutter canceller performs cancellation for the Tx leakage power (from the QC) and the stationary background reflection clutter to enhance the detection sensitivity of the weak vital signals. The integration of the 60-GHz RF sensor consists of the voltage-controlled oscillator, divided-by-2 frequency divider, power amplifier, QC, clutter canceller (consisting of variable-gain amplifier and 360° phase shifter), low-noise amplifier, in-phase/quadrature-phase sub-harmonic mixer, and three couplers.

A fully integrated 60-GHz CMOS direct-conversion Doppler radar RF sensor is fabricated in 90-nm CMOS with a chip size of 2 × 2mm^2 [56]. The total power consumption is 217 mW. Figure 8.10 illustrates the probe-station-based measurement setups for the developed vital-sign Doppler radar sensor chip. The spectrum analyser (Agilent E4440A), signal generators (Agilent E8257D), and oscilloscopes (Agilent DSO9000A) are used to measure the total Tx power, the total Rx gain, the QC isolation, as well as the clutter cancellation performance, respectively. In human vital-signs test, the radar sensor chip is connected by a 7 dB loss V-band cable and the RF probe to a 60 GHz patch-array antenna (17 dBi gain). The radiating 60 GHz CW from the antenna incidents on the human subject and the reflected wave to the antenna is phase modulated by vital signs. The received signal through the QC and amplified by the LNA is down-converted by the SHM. The down-converted in-phase (I)- and quadrature (Q)-channel vital signals are then amplified by the operation amplifier (accompanied by a 5 Hz low-pass filter) and processed by LabVIEW with the NI 6009 multifunction data acquisition (DAQ) in sequence. I and Q two-channel received signals use Complex Signal Demodulation (CSD) to achieve a good frequency spectrum of vital sign.

A 60 GHz 17-dBi patch-array antenna with a size of 4 × 2.2 cm^2 on an RT/Duroid substrate of thickness 0.127 mm is used. Figure 8.11 shows a photograph of the measurement scenario for the human vital-signs detection of the

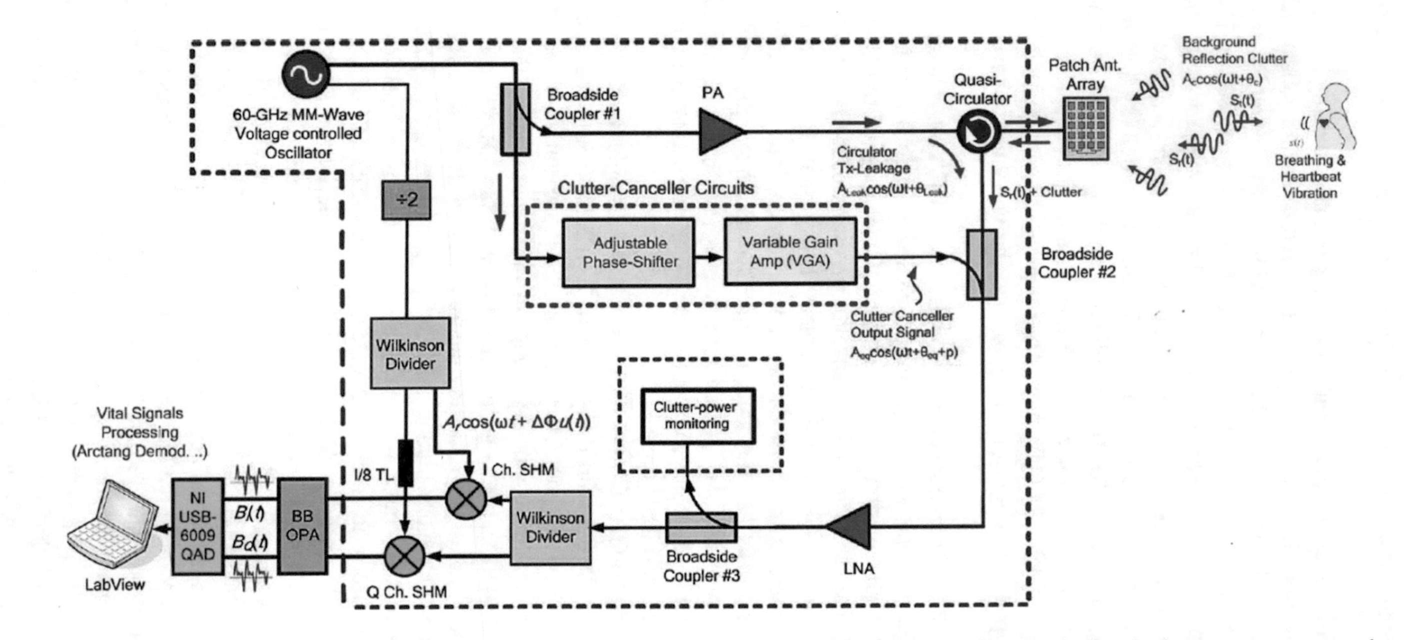

Fig. 8.9 Block diagram of 60-GHz CMOS direct-conversion Doppler radar RF sensor with clutter canceller circuits for single-antenna noncontact human vital-signs detection (Reprinted with permission from IEEE [56])

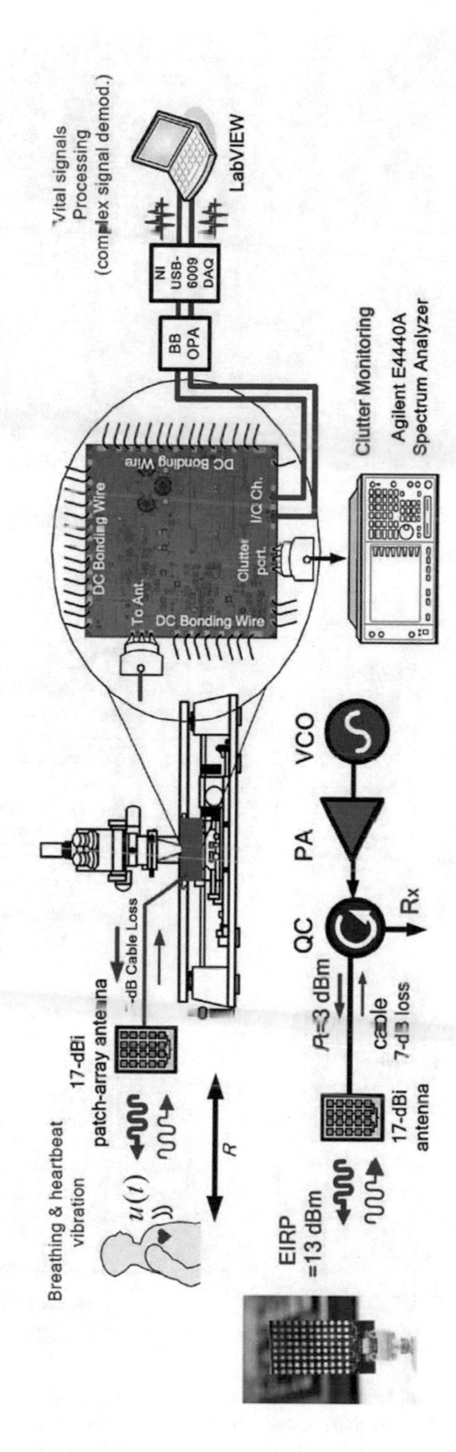

Fig. 8.10 Probe-station-based measurement setup for the radar RF sensor chip. Human vital-signs detection (Reprinted with permission from IEEE [56])

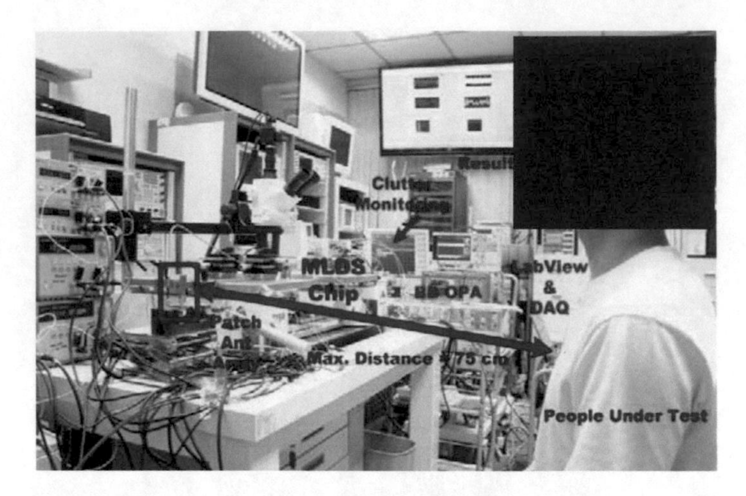

Fig. 8.11 Experimental set up for vital-sign monitoring at 60 GHz (Reprinted with permission from IEEE [56])

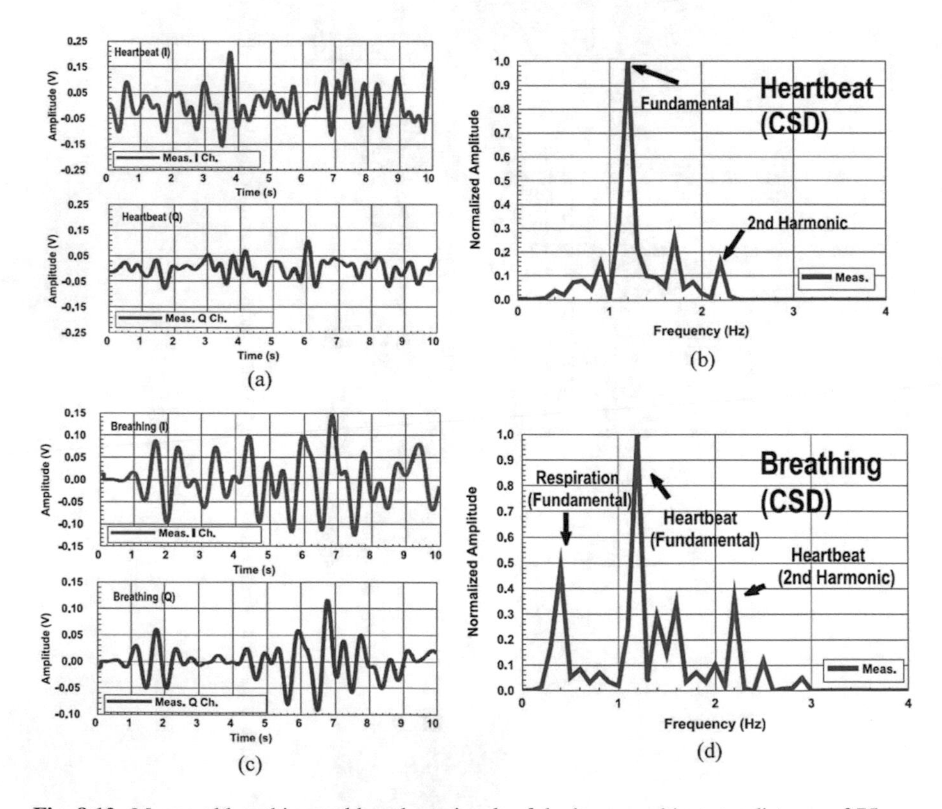

Fig. 8.12 Measured breathing and heartbeat signals of the human subject at a distance of 75 cm: **a** time-domain I/Q signal of the heartbeat signal, **b** CSD frequency spectrum of the heartbeat signal, **c** time-domain I/Q signal of the breathing signal, and **d** CSD frequency spectrum of the breathing signal (Reprinted with permission from IEEE [56])

adar RF sensor chip. For the human vital-signs detection test, Fig. 8.12a–d shows the time-domain waveforms and frequency CSD spectrums of the measured breathing and heartbeat of a human subject at a distance of 75 cm. The measured heartbeat and respiration frequency are around 1.1 Hz (66 beats/min) and 0.4 Hz (24 breaths/min), respectively.

A V-band logarithmic power detector (PD) and the integration with a 60 GHz CMOS vital-signs Doppler radar is proposed in [57]. The PD adopts the successive detection logarithmic amplifier (SDLA) topology and achieves a high dynamic range by replacing the limited amplifiers with millimetre-wave (mmW) linear amplifiers. By feeding the PD output voltage to the A/D pin of the microcontroller unit (MCU), the automatic clutter cancellation function for the vital-signs radar can be performed rapidly. From the measurement results, the PD exhibits a dynamic range and logarithmic errors higher than 35 dB and within $\pm$ 2 dB from 50 to 62 GHz. The whole integrated radar sensor chip is fabricated by a 90-nm CMOS process with a chip size of 2 $\times$ 2.34 mm^2 and a dc power consuming 243 mW. The radar chip and a 60 GHz 17 dBi patch-array antenna are integrated by bond wire interconnection on a compact single carrier board for experimental test [58]. The interconnection has the form of a simple L-C-L structure configured as a T-matching network with bond wires in parallel and a short microstrip line. Open-ended stubs and additional transmission lines (TLs) are also deployed to compensate for imperfect ground connection and improve the impedance matching bandwidth. The total loss of the bond wire interconnection is 3.1 dB, which is less than half of the cable connection (7 dB). Hence, the detection range can be extended from 75 cm to, for example, 120 cm and beyond. By incorporating the MCU to the radar chip board, the fast automatic clutter cancelling can achieve more than 25-dB clutter cancellation. Moreover, it shows a successful vital-signs detection for a distance more than 1.2 m.

A high gain (20.35 dBi), wide-band (3.78 GHz), leaky-wave antenna (LWA) design with beam scanning capabilities (12°) is presented for remote vital sign monitoring (RVSM) utilizing the Doppler radar technique at 60 GHz-band frequencies [59]. Since a narrow frequency band is required for Doppler radar in vital sign monitoring, the antenna beam scanning is achieved by tuning the operating frequency across 62–65 GHz band. A low loss and high efficiency wideband high gain (>20 dBi) beam scanning microstrip antenna array is presented for remote vital sign monitoring utilizing Doppler radar technique at 60 GHz-band [60]. A 32-element (8 $\times$ 4) extended size patch microstrip antenna array is designed at 60 GHz frequency band on RT/D5880 substrate material (ε_r = 2.2) with thickness h = 0.127 mm and copper cladding 17.5 μm. The antenna exhibited about 14° beam scanning within the operating frequency band. Successful remote health monitoring is carried out to detect breath rate (BR) and HR of the human subject using the proposed antennas in [59] and [60].

MIMO array is proposed to form an optimal antenna pattern for tracking a person under test during measurement [61]. The present study uses a UWB mmW radar (central frequency of 60.5 GHz and bandwidth of 1.25 GHz), with a range resolution of 12.0 cm that is sufficiently high to separate echoes from body parts, such as the torso, head, and limbs. The total transmission power was less than 13 dBm, and the array elements were a vertically polarized open-ended rectangular waveguide having an aperture of $3.5 \times 1.7\ \text{mm}^2$ with a single mode TE_{10} at 60.5 GHz. The antenna element gain was 3.9 dBi and the element beam width was $\pm$ 48° (1.7 rad), which covers an area of 3.4 m at 2.0 m from the antennas. Two Tx and four Rx elements constituted a MIMO radar system. The proposed technique was applied to real radar data recorded for a sleeping participant. The Maximum ratio combining (MRC) technique improved the SNR ratio by 8 dB relative to the maximum-power signal channel and the accuracy in measuring the heart rate by 18%.

8.5 60 GHz Activity Monitoring

Movement identification and activity monitoring using mmWave 60 GHz antennas has been explored in this section. Various state-of-the art techniques and novel algorithms are listed below that help in detection, monitoring and identification of the human subject and various physical activities performed using mmWave radar technology and machine learning algorithms to enhance the robustness of the system.

A real-time activity monitoring system is presented by Zeng et al. [62] to track a moving person and conduct activity analysis using 60 GHz mmWave radios. The system was implemented using 60 GHz development platform as shown in Fig. 8.13 by Vubiq [6]. The Tx and Rx are placed at the corners of the room and perform 90° scanning of the room. Figure 8.14a shows the heatmap of the absolute RSS difference values. Figure 8.14b shows the RSS profile when a person is walking along the walking trace shown in Fig. 8.13b. The platform provides a 60 GHz RF front-end and a waveguide module. Horn antenna with 3-dB beam-width of 12° and 24 dBi gain on both Tx and Rx sides was used in this study. A fine resolution programmable rotator is used to conduct the beam steering and rotating. Received signal strength (RSS) profile is constructed without and with the presence of the human subject in the room. The presence of human subject will lead to reflection changes, thus making the human subject detectable. By analysing the RSS profile, the human subject can be tracked and count the steps of walking activity.

A methodology for estimating the velocity of a person walking in indoor environments by using mmWave signals is proposed in [63] and numerical results are reported considering 60 GHz 2×2 MIMO technology. The time-variant Doppler frequencies are estimated by fitting the spectrogram of the complex channel gain of a non-stationary indoor channel model to the spectrogram obtained from the received mmWave signals. The velocity of the moving person is then deduced from estimated time-varying Doppler frequencies. Indoor localization and tracking, using radars technology based on phased-array transceivers at 60 GHz is studied in [64]. The radar sensor IWR6843 SoC is based on a Frequency Modulated

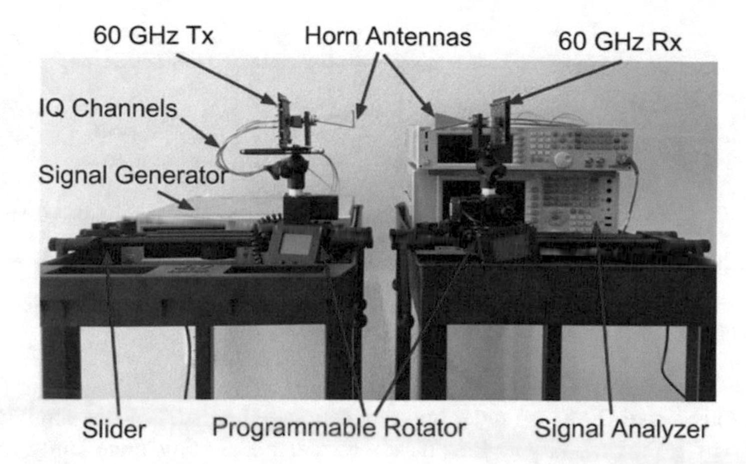

Fig. 8.13 60 GHz transmitter and receiver system setup (Reprinted with permission from IEEE [62])

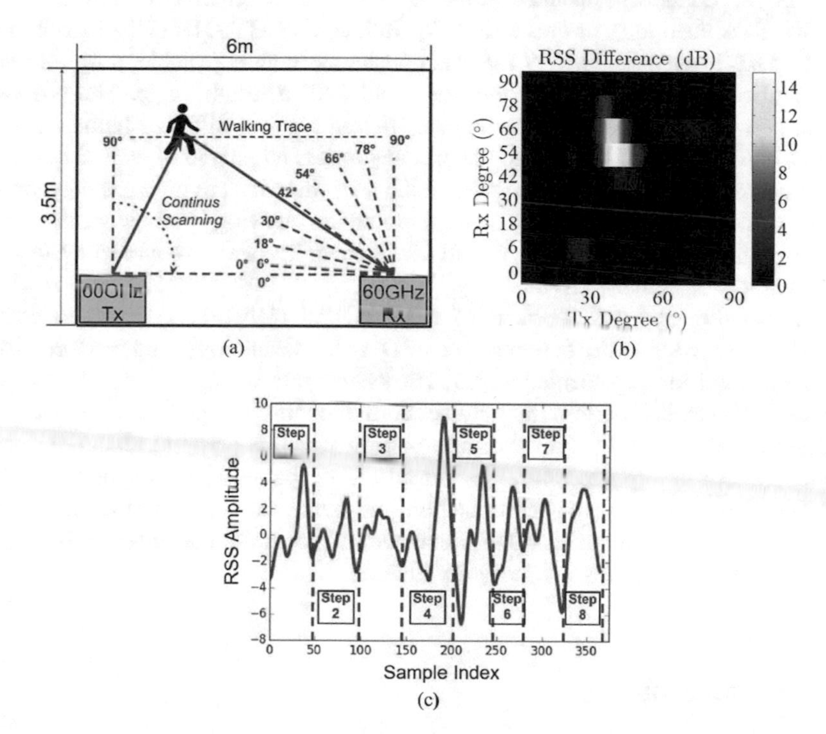

Fig. 8.14 **a** The layout of the experiment for human finding and walking analysis. **b** Heatmap profile for initial human finding process. **c** RSS profile for walking activity (Reprinted with permission from IEEE [62])

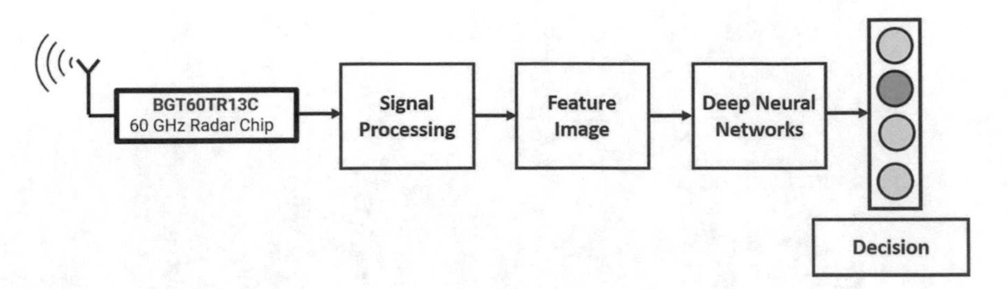

Fig. 8.15 Block diagram of the proposed architecture for activity classification using 60 GHz radar and deep neural networks (After [65], ACM Digital library)

Continuous Wave (FMCW) radar and can estimate the distance, the angle and the velocity of a moving target. Results show that positioning uncertainty is about 30 cm under dynamic conditions and the impact of parameters such as target distance, environment, transition, and allocation start on the localization accuracy.

A novel Euclidean distance softmax layer for 60-GHz radar-based human activity classification is proposed in [65]. Infineon's BGT60TR13C radar chip-set (57–64 GHz) based on FMCW radar is considered with adjustable chirp duration. The field of view covers a 70° elevation and 120° azimuth angle. The proposed neural network architecture is designed to determine four different human activity classes based on their DSs. A deep neural network is employed to learn the activity labels from the Doppler spectrogram (DS) of different activities, namely empty room, walking, standing idle, arm movement, waving and working on a laptop. Figure 8.15 depicts the overall block diagram of the proposed system for classification of human activities.

Investigation and development of Deep Neural Networks (DNNs) to extract temporal and volumetric features from 3D radar technology and perform hand gesture recognition is presented in [66].The system consists of two main parts: (1) a vertical front-facing portion that holds both the 60-GHz phased array and the RealSense camera; and (2) a horizontal portion which consists of electronics such as the frequency ramp generator, the 50-MHz ADC, passive RF components and an Arduino (MCU) for time synchronization and beam steering control. The overall radar system and the associated DNN achieved a recognition accuracy of 93% on a set of 9 different gestures involving two hands.

8.6 Conclusion

UWB and mmWave radar technology are promising technologies for non-contact physiological parameters sensing and activity monitoring suitable for a variety of medical applications, rescue and military operations and general well-being. Various signal processing techniques have been explored with description regarding the state-of-the art methodologies which aid in enhancing the sensing capabilities of the radar technology. Compact antenna prototypes and transceiver modules for UWB and mmWave radar have been presented that are suitable for monitoring, classification, and detection aspects. Machine learning algorithms such as SVM, k-NN, DCNN, Random Forests, Decision Tress with sophisticated statistical analysis techniques have been applied to the signal data measured from the radar setup. The classification performance of various classifiers has been reported for different types of activity monitoring scenarios. Among physiological parameters respiration rate and heart rate have been successfully monitored and noise reduction techniques have been applied to get accurate estimate of the both the vital parameters. Through-wall detection and scenarios related to emergency rescue/search operations have also been discussed in this chapter. Results related to classification and recognition of various activities such as body movements, gait analysis, sleep monitoring, fall detection posture and gesture recognition, and daily activities assessment through IR-UWB and mmWave radar have been reported.

References

1. Fioranelli F, Le Kernec J, Shah SA (2019) Radar for health care: recognizing human activities and monitoring vital signs. IEEE Potentials 38(4):16–23
2. Pisa S, Pittella E, Piuzzi E (2016) A survey of radar systems for medical applications. IEEE Aerosp Electron Syst Mag 31(11):64–81
3. Kebe M, Gadhafi R, Mohammad B, Sanduleanu M, Saleh H, Al-Qutayri M (2020) Human vital signs detection methods and potential using radars: a review. Sensors 20:1454
4. Chauhan SS, Basu A, Abegaonkar MP, Koul SK (2021) Through the wall human subject localization and respiration rate detection using multichannel Doppler radar. IEEE Sens J 21 (2):1510–1518
5. Duan Z, Liang J (2019) Non-contact detection of vital signs using a UWB radar sensor. IEEE Access 7:36888–36895
6. Wang H, Cheng J, Kao J, Huang T (2014) Review on microwave/millimeter-wave systems for vital sign detection. IEEE topical conference on wireless sensors and sensor networks (WiSNet), Newport Beach, CA, pp19–21
7. Choi JW, Yim DH, Cho SH (2017) People counting based on an IR-UWB radar sensor. IEEE Sens J 17(17):5717–5727
8. Tumalskis M (2016) Application of UWB radar for human respiration monitoring. Biomed Eng 19:39–44
9. Cho HS, Park J, Lyu HK (2016) Robust heart rate detection method using UWB impulse radar. In: Proceedings of the IEEE 2016 international conference on information and communication technology convergence (ICTC), Jeju, Korea, 19–21 Oct 2016
10. Nguyen VH, Pyun JY (2015) Location detection and tracking of moving targets by a 2D IR-UWB radar system. Sensors 15:6740–6762
11. Mabrouk M et al (2016) Human breathing rate estimation from radar returns using harmonically related filters. J Sens 1–7
12. Kao TJ, Yan Y, Shen T, Chen AY, Lin J (2013) Design and analysis of a 60-GHz CMOS Doppler micro-radar system-in-package for vital-sign and vibration detection. IEEE Trans Microw Theory Tech 61(4):1649–1659
13. Lazaro A, Girbau D, Villarino R (2010) Analysis of vital signs monitoring using an IR-UWB radar. Prog Electromagnet Res 100:265–284

14. Li W, Lv H, Lu G, Zhang Y, Jing X, Li S et al (2013) A new method for non-line-of-sight vital sign monitoring based on developed adaptive line enhancer using low centre frequency UWB radar. Prog Electromagn Res 133:535–554
15. Khan F, Ghaffar A, Khan N, Cho SH (2020) An overview of signal processing techniques for remote health monitoring using impulse radio UWB transceiver 20(9):2479
16. Xu Y, Dai S, Wu S, Chen J, Fang G (2011) Vital sign detection method based on multiple higher order cumulant for ultrawideband radar. IEEE Trans Geosci Remote Sens 50:1254–1265
17. Zito D et al (2011) SoC CMOS UWB pulse radar sensor for contactless respiratory rate monitoring. IEEE Trans Biomed Circuits Syst 5(6):503–510
18. Schleicher B, Nasr I, Trasser A, Schumacher H (2013) IR-UWB radar demonstrator for ultra-fine movement detection and vital-sign monitoring. IEEE Trans Microw Theory Tech 61 (5):2076–2085
19. Nguyen V, Javaid AQ, Weitnauer MA (2013) Harmonic path (HAPA) algorithm for non-contact vital signs monitoring with IR-UWB radar. In: Proceedings of the IEEE biomedical circuits and systems conference (BioCAS), Rotterdam, The Netherlands, 31 Oct–2 Nov 2013
20. Nguyen V, Javaid AQ, Weitnauer MA (2014) Spectrum-averaged harmonic path (SHAPA) algorithm for non-contact vital sign monitoring with ultra-wideband (UWB) radar. In: Proceedings of the IEEE 2014 36th annual international conference of the IEEE engineering in medicine and biology society, Chicago, IL, USA, 26–30 Aug 2014
21. Wisland DT, Granhaug K, Pleym JR, Andersen N, Støa S, Hjortland HA (2016) Remote monitoring of vital signs using a CMOS UWB radar transceiver. In: Proceedings of the 2016 14th IEEE international new circuits and systems conference (NEWCAS), Vancouver, BC, Canada, 26–29 June 2016
22. Ren L, Wang H, Naishadham K, Kilic O, Fathy AE (2016) Phase-based methods for heart rate detection using UWB impulse Doppler radar. IEEE Trans Microwave Theory Technol 64:3319–3331
23. Li J, Liu L, Zeng Z, Liu F (2014) Advanced signal processing for vital sign extraction with applications in UWB radar detection of trapped victims in complex environments. IEEE J Sel Topics Appl Earth Observations Remote Sens 7(3):783–791
24. Yan J, Zhao H, Li Y, Sun L, Hong H, Zhu X (2016) Through-the-wall human respiration detection using impulse ultra-wide-band radar. In: Proceedings of the IEEE topical conference on biomedical wireless technologies, networks, and sensing systems (BioWireleSS), Austin, TX, USA, 24–27 Jan 2016
25. Hu X, Jin T (2016) Short-range vital signs sensing based on EEMD and CWT using IR-UWB radar. Sensors 2016(16):2025
26. Leem SK, Khan F, Cho SH (2017) Vital sign monitoring and mobile phone usage detection using IR-UWB radar for intended use in car crash prevention. Sensors 17:1240
27. Khan F, Cho SH (2017) A detailed algorithm for vital sign monitoring of a stationary/ non-stationary human through IR-UWB radar. Sensors 17(2):290
28. El-Bardan R, Malaviya D, di Rienzo A (2017) On the estimation of respiration and heart rates via an IR-UWB radar: an algorithmic perspective. In: Proceedings of the ieee international conference on microwaves, antennas, communications and electronic systems (COMCAS), Tel-Aviv, Israel, 13–15 Nov 2017
29. Liang X, Wang Y, Wu S, Gulliver TA (2018) Experimental study of wireless monitoring of human respiratory movements using UWB impulse radar systems. Sensors (Basel) 18(9):3065
30. Shyu KK, Chiu LJ, Lee PL, Tung TH, Yang SH (2018) Detection of breathing and heart rates in UWB radar sensor data using FVPIEF-based two-layer EEMD. IEEE Sens J 19:774–784
31. Zhang Z, Li Y, Mouthaan K, Lian Y (2018) A miniature mode reconfigurable inductorless IR-UWB transmitter-receiver for wireless short-range communication and vital-sign sensing. IEEE J Emerg Select Topics Circ Syst 8(2):294–305

32. Shen H et al. Respiration and heartbeat rates measurement based on autocorrelation using IR-UWB radar. In: IEEE transactions on circuits and systems II: express briefs 65(10):1470–1474
33. Shikhsarmast FM, Lyu T, Liang X, Zhang H, Gulliver TA (2019) Random-Noise denoising and clutter elimination of human respiration movements based on an improved time window selection algorithm using wavelet transform. Sensors 2019(19):95
34. Zhang Y, Li X, Qi R, Qi Z, Zhu H (2020) Harmonic multiple loop detection (HMLD) algorithm for not-contact vital sign monitoring based on ultra-wideband (UWB) radar. IEEE Access 8:38786–38793
35. Ren L, Koo YS, Wang H, Wang Y, Liu Q, Fathy AE (2015) Noncontact multiple heartbeats detection and subject localization using UWB impulse Doppler radar. IEEE Microwave Wirel Compon Lett 25(10):690–692
36. Shang X, Liu J, Li J (2020) Multiple object localization and vital sign monitoring using IR-UWB MIMO radar. IEEE Trans Aerosp Electron Syst 56(6):4437–4450
37. Ding C et al (2018) Non-contact human motion recognition based on UWB radar. IEEE J Emerg Select Topics Circ Syst 8(2):306–315
38. Sharma S, Mohammadmoradi H, Heydariaan M, Gnawali O (2019) Device-free activity recognition using ultra-wideband radios. In: International conference on computing, networking and communications (ICNC), Honolulu, III, USA, pp1029–1033
39. Chen W et al (2019) Non-contact human activity classification using DCNN based on UWB radar. In: 2019 IEEE MTT-S international microwave biomedical conference (IMBioC), Nanjing, China, pp 1–4
40. Qi R, Li X, Zhang Y, Li Y (2020) Multi-classification algorithm for human motion recognition based on IR-UWB radar. IEEE Sens J 20(21):12848–12858
41. Qi F et al (2019) Position-information-indexed classifier for improved through-wall detection and classification of human activities using UWB bio-radar. IEEE Antennas Wirel Propag Lett 18(3):437–441
42. Qi F, Lv H, Wang J, Fathy AE (2020) Quantitative evaluation of channel micro-Doppler capacity for MIMO UWB radar human activity signals based on time-frequency signatures. IEEE Trans Geosci Remote Sens 58(9):6138–6151
43. Skaria S, Al-Hourani A, Evans RJ (2020) Deep-learning methods for hand-gesture recognition using ultra-wideband radar. IEEE Access 8:203580–203590
44. Khan F, Leem SK (2017) Cho SH hand-based gesture recognition for vehicular applications using IR-UWB radar. Sensors 2017(17):833
45. Park J, Cho SH (2016) IR-UWB radar sensor for human gesture recognition by using machine learning. IEEE 18th international conference on high performance computing and communications; IEEE 14th international conference on smart city; IEEE 2nd international conference on data science and systems (HPCC/SmartCity/DSS), Sydney, NSW, pp 1246–1249
46. Wang Y, Fathy AE (2011) Micro-Doppler signatures for intelligent human gait recognition using a UWB impulse radar. In: IEEE international symposium on antennas and propagation (APSURSI), Spokane, WA, pp 2103–2106
47. Koo YS, Ren L, Wang Y, Fathy AE (2013) UWB micro Doppler radar for human Gait analysis, tracking more than one person, and vital sign detection of moving persons. IEEE MTT-S international microwave symposium digest (MTT), Seattle, WA, pp 1–4
48. Piriyajitakonkij M et al (2021) SleepPoseNet: multi-view learning for sleep postural transition recognition using UWB. IEEE J Biomed Health Inform 25(4):1305–1314
49. Kang S et al (2018) Comparison of sleep parameter with actigraphy and IR-UWB sensor based on Polysomnography. In: International conference on network infrastructure and digital content (IC-NIDC), Guiyang, pp 320–323
50. Yim D, Lee WH, Kim JI, Kim K, Ahn DH, Lim YH, Cho SH, Park HK, Cho SH (2019) Quantified activity measurement for medical use in movement disorders through IR-UWB radar sensor. Sensors 19:688

51. Klavestad S, Assres G, Fagernes S, Grønli TM (2020) Monitoring activities of daily living using UWB radar technology: a contactless approach. IoT 2020(1):320–336
52. Mokhtari G, Zhang Q, Fazlollahi A (2017) Non-wearable UWB sensor to detect falls in smart home environment. In: 2017 IEEE International conference on pervasive computing and communications workshops (PerCom Workshops), Kona, HI, pp 274–278
53. Lim S, Lee S, Jung J, Kim S (2020) Detection and localization of people inside vehicle using impulse radio ultra-wideband radar sensor. IEEE Sens J 20(7):3892–3901
54. Chuang H, Kuo H, Lin F, Huang T, Kuo C, Ou Y (2012) 60-GHz millimeter-wave life detection system (MLDS) for noncontact human vital-signal monitoring. IEEE Sens J 12 (3):602–609
55. Shen T, Kao TJ, Huang T, Tu J, Lin J, Wu R (2012) Antenna design of 60-GHz micro-radar system-in-package for noncontact vital sign detection. IEEE Antennas Wirel Propag Lett 11:1702–1705
56. Kuo H et al (2016) A fully integrated 60-GHz CMOS direct-conversion doppler radar RF sensor with clutter canceller for single-antenna noncontact human vital-signs detection. IEEE Trans Microw Theory Tech 64(4):1018–1028
57. Chou C, Lai W, Hsiao Y, Chuang H (2018) 60-GHz CMOS doppler radar sensor with integrated V-band power detector for clutter monitoring and automatic clutter-cancellation in noncontact vital-signs sensing. IEEE Trans Microw Theory Tech 66(3):1635–1643
58. Chan C, Chou C, Chuang H (2018) Integrated packaging design of low-cost bondwire interconnection for 60-GHz CMOS vital-signs radar sensor chip with millimeter-wave planar antenna. IEEE Trans Compon Packag Manuf Technol 8(2):177–185
59. Rabbani MS, Churm J, Feresidis A (2019) Antenna development for multi-functional wireless health monitoring sensor. In: Antennas and propagation conference 2019 (APC-2019), Birmingham, UK, pp 1–4
60. Rabbani MS, Feresidis A (2019) Beam steerable antenna development for wireless health monitoring. 2019 UK/ China Emerging Technologies (UCET), Glasgow, United Kingdom, pp 1–2
61. Sakamoto T (2019) Noncontact measurement of human vital signs during sleep using low-power millimeter-wave ultrawideband MIMO array radar. In: 2019 IEEE MTT-S international microwave biomedical conference (IMBioC), Nanjing, China, pp 1–4
62. Zeng Y, Pathak PH, Yang Z, Mohapatra P (2016) Poster abstract: human tracking and activity monitoring using 60 GHz mmWave. In: 2016 15th ACM/IEEE international conference on information processing in sensor networks (IPSN), Vienna, pp 1–2
63. Hicheri R, Pätzold M, Youssef N (2018) Estimation of the velocity of a walking person in indoor environments from mmWave signals. In: 2018 IEEE Globecom workshops (GC Wkshps), Abu Dhabi, United Arab Emirates, pp 1–7
64. Antonucci A et al (2019) Performance analysis of a 60-GHz radar for indoor positioning and tracking. In: 2019 International conference on indoor positioning and indoor navigation (IPIN), Pisa, Italy, pp 1–7
65. Stadelmayer T, Stadelmayer M, Santra A, Weigel R, Lurz F (2020) Human activity classification using mmWave FMCW radar by improved representation learning publication: mmNets'20. In: Proceedings of the 4th ACM workshop on millimeter-wave networks and sensing systems, pp 1–6
66. Tzadok A, Valdes-Garcia A, Pepeljugoski P, Plouchart JO, Yeck M, Liu H (2020) AI-driven event recognition with a real-time 3D 60-GHz radar system. In: IEEE/MTT-S international microwave symposium (IMS), Los Angeles, CA, USA, pp 795–798

Chapter 9
UWB Radar Technology for Imaging Applications

9.1 Introduction

Medical imaging is an important tool for visualizing the interior of the body for clinical analysis and identifying any abnormality usually in a non-invasive way. Medical imaging has been widely used in all phases of cancer management, identifying bone fracture, and tumors, to name a few applications [1, 2]. Early detection of cancer reduces the significant health risk and providing the right treatment at early stage. Different medical imaging systems such as X-rays, magnetic resonance imaging (MRI), computed tomography (CT)-scan, Positron emission tomography, ultrasound (US) have been developed for detecting tumors in human organs [3, 4].

X-rays is the most used method for early screening of cancer, as it is relatively inexpensive. However, X-rays is limited in sensitivity, and can fail to detect deep-lying cancerous tissues or lead to a false detection [5]. Additionally, the ionization initiated by X-rays can be harmful and have several side effects. Magnetic resonance imaging (MRI) provides very high resolution of images, but it is very expensive and time consuming. This makes it an inappropriate option for early-stage screening and/or regular monitoring. CT-scan is a faster imaging modality. However, it uses X-rays, with a much higher dosage compared to normal X-rays, and is unsafe for frequent screening [6, 7]. Even though CT can provide high spatial resolution, it does not provide high soft tissue contrast, which is a key element for detecting tumors.

The high cost, complexity, and safety aspects of current methods highlights the necessity of microwave imaging, as an early diagnostic tool as well as monitoring tumor changes during chemo/radiotherapy [8–10]. Therefore, microwave imaging (MI) is now a promising method for breast tumor detection because of non-ionizing radiation effect [10]. Microwave imaging is a technique that maps the distribution of electrical property in objects. The basic principle of microwave imaging is based on the varied electrical properties of different tissues, such as the dielectric constant.

S. K. Koul and R. Bharadwaj, *Wearable Antennas and Body Centric Communication*, Lecture Notes in Electrical Engineering 787,
https://doi.org/10.1007/978-981-16-3973-9_9

The microwave antennas can easily distinguish the small signal fluctuations from the variations of the electrical properties of the human tissues [10]. Ultra-wideband (UWB) microwave imaging technology has been extensively explored for detection and diagnosis in a wide range of applications [11, 12]. The UWB characteristic provides high resolution and accurate localization of a target for imaging. The wideband characteristics provide high resolution fast data acquisition and detection, low microwave energy pulse and accurate localization of a target for imaging which are important aspects for electromagnetic medical applications [11, 12].

Applications such as respiration and heart-beat detection, imaging brain for strokes and cerebral edema, breast cancer, bone imaging, heart imaging, and joint tissues, have been investigated [10, 11]. The working principle of the Microwave Imaging (MWI) is based on the difference between the electrical properties (relative permittivity and conductivity) of different tissues. Several research works and initial clinical trials have been reported for microwave imaging systems, especially for breast cancer and brain stroke detection [13, 14].

Other UWB imaging applications deal with through wall imaging for applications related to emergency and rescue, material detection, security, and defence fields. The detection, sensing and localization of the human subjects in such scenarios is an important aspect. UWB radar systems allows detection and tracking of moving targets with an advantage in critical environments or under hindered conditions [15]. UWB radars operating in the frequency range up to 4 GHz offer characteristic of a good penetration of emitted electromagnetic waves through various materials, such as wood, brick, concrete, plastic, rock, ground, snow, etc. Therefore, such radars can detect moving person by measuring changes in the impulse response of complex monitored environments [16]. UWB technology has numerous advantages, such as high detection target range accuracy, good penetrating wall characteristics, improved immunity to external electromagnetic radiation and noise, and robustness to multipath reflections. Compared with other radar systems, the remarkable characteristic of UWB radar is the high spatial resolution because of employment of ultra-narrow pulses [17, 18]. It not only allows the detection of closely positioned targets but also provides information about the shape and the material content of targets.

In this chapter various research studies in simulation and experimental form related to UWB imaging technology for different medical applications have been reported. Several antenna designs, development of image reconstruction algorithms, and providing overview of the complete microwave imaging systems for simple and less complex imaging procedure for medical applications are presented. This chapter also presents an overview of the through wall UWB radar for detection and localizing the human subject in complex and cluttered environments. Current trends and major advancement in the field of UWB imaging are discussed and analysed in detail.

9.2 UWB Radar for Medical Imaging Applications

Microwave imaging is an emerging diagnostic branch that has great potential for the implementation of transportable and low-cost devices [8, 9]. MWI is based on the observation that different tissues or the same tissues but in different functional conditions (i.e., cancerous, and healthy tissue) have different dielectric properties in the microwave band. Two main MWI approaches exist: (1) microwave tomography; (2) UWB radar imaging. Microwave tomography determines morphological (location, size, and shape) and electromagnetic characteristics (permittivity, conductivity, and magnetic permeability) of any abnormal examined tissue [19]. The output is the reconstruction of the complete dielectric profile of the crossed tissues. The tissue reconstruction requires the solution of a very complex mathematical inverse problem. Moreover, the tomography approach requires a classification process, and hence, a training phase to construct a database with many clinical cases and a large amount of a priori and ad-hoc information. In UWB radar imaging, broadband pulses are emitted from antennas properly located around the object to be analysed, non-necessarily in contact with the object and the backscattered signals are collected by the same antennas [20]. Arrival times and amplitudes of the backscattered signals are processed by a beamforming algorithm to locate the scattering points, i.e., the points in which there is a transition from one type of tissue to another. The output is a map of the backscattering energy, i.e., an image where the pixels with higher energy represent potential "abnormal" areas [21]. The microwave imaging system requires to be low cost, compact, portable, and efficient. It should also be able to successfully measure weak scattering fields and provide 3-D computationally efficient imaging. There is also a need to efficiently couple the microwave power to the biological tissues, selection of proper frequency for good resolution and penetration depth. Applications of UWB radar imaging are presented in Fig. 9.1.

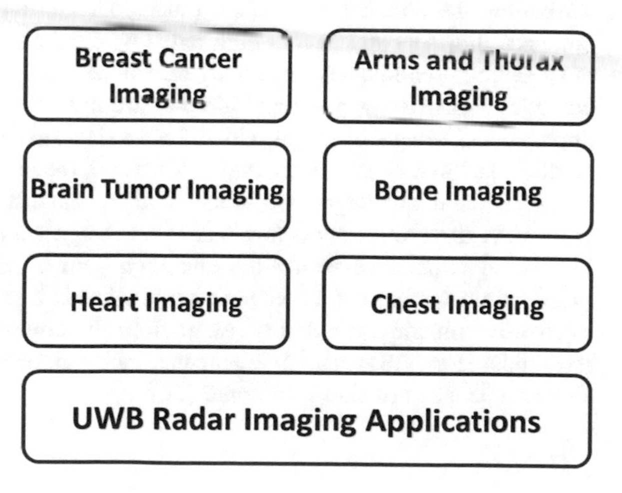

Fig. 9.1 UWB radar imaging for medical applications

In the UWB imaging systems, a very narrow pulse is transmitted from a UWB antenna to penetrate the body. As the pulse propagates through the various tissues, reflections and scattering occur at the interfaces. A particular interest is in the scattered signal from a small size-tissue representing a tumor. The reflected and scattered signals can be received using an UWB antenna, or array of antennas, and used to map different layers of the body. The working principle of microwave imaging is to use backscattering signals to reconstruct images that can detect the presence of objects. The received signal is stored, and the corresponding signal processing algorithm allows for image reconstruction [22].

9.2.1 Antenna Design Requirements

In the microwave imaging system, the antenna is an important element for the final image quality and performance of the imaging system. For an accurate imaging system with high resolution and dynamic range, the transmitting/receiving UWB antenna should be planar, compact in size, and directive with high-radiation efficiency and distortion less pulse transmission/reception [22].

The efficient penetration of the electromagnetic waves into the material under test and the spatial high resolved registration of the reflected signals are crucial tasks of the measurement setup. The main purpose of the antenna is to categorize the variance in dielectric properties between malignant (tumor object) and the healthy breast tissue. High gain, lower resonant frequency, and directive radiation antenna properties are desirable for microwave imaging [23, 24]. An efficient antenna array design concerning biomedical UWB imaging purposes comprises also shape and duration of the signal impulses, fidelity, and physical dimensions of the antenna. Wearable UWB antennas that touch the body are subject to constraints like compact, lightweight, biocompatible and should be preferably flexible.

For the development of an effective MWI system with dynamic range and high resolution, the antenna utilized for transmitter and receiver should be wideband, compact, directive and having high radiation efficiency [24]. The Vivaldi antennas are a suitable candidate for its highly directional radiation beam, broadband nature, compact dimensions, and high gain and are used for several microwave imaging applications. Several antennas with different sizes, gains, efficiencies, and reflection coefficients have been recommended for use in medical imaging systems [24, 25]. Some of them are large, and some of them have a low gain or low radiation efficiency. However, very compact antenna design with improved antenna property like broad impedance bandwidth and high gain is prefered. A lightweight, low profile form factor of this antenna also makes it a preferred antenna system for microwave imaging paradigms. Apart from the commonly used Vivaldi antenna, Pyramidal Horn Antenna, CPW antenna, metamaterials and EBG antenna, and the slotted antennas are also considered [26].

9.2.2 Breast Cancer Detection: State of the Art Techniques and Algorithms

9.2.2.1 Antenna Arrays and Phantoms for Breast Cancer Detection

An UWB microwave system for breast cancer detection is presented in [27] which consists of a hemispherical 16 element real-aperture stacked antenna array designed for a multi-static radar-based detection system. The cavity has planar dimensions of 23×29 mm^2 and is 17 mm long. To absorb the back radiation of the antenna and avoid any resonances, the cavity was lined with the broadband absorbing material (Eccosorb FGM-40 from Emerson and Cuming). On the front face of the antenna a short 5 mm metallic screen was included to decrease the coupling between adjacent array elements. The resulting antenna array is formed around lower part of a 78 mm radius sphere, in four rows of four antennas. The work has been validated through realistic three-dimensional (3D) breast phantom and real breast cancer patients during initial clinical trials.

The antennas are immersed in a matching liquid with material properties (ε_r = 9.5 and attenuation of 1.2 dB/cm at 6 GHz) like that of normal breast-fat, to reduce reflections from the skin and for a more compact antenna design. A 2 mm thick curved skin phantom was developed which is part of the 58 mm radius hemisphere. The skin phantom is fitted into the array and lies 20 mm above the antenna elements. The material is dispersive and at 6 GHz it has a relative dielectric constant of 30 and attenuation of 16 dB/cm. A tumor phantom material with a relative dielectric constant close to 50 and conductivity (7 S/m at 6 GHz) was developed. The contrast between dielectric properties of breast fat and tumor phantom materials is around 1:5. Images are formed using two different beamforming algorithms and the performance of these algorithms is compared through numerical simulation and experimental results. Detection of 4 and 6 mm diameter spherical tumors in the curved breast phantom has been made feasible through the proposed method.

Experimental work on microwave breast cancer imaging using inhomogeneous breast phantoms is presented in [28]. A 31-antenna array operating in 3–10 GHz is used in imaging experiments. Three different breast phantoms are considered, with the contrast between spherical phantom tumors and surrounding materials ranging from 5:1 to 1.6:1 which is challenging for imaging systems. A modified version of the delay-and-sum (DAS) algorithm is proposed for the imaging experiments which considers a new weighting factor and the coherence factor. The algorithm is effective in reducing clutter, providing better images by improving the peak clutter-to-target energy ratio by 3.1 dB.

In the measurement setup, the array relates to coaxial cables to a custom-built network of electromechanical switches and further to a vector network analyser (VNA; Rohde&Schwarz ZVB20), which performs the radar measurement in the frequency domain (Fig. 9.2a). The measured data are transformed into the time domain. Recorded radar signals serve as an input to the image formation (beamforming) algorithm. During measurements, antennas are immersed in a matching liquid to reduce reflections from the skin and for a more compact antenna design.

The modifications to the phantom presented in this work include additional inhomogeneous, high dielectric constant materials, representing the dense breast tissues in comparison to [27]. Each breast phantom is composed of several different dense tissue materials, with dielectric constant ranging from 10 to 30. Different permittivity values were achieved by mixing TX151 and polythene powder with different amounts of water. The skin layer is 2 mm thick; it is a part of an 86-mm-radius hemisphere. All developed phantom materials are dispersive with frequency-dependent characteristics like those presented in [27].

The breast phantom which is closest to the MRI breast image is shown in Fig. 9.2b. The interior was composed of two inhomogeneous dense materials with $\varepsilon_r = 20$ and $\varepsilon_r = 30$ (at 3 GHz), respectively. Then, the 10 mm spherical phantom tumor was placed about 10 mm away from those dense materials. Imaging results for the most complex phantom is shown Fig. 9.2c using beam forming algorithm

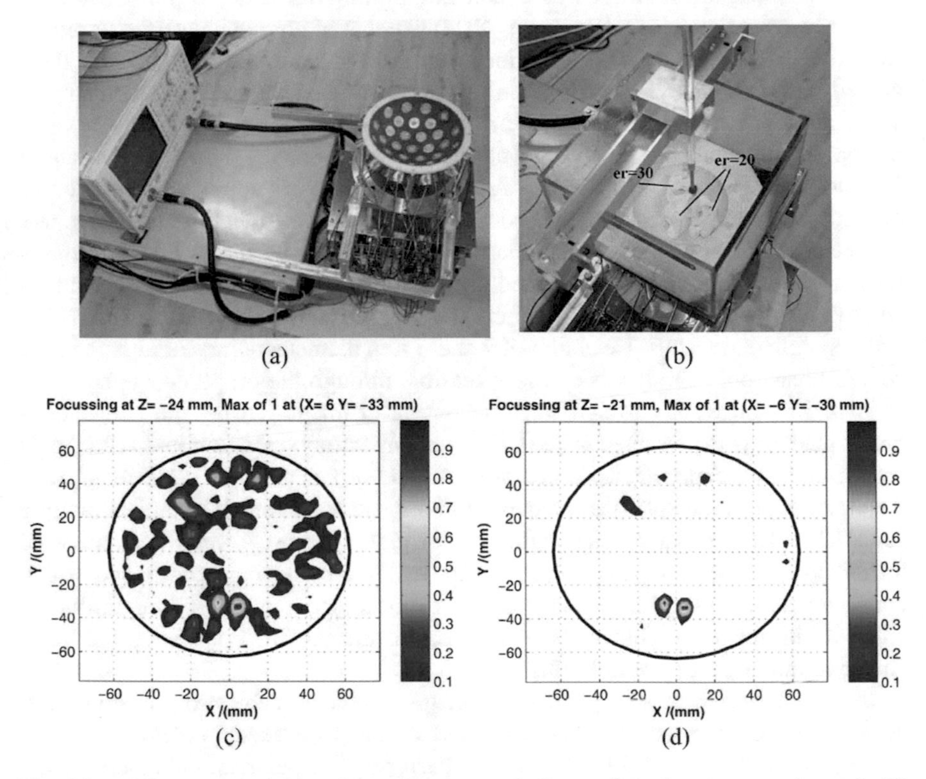

Fig. 9.2 **a** Microwave radar-based imaging system. **b** Setup of the imaging experiment 3. 2D imaging results for experiment 3. **c** Results when CF = 1 (coherence quality weighting not applied). **d** Results after the CF weighting was applied (Reprinted with permission from IEEE [28])

and significantly better results were observed when coherence factor (CF) weighting is applied as observed in Fig. 9.2d. The results show accurate detection of the tumor even in a high clutter level caused due to the use of heterogeneous phantoms.

A 60-element antenna array for breast cancer detection at 4–8 GHz with the aim of improving immunity to clutter and decrease of scan times is reported in [29]. New connector and cable solutions were explored to minimize the element size. The previous 31-element array [28] accomplished the necessary 465 pairs of unique S-parameter measurements in 90 s, the new 60-element array completes 1770 such measurements in only 10 s.

A modified imaging method that presents reduced clutter images using inhomogeneous (complex) breast phantom is presented in [30]. A contrast-enhanced imaging scenario is assumed in the investigations. To approximate dense heterogeneous breast tissues, three randomly shaped scatters with high dielectric constant are placed in the breast phantom. Two tumor-like inclusions with different dielectric constants were used to approximate a contrast-enhanced imaging scenario. Through differential imaging, the tumor can be localized successfully with the 31-element antenna array. The improvement in the image has been achieved by reducing the number of antennas used in focusing from 31 to 20, 15 and 10 elements, selecting only the antennas that are nearest to the tumor location. Results show that the level of clutter energy is reduced to 1/14 of its original value.

An experimental system for microwave breast cancer detection that uses multi-static radar is presented in [31]. The system operates in time-domain by transmitting a short-duration pulse and collecting the signals scattered within the breast. A radome is the interface between the system and the breast. It is a hemispherical dielectric bowl in which the breast is placed and its exterior surface containing slots that house the 16 antennas. A short-duration generic pulse is reshaped for the desired frequency content; then, it is fed to the antenna array; the emitted microwave pulse scatters within the breast and the signals collected by the antennas are recorded by an oscilloscope.

The models used to test the system are shaped and sized as a typical B-cup breast. They contain distinct fat, skin, and tumor phantoms each approximating the dielectric properties of the corresponding breast tissues over the microwave frequency range of interest. The realistically shaped phantoms measure 17.5×6.5 cm^2 at maximum cross section, with a height of up to 8 cm. They are composed of an uneven, textured 2 mm skin layer filled with fat-mimicking material and tumors of given shape/size placed at any desired location.

Two different breast phantoms are examined in this work. Each phantom contains a roughly spherical tumor with a radius of less than 1 cm representing the limiting case for *Stage 1* breast cancer, an optimal stage for detection and successful treatment. The tumors are cantered at depths between 1 and 3 cm above the chest wall. Images are generated using the delay-multiply-and-sum (DMAS) algorithm.

A single- and dual-polarization antennas for UWB breast cancer detection systems using an inhomogeneous multilayer model of the human breast is presented in [32]. Antennas made from flexible materials are more easily adapted to wearable applications. Miniaturized flexible monopole and spiral antennas on a 50-μm bio-compatible Kapton polyimide substrate with ε_r of 3.5 are designed, using the high-frequency structure simulator (HFSS), to be in contact with biological breast tissues. A superstrate layer identical to the substrate layers covers the metal antenna traces to isolate the antenna from the biological tissues. The proposed antennas are 20×20 mm^2 in size and are designed to operate in a frequency range of 2–4 GHz. Two flexible low cost, light weight, conformal 4×4 UWB antenna arrays (single and dual polarization), in a format like that of a bra were developed for a radar-based breast cancer detection system.

The breast as a communication media is modelled in HFSS by multi-layered homogenous biological tissues and each biological tissue is defined as a dispersive dielectric in a homogeneous medium [33]. A multilayer model that is used to design the antenna array includes skin, fat, gland, and muscle is developed. Measurements were also carried out to validate the model and the antenna design. The homogenous phantom is of stable rubber construction and consists of a thin-skin-mimicking layer filled with adipose-mimicking material. The dielectric properties are matched to those of the actual fat and skin tissues.

The wearable antenna array prototype proposed in for breast imaging uses flexible monopole antennas, are intended to contact the skin directly. This eliminates the need for an immersion medium and allows for knowledge of the skin relative to the antennas. The radiation for the proposed monopole antenna is not end-fire, but broadside. This difference in radiation pattern allows the wearable prototype, where the antennas are now tangential to the skin surface, to eliminate the need for the ceramic radome that held the antennas perpendicularly to the breast surface in the table-based prototype. The wearable prototype is composed of a 16-element antenna array, embedded inside a bra with contacting the skin is presented in [34]. The antennas are distributed asymmetrically around the bra surface, to reduce imaging artifacts that can be induced by a symmetric array. The antenna connectors protrude out of the bra, where they are connected to cables that are in turn connected to the switching matrix and UWB imaging system. Photographs of the antenna and the array from the table-based prototype are shown in Fig. 9.3a–d, and the flexible array within the wearable prototype is shown in Fig. 9.3e.

9.2.2.2 UWB Breast Cancer Imaging and Neural Networks

An experimental early breast cancer detection system in terms of heterogeneous breast phantom is proposed in [35]. The system consists of UWB transceivers and Neural Network (NN) based Pattern Recognition (PR) software for imaging. The materials to construct a simple breast phantom with tumor are: (i) A mixture of petroleum jelly, soy oil, wheat flour and water as heterogeneous tissue; (ii) A particular glass as skin; and (iii) A specific mixture of water and wheat flour as cancer-tissue. All the materials and their mixtures are considered according to the ratio of the dielectric properties of the breast tissues [36]. The UWB signals are transmitted from one side of the breast phantom and received from opposite side

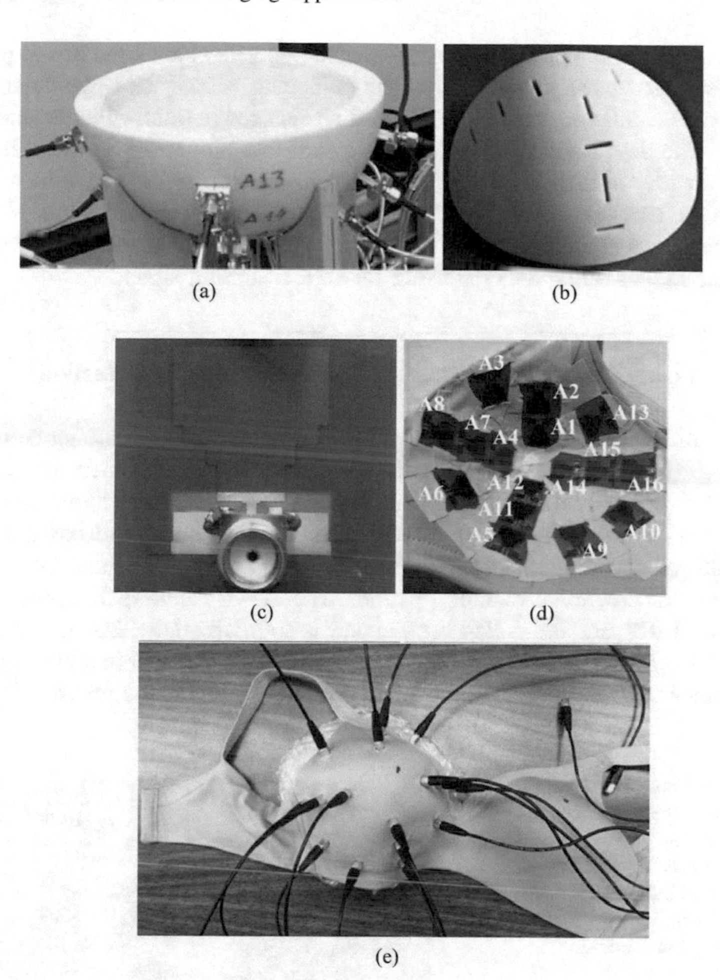

Fig. 9.3 Components of the table-based prototype. **a** Antenna array and radome; **b** bottom view of radome with antenna slots visible. The wearable prototype; **c** close-up photograph of the connectorized monopole antenna; **d** photograph of the antenna array inside the bra-cup, with antenna numbers marked; **e** photograph of the outside of the bra showing the SMA cables that connect to the antennas (the bra is sitting on a breast model) (Reprinted with permission from IEEE [34])

diagonally. By using discrete cosine transform (DCT) of the received signals, a Neural Network (NN) is trained, tested, and interfaced with the UWB transceiver to form the complete system. The achieved detection rate of cancer cell's existence, size and location are approximately 100%, 93.1% and 93.3% respectively.

A low cost and non-invasive breast cancer detection system for early detection consisting of UWB antennas fabricated on Zelt-fabric and software which consists of a NN module [37]. The received signals are fed into the NN module for further processing. The heterogeneous breast phantom is placed in the center and a pair of

UWB antennas was placed diagonally on the opposite side of the breast phantom. Heterogeneous breast phantom contains skin, fatty tissue, and glandular tissues. K-fold cross validation-based feed forward NN is used to train, validate, and test the features. The developed feed forward NN with two hidden layers with 20 hidden neurons in first hidden layer and 1 hidden neuron in the second hidden layer, 4 nodes in the input layer and 4 nodes in the output layer. The system can screen the breast cancer with average detection performance of 87.55% using backward scattering signals while 84.17% using forward scattering signal.

9.2.2.3 Vivaldi Antennas for Microwave Breast Tumor Detection

A breast phantom measurement system is presented using a side slotted Vivaldi antenna [38] shown in Fig. 9.4a, b. The radiating fins are modified by etching six side slots to enhance the electrical length and produce stronger directive radiation with higher gain. The size of the antenna is $8.8 \times 7.5\ \text{cm}^2$ and has a fractional bandwidth of approximately 127% from 1.54 to 7 GHz for return loss less than 10 dB with a directional radiation pattern. The average gain of the proposed prototype is 8.5 dBi, and the radiation efficiency is approximately 92% on average over the operating bandwidth. The antenna is used as the transceiver in a breast phantom measurement system to detect unwanted tumor cells inside the breast.

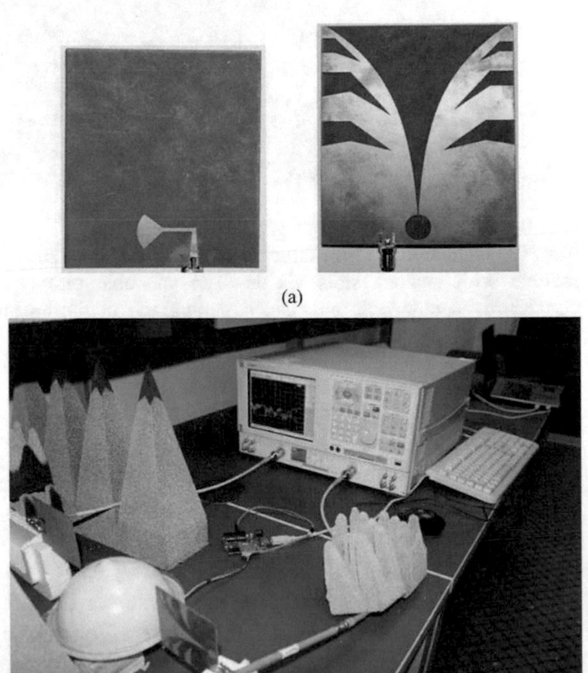

Fig. 9.4 **a** Vivaldi antenna design and **b** the experimental setup for breast cancer imaging consisting of the antennas and the breast phantom (Reprinted with permission from IEEE [38])

The size of the breast phantom considered is approximately 16 × 8 cm^2, with the standard dielectric constant of a human breast and a target tumor of 10 mm diameter inside. The phantom contains four layers, namely, the skin layer, the breast tissue layer or fat, the tumor and the normal air layer. Tumorous cells normally have a high-water content and a dielectric constant that is higher than that of low water tissues, such as fat. The proposed antennas are placed face to face at 18 cm, and each of them is 6.5 mm from the breast skin layer. The breast phantom is placed on the rotation platform, and two antennas are placed beside the phantom and are connected to the VNA (Agilent N5227A). The VNA is connected to a personal computer (PC) via a General Purpose Interface Bus (GPIB) port, and data are transmitted to PC through this setup. The experimental setup is shown in Fig. 9.4b. Finally, the effects of the normal breast tissue and tumor on both forward and backscattering signals are studied and processed.

A system for microwave breast tumor detection is presented in [39] using iteratively corrected coherence factor delay and sum (CF-DAS) algorithm to produce stable and accurate images. Several low-cost lab-based homogenous and heterogeneous phantoms containing the dielectric property of the human breast and tumor tissue are prepared to test the system efficiency. The geometric design and fabricated Vivaldi antenna are presented in [39] where FR4 is used as the substrate material with 1.6 mm thickness, the relative permittivity of 4.4 and loss tangent of 0.02. The antenna covers the dimension of 40 × 40 × 1.6 mm^3 with modified patch and ground, an elliptical notch for increasing the directivity of the radiating elements and a 50-Ω microstrip feed line. The bandwidth of about 8.4 GHz (125.92%) ranging from 2.5 to above 11 GHz frequency ranges is achieved. The average realized gain is 6 dBi together with a maximum peak of 7.2 dBi at 8.3 GHz.

The designed imaging system comprises an array of nine improved modified antipodal Vivaldi antennas that can work across the UWB band (2.5 to 11 GHz). A suitable SP8T device is used to enable the eight receiver antennas to 50 rotated positions to send reflected microwave signals whereas the reflected backscattered signals are recorded by VNA using the MATLAB based software architecture. Laboratory-based breast phantom that emulates the dielectric properties of real breast tissues with tumor tissue is fabricated and measured to test the validity of the imaging system. After collecting the data, an iteratively enhanced CF-DAS algorithm is used to detect the tumors inside the phantoms. The proposed method achieves more than 10 dB improvement over conventional CF-DAS in terms of signal-to-mean ratio (SMR) in this study.

9.2.2.4 Microwave Breast Imaging Using Time-Domain UWB CSAR Technique

A combination of time-domain UWB radar technique with CSAR processing to develop a high-resolution imaging method for breast cancer detection is proposed in [40]. CSAR is a modified version of the Global Back Projection (GBP) in the time domain, for circular data acquisition. A sectional CSAR image reconstruction is considered to reduce the errors due to the difference in permittivity of various tissues in the breast. This method is very fast and easy to implement, compared to frequency domain tomographic methods. The method can be suitable for early diagnosis and treatment monitoring during chemotherapy or radiotherapy.

Advanced Phantom: 3D Printed Phantom

The human breast consists of three different tissues: adipose, glandular, and fibro-connective tissues and the percentages of each of the tissues can vary for everyone. The advanced breast phantom is built based on MRI of real patient using 3D printing of acrylonitrile butadiene styrene (ABS) plastic ($\varepsilon_r = 2.25$). The breast is 30 cm in circumference and projects outwards 7 cm and the tumor is a cube of sides of 1 cm as seen in Fig. 9.5a, b. The interior distribution of the advanced phantom is filled with solutions emulating glandular tissues and tumor to accurately mimic the human breast. The solution consists of polyethylene glycol mono phenyl ether (Triton X-100) and deionized water. The solutions emulating the breast tissues or tumor are different ratios of deionized water and Triton X-100. The dielectric properties of the emulated breast phantom solutions were verified using a dielectric probe (85070*E*) from Keysight Technologies and are plotted in Fig. 9.5c, d.

The advanced phantom contains a 1 cm^3 emulated tumor shown in Fig. 9.6a. Figure 9.6c shows the positive image reconstructed using the average group velocity in the medium. The image is blurred, the diameter of the fat, and the glandular tissues also changed, due to the difference between actual and the average group velocities. To solve this problem the sectional image reconstruction method that uses different group velocities is applied. The major reflections, at the matching liquid/fat interface and the fat/glandular interface are presented in Fig. 9.6b. Figure 9.6d shows the sectional reconstruction of the image that is vivid and demonstrates the dimensions of phantom, shape, and location of the tumor correctly. The phantom was scanned using a 3 T MRI system (Fig. 9.6e) to evaluate the ability of the UWB-CSAR technique.

9.2.2.5 Antenna Designs Comparison for Breast Tumor Detection

In this work, additional antenna factors which can affect the effectiveness of skin artefact removal and image reconstruction algorithms that are quite relevant in the context of microwave imaging are considered [41]. Strategies are proposed to minimize or compensate its adverse effects on the quality of the inverted image. Three antenna topologies are considered for imaging the breast namely, Balanced Antipodal Vivaldi Antenna (BAVA), Planar Monopole Antenna planar monopole, and an exponentially tapered slot-based antenna (XETS). Angular characterization of the “phase center” is an effective technique to take advantage of prior knowledge about the antenna and improve the image focusing. Results show very good focusing by all three antennas, by applying calibration strategies to significantly improve the image quality.

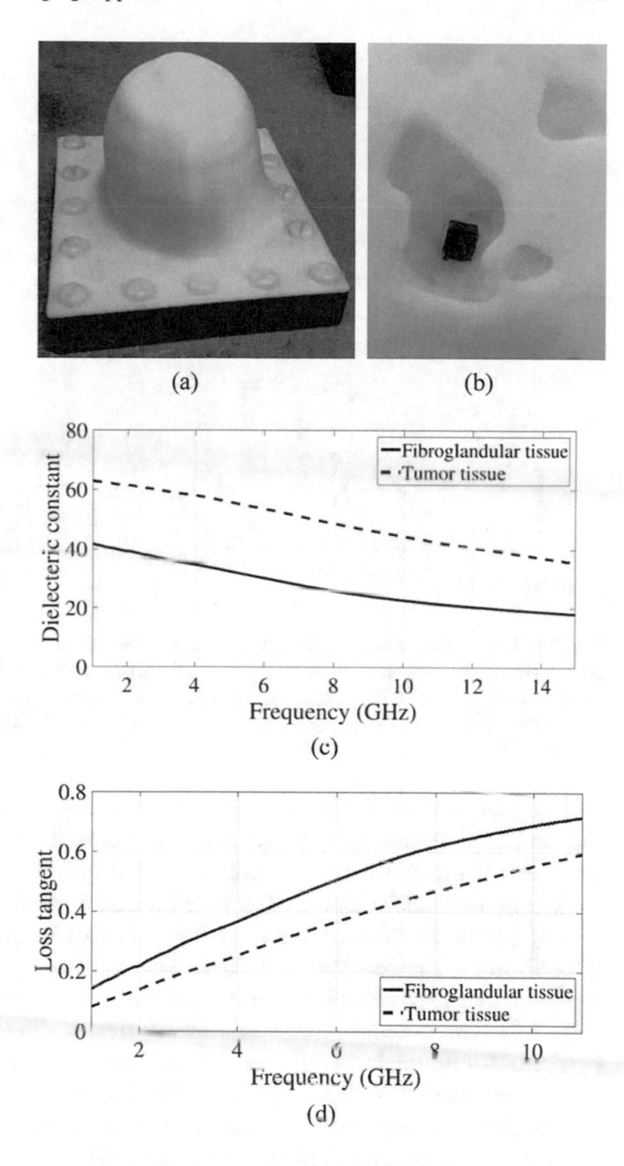

Fig. 9.5 Breast phantoms to perform measurement scenarios: **a** 3D printed phantom, **b** emulated tumor using a 1 cm^3 plastic container filled with mimicking solution, **c** Measured dielectric constant, **d** loss tangent, versus frequency for mimicking solutions used for breast phantom (Reprinted with permission from IEEE [40])

9.2.2.6 Optimized UWB Monopole Antenna Design for Microwave Imaging

A cost-efficient design, optimization, and physical implementation of a compact slotted ultra-wideband (UWB) monopole antenna for body-centric imaging applications is reported in [42]. A parallel surrogate model assisted hybrid differential evolution for antenna optimization (PSADEA) is proposed to optimize the design. The antenna is fabricated on FR4 substrate and provides a good reflection coefficient (S_{11} -10 dB) in the UWB frequency band from 3.1 to 10.6 GHz for free space

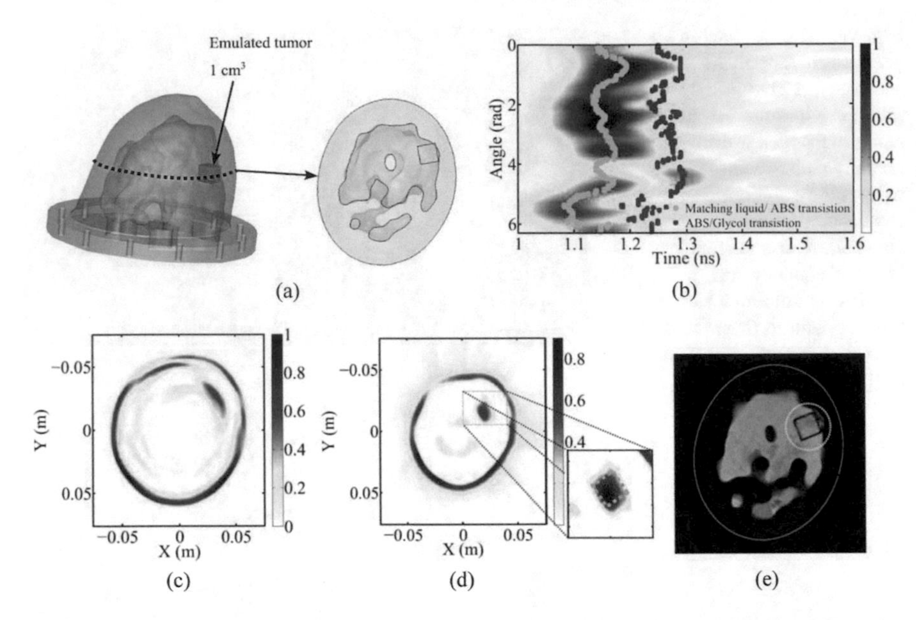

Fig. 9.6 Cancerous advanced phantom; **a** transparent phantom in different views (the cube region is the simulated tumorous tissue). **b** Plot of transition into different medium. Reconstructed positive images using **c** the average group velocity in the medium **d** sectional reconstruction **e** MRI image for comparison (Reprinted with permission from IEEE [40])

and in proximity of the body or breast mimicking tissues (phantoms). The breast mimicking phantom was characterised by electrical properties including conductivity $\sigma = 0.4$ S/m and relative permittivity of $\varepsilon_r = 9$ and the tumor prepared from scattering material including 10 g of wheat flour mixed with 5.5 g of water.

The combined mixture has a relative permittivity of 23 with a conductivity of 2.57 S/m at a frequency of 4.7 GHz. This mixture is used in this experiment to represent and place inside the breast phantom oil for the experimentation analysis (Fig. 9.7). The circular tube is mimicking the skin being replaced by a Plexiglas barrier of 2 mm thickness having dielectric constant varying between 2.39 to 2.59, with corresponding conductivity of 0.009–0.007 S/m. Image reconstruction algorithms like delay-multiply-and-sum (DMAS) modified weighted- delay-and-sum (MWDAS) are applied to obtain the images after signal processing.

9.2.2.7 Various Antenna Designs for UWB Imaging

An ultra-wideband (UWB) printed monopole antenna (PMA) for use in a circular cylindrical microwave imaging system is presented in [43]. The proposed antenna consists of a square radiating patch and a ground plane with a pair of E-shaped slots and a rectangular slot with a pair of horizontal T-shaped strips protruded inside the slot, which provides a wide usable fractional bandwidth of more than 120%

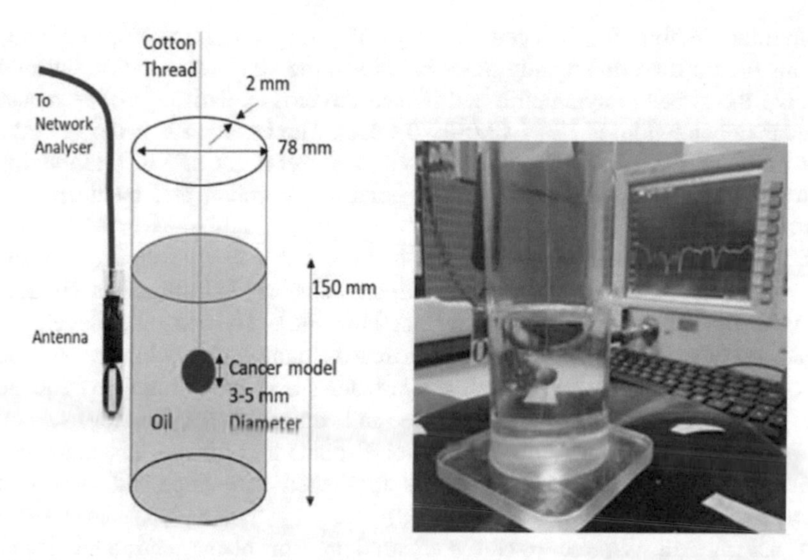

Fig. 9.7 Proposed antenna placed in close (varying) distances to the breast mimicking phantom (Reprinted with permission from IEEE [42])

(2.97–12.83 GHz). By cutting two modified E-shaped slots with variable dimensions on the ground plane corners and by inserting two T-shaped strips inside the rectangular slot located in the ground plane center, additional resonances are excited, and hence much wider impedance bandwidth can be produced, especially at the higher band. The proposed antenna has an ordinary square radiating patch, therefore displays a good omnidirectional radiation pattern even at higher frequencies, and its radiation efficiency is greater than 86% across the entire radiating band. The designed antenna has a small size of $12 \times 18\ \text{mm}^2$. Simulated and measured results are presented to validate the usefulness of the proposed antenna structure for circular cylindrical microwave imaging system.

A compact 4×4 planar UWB antenna array with the total size of $44 \times 52.4\ \text{mm}^2$ was developed for radar-based breast cancer detection system [44]. The center frequency and the bandwidth of the antenna were 6 and 12.5 GHz, respectively. The antenna consists of a square slot set in a ground plane on one side of a Duroid RT 6010LM substrate with a relative permittivity of 10.2. A stack of dielectric materials 40 mm thick was inserted under the antenna as skin and adipose phantoms to optimize the dimensions of the antennas. The measured 10-dB bandwidth of the antenna was 12.5 GHz (3.5–15 GHz). A quasi-three-dimensional confocal imaging was performed using the developed breast phantoms. A 2 mm tumor phantom in an inhomogeneous structure with a glandular phantom was easily detected and the two separate tumor phantoms, which were located at the depth of 23 mm with the spacing of 10 mm.

A dual-polarized UWB slot antenna is presented in [45] in which the slot has an overall shape of a square ring whose four arms are identical and stepped rectangular. While two adjacent arms of the ring are microstrip-fed and are responsible for the orthogonally polarized fields, the other two arms are introduced to make the structure symmetric and reduce the cross polarization. The measured impedance bandwidth is 120% (3.0–12.0 GHz). The isolation is better than 20 dB over most of the band. The antenna will be useful for UWB multiple-input–multiple- output (MIMO) transmission, polarimetric radar, and medical imaging applications.

A small modified antipodal Vivaldi antenna (MAVA) with rectangular slits at sun-shaped configuration and a half elliptical-shaped dielectric lens (HEDL) is designed for microwave imaging applications [46]. The slits are developed to extend low end of frequency band and increase antenna gain at lower frequencies, while the lens is applied to this antenna to feature high gain at higher frequencies, high front-to-back ratio, low side-lobe and cross-polarization levels, narrow half-power beamwidth, and modification on E-plane tilt of beam. The antenna with small size of $30 \times 55 \times 0.508$ mm^3 is fabricated and employed as a part of the microwave imaging system (Fig. 9.8a).

A nature fern inspired fractal leaf structure for planar antipodal design is implemented in [47] which operates from 1.3 to 20 GHz. The lower operating frequency of this antenna is reduced by 19% with the second iteration as compared to the first iteration of fractal leaf structure. The prototype antenna is fabricated and tested in frequency as well as in time domains to obtain various transfer characteristics along with common antenna parameters. Experimental results show that good wideband feature, stable radiation pattern, and promising group delay of less than 1 ns signatures are obtained, which agree well with the simulated data. The miniaturized proposed antenna structure becomes an attractive choice in microwave imaging applications because of its ultrawide fractional bandwidth at 175%, high directive gain of 10 dBi, and a good fidelity factor above (>90%).

An UWB metamaterial slab (meta-slab) covered antipodal Vivaldi antenna (AVA) operating from 3.68 to 43.5 GHz with high-gain and stable radiation patterns is presented in [48] suitable for imaging applications (Fig. 9.8b). The meta-slab exploiting the dispersive nature of meta-material exhibits an effective permittivity function higher than the one of the bare dielectric slabs. The high permittivity meta-slab sucks energy from the tapered slot and the flare termination of the Vivaldi antenna and transmits the energy to the end fire direction. Measured far-field results show the proposed antenna provides boresight gain >10 dBi over the 10–20 GHz, >15 dBi over the 20–32 GHz, and >17 dBi over the 32–40 GHz.

The omnidirectional planar UWB antenna for imaging purposes is fabricated using PCB technique on FR4 substrate with a permittivity ε_r of 4.3, loss tangent $\tan\delta = 0.025$ and substrate thickness of 1.58 mm [49]. The metallic reflector is made of brass and the planar antenna is held and fixed inside the reflector using plastic screws at the four corners of the omnidirectional antenna. The −6 dB matching bandwidth is 135% from 1.5 to 7.7 GHz, and the −10 dB matching bandwidth is 100% from 2.5 to 7.7 GHz, for the final prototype.

An UWB active slot antenna for tissue sensing arrays is developed and measured to operate from 3 to 8 GHz for breast imaging [50]. It integrates a low-noise

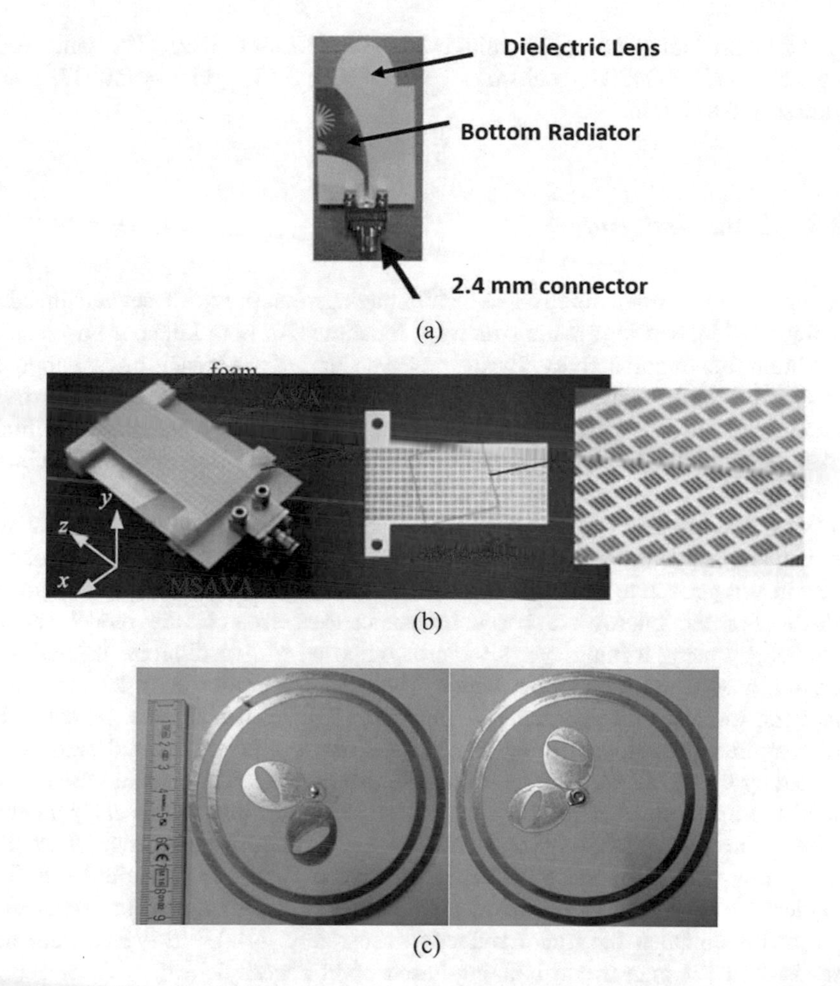

Fig. 9.8 **a** Photograph of the fabricated MAVA-HEDL (Reprinted with permission from IEEE [46]). **b** Photograph of the proposed antenna MSAVA consisting of a conventional AVA and two meta-slabs (Reprinted with permission from IEEE [48]). **c** UWB CP crossed dipole fabricated antennas: top view and bottom view (Reprinted with permission from IEEE [51])

amplifier (LNA) with a printed-slot antenna to achieve gain enhancement of about 20 dB. The integration of a LNA with the antenna significantly improves the system signal-to-noise ratio (SNR), which is one of the greatest challenges in microwave tissue imaging.

A UWB circular polarized (CP) crossed dipole antenna was designed and fabricated for imaging applications [51]. The presented antenna comprises of two orthogonal dipoles with elliptical arms vacated by rotated elliptical slot as shown in Fig. 9.8c. The slots and a modified feeding configuration enhance both impedance and axial ratio bandwidth. The antenna has 4.75:1 impedance bandwidth

(1.6–7.6 GHz) and 4.1:1 axial ratio bandwidth (1.8–7.1 GHz). The antenna is designed on RO4003 substrate with $\varepsilon_r = 3.55$, $\tan\delta = 0.0027$, and thickness = 0.817 mm.

9.2.3 Brain Imaging

Imaging of brain, often referred as neuroimaging, comprises of several imaging techniques addressed to provide structural, functional, or both kinds of information about brain tissues and activity. Brain strokes occur due to a loss of blood supply to a part of the brain caused by a blood clot (ischemic stroke) or a haemorrhage (haemorrhagic stroke) [52]. Figure 9.9 shows illustration of a haemorrhagic brain stroke. Both kinds of stroke present similar symptoms but radically different treatment which must be administered as fast as possible.

Microwave imaging systems have the potential of being relatively low-cost and portable, and hence very suitable for fast and accurate diagnostics of brain stroke or cancer in comparison to most of the imaging technologies that are high cost and not portable. For the microwave based human cancer research, the model can be simplified by using a multilayer structure composed of five different layers, skin, fat, skull, grey matter and white matter. The setup for brain tumor detection contains a transmitter, a receiver, and the phantom [53]. An UWB pulse generated by the transmitter propagates via air to the head and the back-scattered signals are recorded by the receivers. Both the transmitter and the receiver antennas are rotated around the brain together to complete a scan. It shows that the receiver antenna receives more energy when pointed at the tumor. Using this knowledge, the tumor location can be determined by rotating the antenna array to scan the head. The analytical signal and peak detection method have been employed to remove the noise and distinguish the tumor reflection accurately. An UWB Vivaldi antenna array is used for microwave imaging-based brain cancer detection. A cancerous

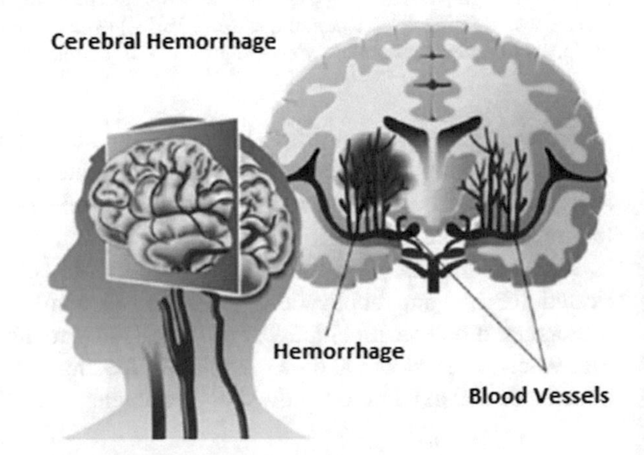

Fig. 9.9 Illustration of haemorrhagic brain stroke (Reprinted with permission from IEEE [52])

brain model which consists of a 5 mm tumor and four layers is simulated with CST Microwave Studio. The input pulse is radiated into brain model and the pulse response is recorded by the receiver antenna. The analytical signal function has been employed to eliminate the noise and the scanning method is applied for microwave imaging. The image processing result shows a good agreement with the actual tumor location in the brain model.

Human head is generally known to be consisting of six layers including: skin, skull, cerebrospinal fluid (CSF), grey matter and white matter. Microwave imaging exploits the difference between the dielectric properties of tissues to detect and localize malfunctions in biological tissues [54]. The electromagnetic (EM) characteristics of each layer of the head versus frequency is presented in this work. Each layer is totally dispersive as in real situation and the EM characteristics of individual layers.

The design of an antipodal exponentially corrugated tapered-slot antenna to be used in a microwave system for brain stroke detection is presented in [55]. The antenna is designed to operate across the band from 1 to 4 GHz that is widely used for head imaging. Tapered corrugations are used to reduce the size of the antenna and to improve its directivity especially at the lower end of the band. The antenna has dimensions of (95 × 120 mm^2) using the high dielectric constant substrate Rogers R03010 (thickness = 1.28 mm, and dielectric constant = 10.6). The results indicate an antenna with a -10 dB return loss bandwidth and better than 10 dB front-to-back ratio across the band from 1.1 to 4 GHz.

A system designed to operate across the band from 1–4 GHz achieves satisfactory imaging results of the brain [52]. A circular array of sixteen directional antennas operating across the band 1–4 GHz is designed for brain imaging. The antenna is designed with corrugations for miniaturization, with dimensions of (11 × 9 cm^2) on the substrate Rogers RO3010 ($\varepsilon_r = 10.2$, and thickness = 1.28 mm). The return loss and antenna prototype is presented in Fig. 9.10a. The system uses the monostatic radar mode of operation. In this case, each of the antenna elements is sequentially used for transmitting and receiving the wideband signals. The scattered/reflected signals at each antenna are recorded in the frequency domain using the VNA which are transformed to time-domain. An imaging algorithm based on the confocal delay-and-sum algorithm is utilized for the post-processing of the determined target response at different time shifts.

A realistic head phantom that includes the main tissues found in the human brain; skin, skull, grey matter, white matter, and the cerebral spinal fluid (CSF) was fabricated. The obtained images are shown in Fig. 9.10b for two different realistic locations of the stroke. As the images portray, the stroke is clearly detected and localized for the two investigated cases.

A low complexity pre-processing algorithm based on the principal component analysis (PCA) processing is presented in [56] to improve the performance of the stroke detection imaging system. The algorithm is compared with state-of-the art algorithms in a monostatic and in a multistatic modes. The DAS algorithm has the attracting feature of having low computational complexity, but it offers limited robustness to artifacts and requires a very high number of antennas. Backscattered

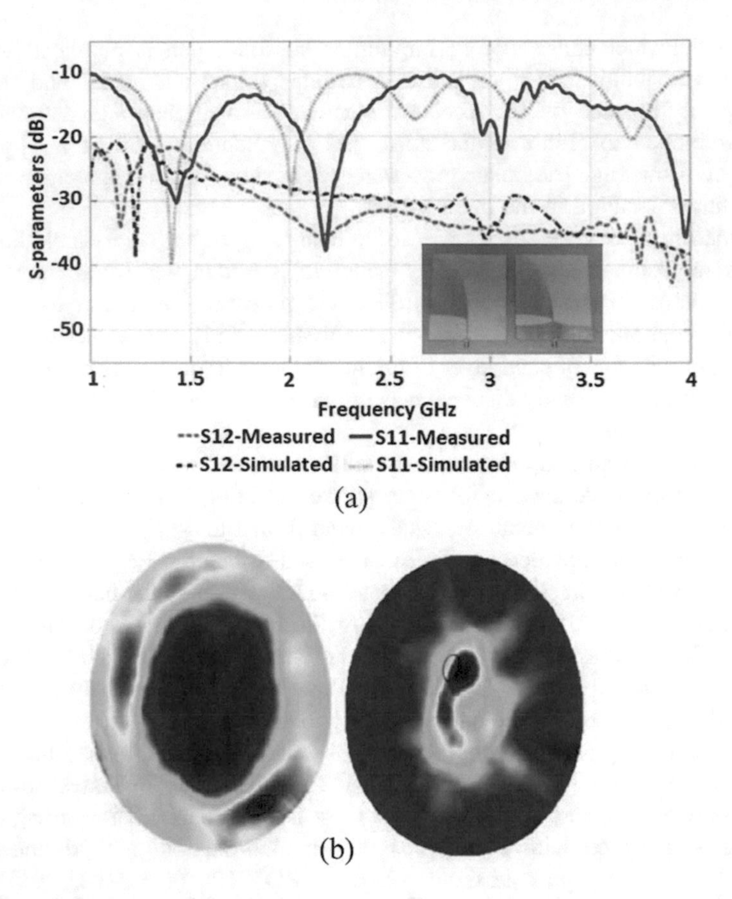

Fig. 9.10 **a** Reflection coefficient and mutual coupling, and time-domain performance of the antenna (inset: top and bottom views of the wideband antenna). **b** Two-dimensional imaging results. The real location of stroke is indicated by an ellipse (Reprinted with permission from IEEE [52])

signals are generated by a FDTD using a 3D head model and eight transmitting/receiving antennas. The performance has been assessed considering a spherical bleeding (haemorrhage) with a diameter of 20 mm at different positions of the scalp.

A coplanar wave guide (CPW) fed ultra-wideband (UWB) Vivaldi antenna is considered for stroke detection [57]. A six layered namely skin, fat, skull, dura, cerebrospinal fluid (CSF), brain (white matter) human head model is designed and simulated with 5 mm tumor. The size of the antenna is of $30 \times 30 \times 0.8$ mm^3. FR4 (lossy) dielectric material having dielectric constant of 4.3 and loss tangent of 0.025 has been used as the substrate. Various radiation properties of the CPW fed UWB Vivaldi antenna have been analysed for both the healthy brain model and tumor induced brain model. The specific absorption rate (SAR) is higher for tumor induced brain model compared to a healthy brain model. A UWB radar system (1–8 GHz) for brain-stroke detection was simulated using a 3D anatomical head phantom with strokes of different sizes and in different positions [58]. The position of the stroke was estimated by processing reflected signals using delay and sum beam-forming algorithm which could accurately detect the location of the stroke. A UWB bow-tie antenna fabricated on Rogers Duroid substrate (4003) with dimension 20 × 40 mm^2 working in 1–8 GHz frequency band was used. The imaging system is composed of 10 multi-static UWB dipole antennas which were directly attached to the head. During the simulation, one antenna is selected as a transmitting antenna and other 9 are receiving antennas. The simulation was performed using EM simulation software SEMCAD which is based on the FDTD algorithm. Design of UWB antenna for brain cancer detection with circular modification patch microstrip antenna with slotted patch, partial ground plane and reflector is reported in [59]. The substrate material of the antenna is Rogers Duroid RT5880 with dielectric constant 2.2 and thickness 1.575 mm. The simulation result show that the antenna offers bandwidth ranging from 3.1 to 10.6 GHz and the radiation pattern is unidirectional.

9.2.4 Time-Lapse Imaging of Human Heart Motion

MWI has been researched especially for stroke imaging and breast cancer imaging. Another field in which radar and MWI techniques may prove successful, is monitoring of the cardiovascular system inside the body. The ability of microwave frequencies to penetrate tissues allows for studying the mechanical functioning of the cardiovascular system. This would complement bioelectric activity measured through electrocardiography (ECG). The recent innovation of a CMOS implementation of a coherent pulsed UWB radar [17], provides the opportunity of in-body radar imaging with extremely small and low-cost electronics. IR-UWB radar allows for sensing at a range of different depths in the body and thus for localization of scattering sources in space. When multiple sensors are combined, 2D or 3D imaging through beamforming can be performed [60]. The theoretical implications of 2D and 3D imaging are discussed, this work focuses on 1D sensing of the cardiovascular system.

Coherent ultra-wideband (UWB) radar-on-chip technology shows great promise for developing portable and low-cost medical imaging and monitoring devices. Cardiovascular signals were recorded using radar-on-chip systems and electrocardiography (ECG). Through ECG-aligned averaging, the arterial pulse wave could be measured at several locations in the body. Pulse arrival time could be determined with high precision, and blood pressure pulse wave propagation through different arteries was demonstrated. In addition, cardiac dynamics were measured from the chest. A portable and low-cost device for long-term monitoring of the cardiovascular system is proposed and provides the fundamentals necessary for developing UWB radar-on-chip imaging systems.

In this work, a radar system for ultra-wideband (UWB) imaging of the human heart is presented [61]. To make the radar waves penetrate the human tissue the antenna is placed very close to the body. The antenna is an array with eight elements, and an antenna switch system connects the radar to the individual elements in sequence to form an image. Successive images are used to build up time-lapse movies of the beating heart. Measurements on a human test subject are presented and the heart motion is estimated at different locations inside the body. The movies show rhythmic motion consistent with the beating heart, and the location and shape of the reflections correspond well with the expected response form the heart wall. The spatial dependent heart motion is compared to ECG recordings, and it is confirmed that heartbeat modulations are seen in the radar data.

By placing the radar antennas in contact with or very close to the human body, the strong reflection from the air-body surface is reduced and the radar can sense waves that penetrate the tissue and are reflected from the heart wall. In this way the radar directly measures the mechanical movement of the heart and can noninvasively provide information for heart diagnosis. With imaging radar, it is possible to simultaneously measure how different parts of the heart are moving. This can give valuable information on the physical state of the heart. In this work the feasibility of radar imaging of the human heart is studied using a simplistic radar system based on a vector network analyser (VNA), a switched array system and a delay and sum (DAS) based beamformer.

Extracting the complex value from the same pixel of every image in the time-lapse data cube of images gives a signal describing the time variation of the signal at the location corresponding to the selected pixel. Modulations observed in the phase of the slow time profile will be strongly influenced by the heart motion and is used as an indicator for the heart motion. A movie was made of the time-lapsed images where a rhythmic motion is visible. The observed motion is most clear around the reflectors. Slow time profiles were extracted at the four different positions marked in Fig. 9.11 which is one of the images from the time lapse data and the heartbeat motion were estimated. The results are shown in Fig. 9.12a–d where the estimates are plotted together with synchronized ECG recordings for three heartbeat cycles.

9.3 Through Wall Imaging

This section is concentrated on overview and latest techniques related to the ultra-wideband through wall imaging (UWB-TWI) radar systems. Various aspects related to design and development of a suitable through wall imaging system at both circuits modelling, signal processing algorithms for imaging and detection, and electromagnetic aspects like through-wall propagation, antenna effects, the overall propagation environment and 3-D radar targets scattering needs have been discussed.

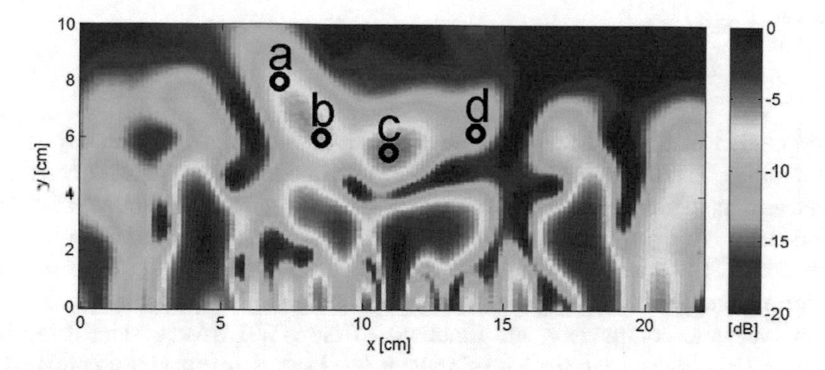

Fig. 9.11 A single image from the time-lapse imaging data. Phase modulation estimation is carried out from the marked points **a–d** (Reprinted with permission from IEEE [61])

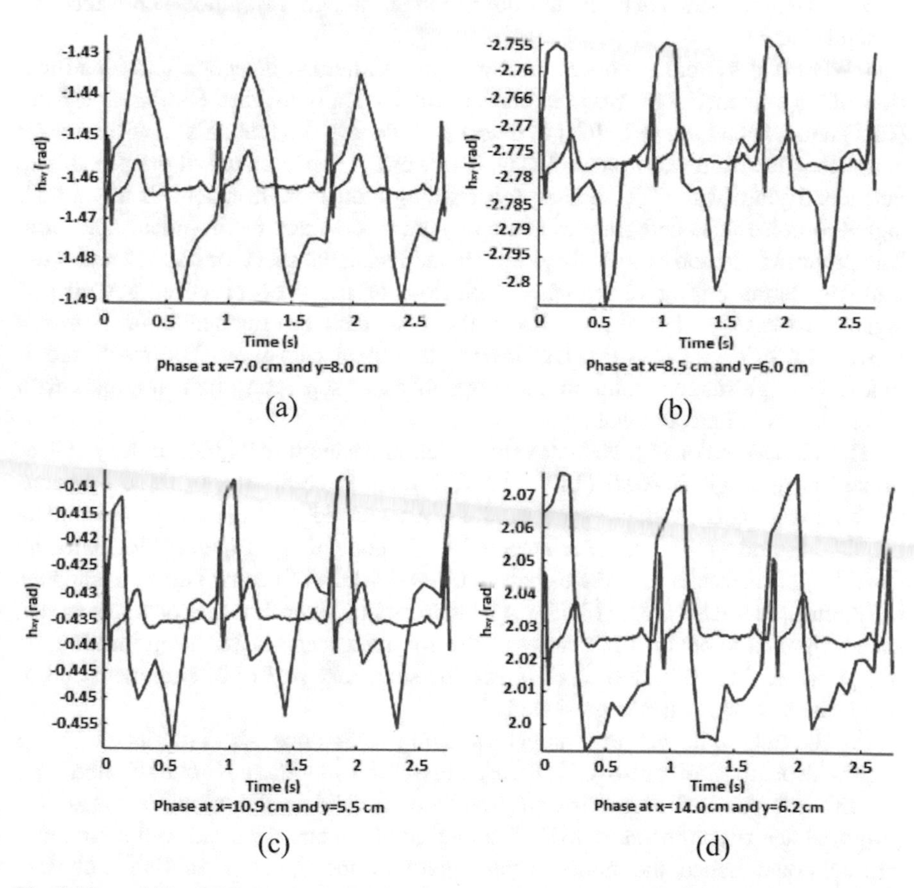

Fig. 9.12 a–d The ECG data is plotted in red and scaled in amplitude to make the plots readable. The given y-axis applies to the radar data only (Reprinted with permission from IEEE [61])

9.3.1 UWB Though Wall System Design Aspects and Imaging Techniques

A simulation platform for localization purposes is proposed to accurately analyse and simulate UWB-TWI SAR systems. Various modules and features considered are electromagnetic simulator to account for the wall presence and target scattering, a wideband back projection imaging algorithm, and different transceiver architectures and 1×8 linear Vivaldi antenna array configurations [17]. The developed platform is capable of time gating to suppress early wall reflections and simulates various discrete components and functions of the UWB SAR system using both linear and nonlinear analysis. The simulator has been experimentally validated for both dielectric and metallic targets using a mock-up human model. The developed simulator can also be used to study various effects related to operation frequency, pulse width, pulse shape, wall dispersion, and carrier leakage. It predicts UWB system performance such as detection range, image resolution, and receiver dynamic range.

UWB radar system is chosen for through-wall human detection and identification of movements. The work is carried out by applying fast Fourier transform (FFT) and *S* transform (ST) [62] to detect and identify the human's time-frequency characteristics from experimental data. The work is concentrated in the processing and identifying of the life signals under strong clutter environment. It has a high signal-to-noise ratio and simpler to implement in complex environment detection. The proposed method is suitable to search and locate the survivor trapped under the building debris during earthquake, explosion, or fire. The proposed set up and signal processing algorithms extract the characteristic frequency of human's movement, calculate the size, and locate the human's position. The result shows that it has high range resolution and helps in better separating human targets with other objects behind the wall.

The FMCW radar is a cost-effective solution for high resolution radar applications. The through-the-wall (TTW) FMCW radar is composed of three interconnected main parts: a microwave module, a switched antenna array and a data acquisition and processing. An array of 16 Exponentially Tapered Slot Antenna (ETSA) [63] is used and operates between 1 and 4 GHz. Spacing between antennas is 7.5 cm. For each 16 data blocks, a beam forming algorithm that operates in real time is applied to obtain a 2D image of the observed scene. After beam forming, an image processing algorithm to eliminate the stationary part of environment and to retain the moving target is presented.

A through-wall moving target imaging technique is introduced using ultra-wideband short pulse (UWB-SP) radar. The modified Kirchhoff migration algorithm is selected as the imaging method to validate the effectiveness of the proposed detection procedure [64]. The motion of the target behind wall is irregular and unknown, hence, the model of the target with the "go-stop" motion is chosen. A background removal strategy without complicated operations is introduced to extract the "go-stop" target signal.

Strategy of Moving Target Imaging

The proposed imaging method for through wall tracking of moving person consists of three phases, namely background subtraction to improve the signal to noise ratio, modified Kirchhoff migration imaging to improve both spatial resolution and signal to clutter noise ratio (SCNR) under near-field conditions and Hilbert processing. The side lobe artifacts produced in the array-based difference images confuse the actual location of the moving person. To improve the SCNR of the array-based difference image, the Hilbert transformation is considered to generate clear image.

A UWB-SP MIMO through wall radar system focusing on the 500 $\sim$ 3.0 GHz, consists of several major subsystems, such as the MIMO antenna array, the data acquisition module, and the system control/processing module. The person undergoing the first and second motion can be detected successfully and accurately localized.

Experimental investigation is carried out to show the potential of the proposed UWB-TWR radar for detection, localization, tracking and identification of human beings in motion through a wall [15]. The antenna is a 2D array of 108 elementary UWB antennas "Double Tapered Slot Antennas" (or "DETSA") of dimensions 75 $\times$ 59 cm^2. The specific element antenna used in the DIAMS array is printed on a "Roger Duroid" substrate (relative permittivity $\varepsilon_r = 2.2$, thickness = 1.27 mm). The exponentially tapered profile of the radiating area is designed to confer a good impedance antenna matching from 1 to 4 GHz.

The radar image stream generated is divided into two parts: The first one is the dynamic sequence, which contains moving targets. The second is the static sequence which contains the inner walls of the scene under surveillance (SUS) and the stationary targets. The partition is done using an algorithm, to consider the characteristics of the radar images. The interior walls of the SUS are highlighted using a discrete Radon transform-based (DRT) algorithm performed on to the static sequence. The dynamic sequence is processed by a CFAR (Constant False Alarm Rate) detector modified to consider the orientation of the target signatures in radar images. A multi-target multi-hypothesis tracking (MHT) is applied on the detected targets, to provide the trajectories of moving targets. Finally, 2D or 3D representations of the detected targets are generated.

An effective method is proposed to accurately detect trapped victims even in low SNCR environments such as in long range and through-wall conditions. Localization is achieved using the short-time Fourier transform (STFT) of the kurtosis and standard deviation of the received signal [65]. Further, an improved arctangent demodulation (AD) technique is presented to estimate human micro-motion frequencies based on a multiple frequency accumulation (FA) method. The trapped victim is then identified in a distance frequency matrix. Results indicate that the proposed approach can effectively suppress static and non-static clutter, linear trend, harmonics, and the product of respiration and heartbeat signals. The experiments were carried out in indoor and outdoor environments as shown in Fig. 9.13a, b with subjects positioned at various distances on the other side of the wall. The estimated ranges obtained using the STFT technique are shown in Fig. 9.13c and for the outdoor scenario in Fig. 9.13d.

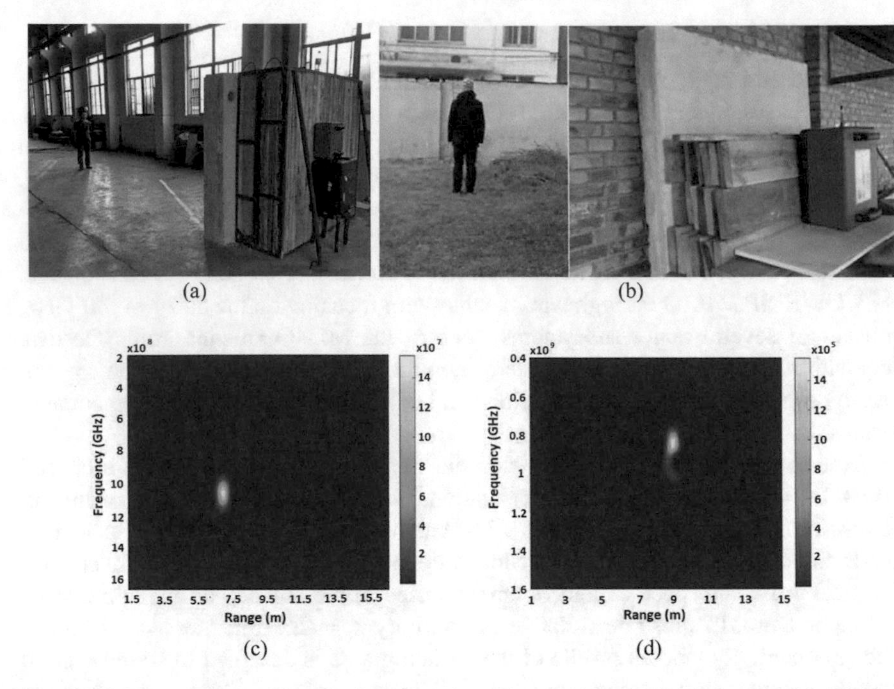

Fig. 9.13 The experimental setup for through wall radar imaging. **a** Indoor and **b** Outdoor environment. **c** The estimated range obtained at a distance from the antenna of 7 m in indoor environment. **d** Range estimation in the outdoor environment with a distance between the subject and radar of 9 m (Reprinted with permission from IEEE [65])

9.3.2 Through Wall Human Sensing and Building Layout Reconstruction Using UWB-MIMO

Layout reconstruction method based on the UWB-TWI radar under one single observation and simultaneously identification of the human subject is proposed in [15]. A coherent processing interval consisting of several successive received echoes in the initial stage is first employed to construct a range-Doppler (RD) spectrum. In the RD spectrum, a series of selected discrete Doppler frequency signals are used to form Doppler back projection (B-P) images. The Doppler B-P image stack and a 3-D constant false alarm rate (CFAR) detector is considered to extract the building layout. The obtained layout is used as auxiliary information which is merged with the simultaneous human identification. The proposed approach can effectively extract the covered layout of multiple walls under one single view and detect presence of the concealed human subject.

Detecting the scene with moving targets via static UWB MIMO radar and obtain the corresponding spatial-Doppler information is the focus of this work. A UWB MIMO TWIR with a 3-m array to sense the concealed human target, where two transmitting elements are at the left and right ends and 11 uniformly spaced

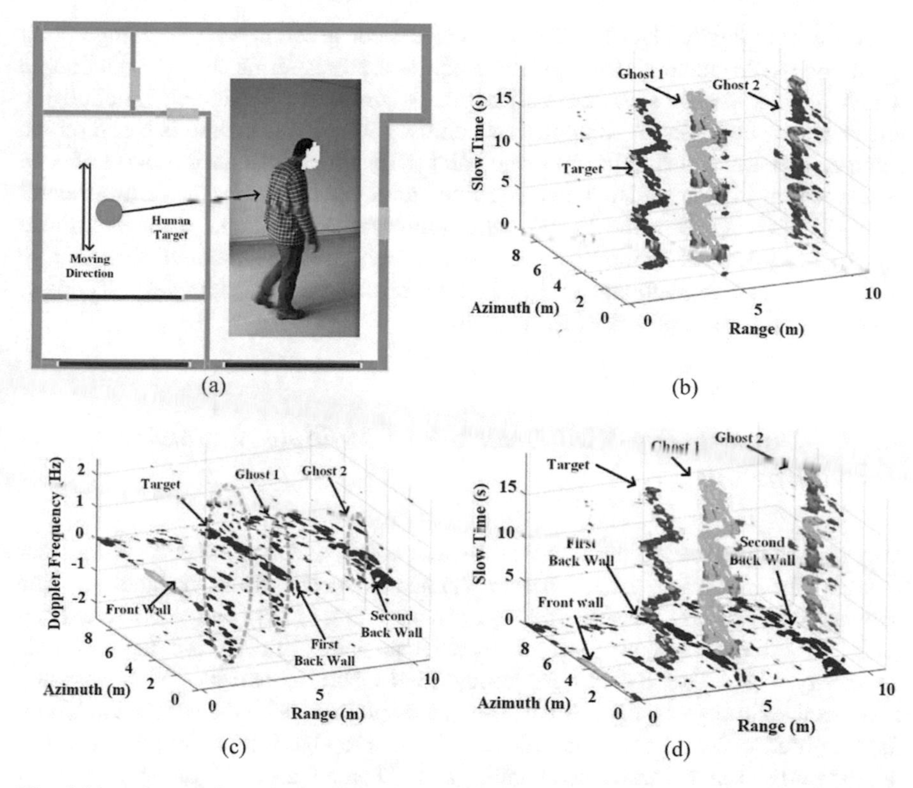

Fig. 9.14 **a** Scene of experiment B. **b** Stack of human indication results in a certain coherent processing interval (CPI). **c** Detection results using the 3-D CFAR in the Doppler B-P image stack. **d** Fusion result of human indication and layout reconstruction (Reprinted with permission from IEEE [15])

receiving elements are in the middle. The radar transmits stepped frequency continuous wave signals in 0.5–2 GHz at a step of 2 MHz with a PRF $fp = 5$ Hz. In experiment A, a person was standing inside the room and waving his arm back and forth. In experiment B, a person was walking inside the room parallel to the front wall, the brick concrete front wall to penetrate has a thickness of 0.28 m (Fig. 9.14a).

Experiment B: Walking Person

Figure 9.14b shows a stack of piled indication results of the walking person, which suffers from much more serious ghost issue. Ghost 1 is even stronger than the walking person, and ghost 2 is also considerable. In Fig. 9.14c, the walking macro motion has a wide and complex Doppler spectrum, almost spanning the entire Doppler band. With the auxiliary information provided by the layout reconstruction obtained by the 3-D CFAR detector applied to the Doppler B-P image stack, distinction between the real target and shadow ghosts is possible based on their locations Fig. 9.14d.

A low complexity algorithm for 3-D image reconstruction with the Range-point migration (RPM) method using a sample-point-based scattering center extraction, in which a golden ratio sampling pattern and a simplified evaluation function are introduced to upgrade the computational efficacy [66]. This method is based on the batch conversion algorithm from range point (RP), which is defined as a set of each observed time delay and antenna location, to a corresponding scattering center point. The results from the FDTD-based numerical simulations, using a realistic human body model, demonstrate that the proposed RPM method significantly accelerates the computational speed without sacrificing the reconstruction accuracy, compared with the original RPM method.

9.3.3 Through-The-Wall Detection of Multiple Stationary Humans

This research proposes a through-wall S-band UWB switched antenna- array radar scheme for detection of stationary human subjects from respiration. The antenna-array radar consists of one transmitting (Tx) and five receiving (Rx) Vivaldi type antennas with high antenna gain (10 dBi) and narrow-angle directivity [18]. The S-band frequency (2–4 GHz) is capable of penetrating non-metal solid objects and detecting human respiration behind a solid wall. Under the proposed radar scheme, the reflected signals are algorithmically pre-processed and filtered to remove unwanted signals, and 3D signal array is converted into 2D array using statistical variance. The images are reconstructed using back-projection algorithm prior to Sinc-filtered refinement. To validate the detection performance of the through-wall UWB radar scheme, simulations are carried out and the experiments are performed with single and multiple real stationary human subjects and a mannequin behind the concrete wall.

The TW-UWB switched-antenna-array radar is shown in Fig. 9.15a, consisting of the Tx and Rx antennas connected to the switch network. Figure 9.15b shows the experimental scene with two stationary real human subjects and a mannequin. Signal processing algorithms are applied on the Rx raw data of each channel. The 2D array is reconstructed using back-projection algorithm (Fig. 9.15c and further refined by the Sinc filter (Fig. 9.15d). The proposed method is capable of distinguishing between both human subjects and the mannequin.

9.3.4 Current State-Of-The-Art Techniques for UWB TWI

A UWB through-wall imaging architecture is developed which can scan the scene automatically by using SPT8 switch that facilitates to obtain 2D images of person with high resolution in cross-range direction [67]. The experiment to see through

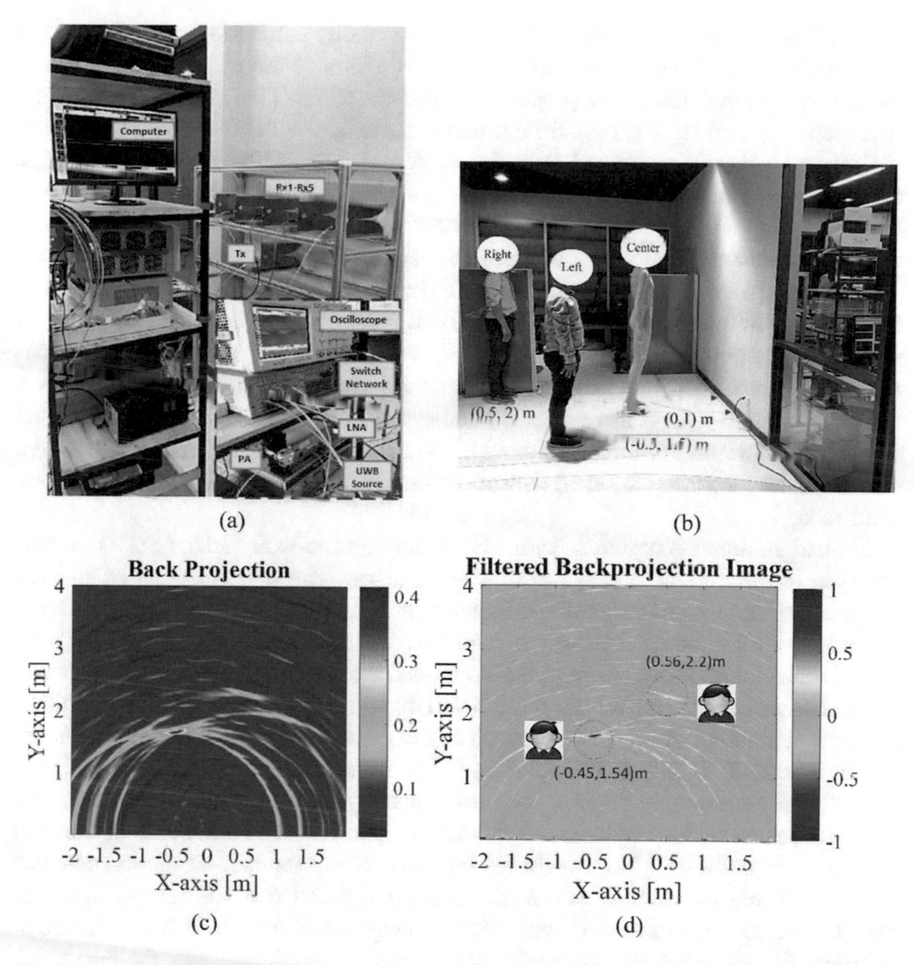

Fig. 9.15 **a** Experimental setup of the switched-antenna-array radar prototype. The space between receiving antennas (Rx1–Rx5) is 30 cm. Through-the-wall detection of two human subjects at (−0.5, 1.5 m) and (0.5, 2 m) and a mannequin at (0 1 m): **b** the experimental scene with two real humans and mannequin, **c** radar imaging by back projection, **d** refined radar imaging using Sinc filtered back projection (After [18], Sensors)

the wall is achieved based on RF components, Field Programmable Gate Array (FPGA) board for UWB signal generation and high sampling oscilloscope (5 Gsa/s) for signal acquisition. Direct acquisition is possible in the time domain of UWB impulse signal unlike the other platforms that operate in frequency such as FMCW and SFCW domain, in which the system should step through several frequencies before information processing. The integration of SPT8 switch makes the system automatically scan the scene and obtain images of the object with high resolution in cross-range, the system architecture is also based on FPGA to generate an extremely narrow pulse of 1 ns duration, RF components and PulsON antennas.

A technique for a super-resolution Doppler velocity estimation algorithm based on Gaussian kernel density estimation, which converts observed range—τ points to Doppler-associated ranges is proposed related to UWB-TWI scenario [68]. This approach also aids in super-resolution range extraction with a compressed sensing (CS) filter, which is combined with the range-point migration (RPM) method for human body imaging associated with micro-Doppler components. 2-D or 3-D numerical simulations, including human body imaging scenario, demonstrate that the proposed method allows both accurate Doppler velocity estimation and human body imaging, which can be updated at the pulse-repetition interval. The 2-D FDTD-based numerical simulation of a UWB-TWI scenario and the experimental validation, assuming free-spacing imaging with the actual Impulse UWB radar system, demonstrated that the proposed method can be used for accurate motion vector estimation with considerably higher temporal resolution. The 3-D GO-based simulation, assuming human-mimicking objects, validates enhancement in the reconstruction accuracy for each part of the human body, even for lower frequency band radar.

Human motion recognition with UWB through-the-wall radar (TWR) is proposed in [69] for view of range profile serialization. The original radar echoes are firstly converted into range profiles. Experimental data with respect to four different behind-the-wall human motions is collected by self-developed UWB TWR to validate the effectiveness of the proposed model. The results show that the proposed model can recognize the human motion serialization and achieve 93% recognition accuracy within the initial 20% duration of the activities (the average durations are 4, 5.5, 3 and 4.5 s), which is important for real-time human motion recognition.

Different movements such as boxing, walking, picking, arm raising/lowering were performed by four volunteers with 1.5 m apart from the other side of the wall along the direction of light-of-sight for the radar. The examples of motion samples have ample interferences as well as the noise energy filled in the range bins other than the zones that appeared with strong energy resulting from the volunteers' motions. To reduce the influence of clutter on the recognition results, the maximum entropy threshold is chosen to determine the threshold E. The sample data can show the characteristics of hidden human motion more clearly, which lays a foundation for the training and testing of sequential recognition model of hidden human motion based on deep learning.

9.4 Conclusion

UWB impulse radar technology has been the subject of significant research because of its simple structure, low-cost, and low-power consumption. Compared with other radar systems, the remarkable characteristic of UWB radar is the high spatial resolution because of employment of ultra-narrow pulses. It not only allows the detection of closely positioned targets but also provides information about the shape and the material content of targets. UWB radar has numerous advantages, such as high detection target range accuracy, good penetrating wall characteristics, improved immunity to external electromagnetic radiation and noise, and robustness to multipath reflections.

In this chapter the capability of using UWB microwave imaging for the detection and localization of breast tumors, heart motion analysis, brain stroke identification, has been explored. State-of-the-art and advancement in numerical and experimental approaches have been reported to obtain accurate detection of the tumor. These include various aspects of compact antenna design for medical imaging applications, simple, portable, and cost-effective UWB radar system set up, novel image reconstruction algorithms to obtain clutter and noise free images. Latest techniques and methodologies related to UWB through wall imaging have also been reported which have the capability to detect single or multiple human subjects behind obstacles in challenging indoor and outdoor environments.

References

1. Guy C, Ffytche D (2005) An introduction to the principles of medical imaging, revised edition, pp 267–294. Imperial College Press
2. Kasban H, El-Bendary MAM, Salama DH (2015) A comparative study of medical imaging techniques. Int J Inf Sci Intell Syst 4(2):37–58
3. Sathish D, Kamath S, Rajagopal KV, Prasad K (2016) Medical imaging techniques and computer aided diagnostic approaches for the detection of breast cancer with an emphasis on thermography—a review. Int J Med Eng Inf 8(3):275–299
4. Barber D (2018) Fundamentals of medical imaging: Paul Suetens. Med Eng Phys 25
5. O'Loughlin D, J O'Halloran M, Moloney BM, Glavin M, Jones E, Elahi MA. Microwave breast imaging: clinical advances and remaining challenges. IEEE Trans Biomed Eng 65 (11):2580–2590
6. Hall EJ, Brenner DJ (2008) Cancer risks from diagnostic radiology. Brit J Radiol 81 (965):362–378
7. Lin X, Chen Y, Gong Z, Seet B, Huang L, Lu Y (2020) Ultrawideb and textile antenna for wearable microwave medical imaging applications. IEEE Trans Antennas Propag 68 (6):4238–4249
8. Nikolova N (2017) Introduction to microwave imaging Cambridge. Cambridge University Press, UK
9. Cheng GG, Zhu Y, Grzesik J (2013) Microwave medical imaging techniques. In: 7th European conference on antennas and propagation (EuCAP), Gothenburg, pp 2669–2673
10. Chandra R, Zhou H, Balasingham I, Narayanan (2015) On the opportunities and challenges in microwave medical sensing and imaging. IEEE Trans Biomed Eng 62(7):1667–1682
11. Pan J (2008) Medical applications of ultra-wideband (UWB)
12. Staderini E (2002) UWB radars in medicine. IEEE Aerosp Electron Syst Mag 17(1):13–18
13. Pancera E (2010) Medical applications of the ultra-wideband technology. In: 2010 Loughborough antennas and propagation conference, Loughborough, pp 52–56
14. Ghosh D, Sahu PK (2016) UWB in healthcare. In: 2016 International conference on electromagnetics in advanced applications (ICEAA), pp 679–682
15. Song Y, Hu J, Chu N, Jin T, Zhang J, Zhou Z (2018) Building layout reconstruction in concealed human target sensing via UWB MIMO through-wall imaging radar. IEEE Geosci Remote Sens Lett 15(8):1199–1203

16. Millot P et al (2015) An UWB through-the-wall radar with 3D imaging, detection and tracking capabilities. In: 2015 European radar conference (EuRAD), Paris, pp 237–240
17. Wang Y, Fathy AE (2012) Advanced system level simulation platform for three-dimensional UWB through-wall imaging SAR using time-domain approach. IEEE Trans Geosci Remote Sens 50(5):1986–2000
18. Rittiplang A, Phasukkit P (2020) 1-Tx/5-Rx through-wall UWB switched-antenna-array radar for detecting stationary humans. Sensors 20(23):6828
19. Grzegorczyk TM, Meaney PM, Kaufman PA, Di Florio-Alexander RM, Paulsen KD (2012) Fast 3-D tomographic microwave imaging for breast cancer detection. IEEE Trans Med Imag 31(8):1584–1592
20. Mahmud MZ, Islam MT, Misran N, Almutairi AF, Cho M (2018) Ultra-wideband (UWB) antenna sensor based microwave breast imaging: a review. Sensors 18(9):2951
21. Alqadami ASM, Nguyen-Trong N, Mohammed B, Stancombe AE, Heitzmann M, Abbosh A (2020) Compact unidirectional conformal antenna based on flexible high-permittivity custom-made substrate for wearable wideband electromagnetic head imaging system. IEEE Trans Antennas Propag 68(1):183–194
22. Abbosh M (2008) Directive antenna for ultrawideband medical imaging systems. Int J Antenna Propag
23. Helbig M, Kmec M, Sachs J, Geyer C, Hilger I, Rimkus G (2012) Aspects of antenna array configuration for UWB breast imaging. In: 6th European conference on antennas and propagation (EUCAP), Prague, 2012, pp 1737–1741
24. Borja B, Tirado JA, Jardón H (2018) An overview of UWB antennas for microwave imaging systems for cancer detection purposes. Prog Electromagn Res B 80:173–198
25. Tangwachirapan S, Thaiwirot W, Akkaraekthalin P (2019) Design of ultra-wideband antipodal Vivaldi antenna with square dielectric lens for microwave imaging applications. In: 7th International electrical engineering congress (iEECON), Hua Hin, Thailand
26. Abbak M, Akıncı MN, Çayören M, Akduman İ (2017) Experimental microwave imaging with a novel corrugated Vivaldi antenna. IEEE Trans Antennas Propag 65(6):3302–3307
27. Klemm M, Craddock IJ, Leendertz JA, Preece A, Benjamin R (2009) Radar-based breast cancer detection using a hemispherical antenna array—experimental results. IEEE Trans Antennas Propag 57(6):1692–1704
28. Klemm M, Leendertz JA, Gibbins D, Craddock IJ, Preece A, Benjamin R (2009) Microwave radar-based breast cancer detection: imaging in inhomogeneous breast phantoms, IEEE Antennas Wirel Propag Lett 8:1349–1352
29. Klemm M et al (2011) Development and testing of a 60-element UWB conformal array for breast cancer imaging. In: Proceedings of the 5th European conference on antennas and propagation (EUCAP), Rome, 2011, pp 3077–3079
30. Abutarboush HF, Klemm M (2013) Signal selection for contrast-enhanced UWB microwave radar imaging with inhomogeneous breast phantoms. IEEE Antennas Wirel Propag Lett 12:1408–1411
31. Porter E, Kirshin E, Santorelli A, Coates M, Popović M (2013) Time-domain multistatic radar system for microwave breast screening. IEEE Antennas Wirel Propag Lett 12:229–232
32. Bahramiabarghouei H, Porter E, Santorelli A, Gosselin B, Popović M, Rusch LA (2016) Flexible 16 antenna array for microwave breast cancer detection. IEEE Trans Biomed Eng 62 (10):2516–2525
33. Gabriel S et al (1996) The dielectric properties of biological tissues: II measurements in the frequency range 10 Hz to 20 GHz. Phys Med Biol 41:2251–2269
34. Porter E, Bahrami H, Santorelli A, Gosselin B, Rusch LA, Popović M (2016) A Wearable microwave antenna array for time-domain breast tumor screening. IEEE Trans Med Imag 35 (6):1501–1509
35. Alshehri S, Jantan A, Raja Abdullah RSA, Mahmud R, Khatun S, Awang Z (2011) A UWB imaging system to detect early breast cancer in heterogeneous breast phantom. In: International Conference on Electrical, Control and Computer Engineering 2011 (InECCE), Pahang, pp 238–242

36. Lazebnik M, Popovic D, McCartney L, Watkins CB, Lindstrom MJ, Harter J et al (2007) A large scale of the ultrawideband microwave dielectric properties of normal, benign and malignant breast tissues obtained from cancer surgeries. Phys Med Biol 52:6093–6115
37. Vijayasarveswari V, Jusoh M, Khatun S, Fakir MM (2017) Scattering performance verification based on UWB imaging and neural network. In: IEEE 13th international colloquium on signal processing and its applications (CSPA), Batu Ferringhi, pp 238–242
38. Islam MT, Mahmud MZ, Misran N, Takada J, Cho M (2017) Microwave breast phantom measurement system with compact side slotted directional antenna. IEEE Access 5:5321–5330
39. Kibria S, Samsuzzaman M, Islam MT, Mahmud MZ, Misran N, Islam MT (2019) Breast phantom imaging using iteratively corrected coherence factor delay and sum. IEEE Access 7:40822–40832
40. Oloumi D, Winter RSC, Kordzadeh A, Boulanger P, Rambabu K (2020) Microwave imaging of breast tumor using time-domain UWB circular-SAR technique. IEEE Trans Med Imag 39 (4):934–943
41. Felício JM, Bioucas-Dias JM, Costa JR, Fernandes CA (2019) Antenna design and near-field characterization for medical microwave imaging applications. IEEE Trans Antennas Propag 67(7):4811–4824
42. Danjuma M, Akinsolu MO, See CH, Abd-Alhameed RA, Liu B (2020) Design and optimization of a slotted monopole antenna for ultra-wide band body centric imaging applications. IEEE J Electromagn RF Microwaves Med Biol 4(2):140–147
43. Ojaroudi N, Ojaroudi M, Ghadimi N (2012) UWB omnidirectional square monopole antenna for use in circular cylindrical microwave imaging systems. IEEE Antennas Wirel Propag Lett 11:1350–1353
44. Sugitani T, Kubota S, Toya A, Xiao X, Kikkawa T (2013) A compact 4 × 4 planar UWB antenna array for 3-D breast cancer detection. IEEE Antennas and wireless propagation letters 12:733–736
45. Krishna RVSR, Kumar R (2016) A dual-polarized square-ring slot antenna for UWB, imaging, and radar applications. IEEE Antennas Wirel Propag Lett 15:195–198
46. Moosazadeh M, Kharkovsky S, Case JT, Samali B (2017) Improved radiation characteristics of small antipodal Vivaldi antenna for microwave and millimeter-wave imaging applications. IEEE Antennas Wirel Propag Lett 16:1961–1964
47. Biswas B, Ghatak R, Poddar DR (2017) A fern fractal leaf inspired wideband antipodal Vivaldi antenna for microwave imaging system. IEEE Trans Antennas Propag 65(11):6126–6129
48. Li X, Zhou H, Gao Z, Wang H, Lv G (2017) Metamaterial slabs covered UWB antipodal Vivaldi antenna. IEEE Antennas Wirel Propag Lett 16:2943–2946
49. Shao W, Edalati A, McCollough TR, McCollough WJ (2018) A time-domain measurement system for UWB microwave imaging. IEEE Trans Microw Theory Tech 66(5):2265–2275
50. Foroutan C, Nikolova NK (2019) UWB active antenna for microwave breast imaging sensing arrays. IEEE Antennas Wirel Propag Lett 18(10):1951–1955
51. Akbarpour A, Chamaani S (2020) Ultrawideband circularly polarized antenna for near-field SAR imaging applications. IEEE Trans Antennas Propag 68(6):4218–4228
52. Abbosh A (2014) Microwave-based system using directional wideband antennas for head imaging. In: 2014 International workshop on antenna technology: small antennas, novel EM structures and materials, and applications (iWAT), Sydney, NSW, pp 292–295
53. Zhang H, Flynn B, Erdogan AT, Arslan T (2012) Microwave imaging for brain tumour detection using an UWB Vivaldi antenna array. In: Loughborough antennas and propagation conference (LAPC), Loughborough, pp 1–4
54. Jalilvand M, Li X, Zwick T (2013) A model approach to the analytical analysis of stroke detection using UWB radar. In: 7th European conference on antennas and propagation (EuCAP), Gothenburg, pp 1555–1559

55. Mohammed B, Abbosh A, Ireland D (2012) Directive wideband antenna for microwave imaging system for brain stroke detection. In: Asia pacific microwave conference proceedings, Kaohsiung, pp 640–642
56. Ricci E, di Domenico S, Cianca E et al (2017) PCA-based artifact removal algorithm for stroke detection using UWB radar imaging. Med Biol Eng Compu 55:909–921
57. Paul LC, Hossain MN, Mowla MM, Mahmud MZ, Azim R, Islam MT (2019) Human brain tumor detection using CPW fed UWB Vivaldi antenna. In: IEEE international conference on biomedical engineering, computer and information technology for health (BECITHCON), Dhaka, Bangladesh, pp 1–6
58. Fiser O, Hruby V, Merunka I, Vrba J and Vrba J (2018) Numerical study of stroke detection using UWB radar. In: progress in electromagnetics research symposium (PIERS-Toyama), Toyama, pp 160–163
59. Widyatama PZ, Nugroho BS, Nur LO (2019) Design of circular modified UWB antenna microstrip for brain cancer detection. In: IEEE Asia Pacific conference on wireless and mobile (APWiMob), BALI, Indonesia
60. Lauteslager T, Tømmer M, Lande TS, Constandinou TG (2019) Coherent UWB radar-on-chip for in-body measurement of cardiovascular dynamics. IEEE Trans Biomed Circuits Syst 13(5):814–824
61. Brovoll S, Berger T, Paichard Y, Aardal Ø, Lande T, Hamran S (2014) Time-lapse imaging of human heart motion with switched array UWB radar. IEEE Trans Biomed Circuits Syst 8 (5):704–715
62. Li J, Zeng Z, Sun J, Liu F (2012) Through-wall detection of human being's movement by UWB radar. IEEE Geosci Remote Sens Lett 9(6):1079–1083
63. Maaref N, Millot P (2012) Array-based UWB FMCW through-the-wall radar. In: Proceedings of the 2012 IEEE international symposium on antennas and propagation, Chicago, IL, pp 1–2
64. Wu S, Tan K, Xu Y, Chen J, Meng S, Fang G (2012) A simple strategy for moving target imaging via an experimental UWB through-wall radar. In: 14th International conference on ground penetrating radar (GPR), Shanghai, pp 961–965
65. Liang X, Zhang H, Fang G, Ye S, Gulliver TA (2017) An improved algorithm for through-wall target detection using ultra-wideband impulse radar. IEEE Access 5:22101–22118
66. Akiyama Y, Kidera S (2019) Low complexity algorithm for range-point migration-based human body imaging for multistatic UWB radars. IEEE Geosci Remote Sens Lett 16(2):216–220
67. Saad M, Maali A, Azzaz MS, Kakouche I (2020) An experimental platform of impulse UWB radar for through-wall imaging based on FPGA. In: 1st International conference on communications, control systems and signal processing (CCSSP), EL OUED, Algeria, pp 198–201
68. Setsu M, Hayashi T, He J, Kidera S (2020) Super-resolution doppler velocity estimation by kernel-based range—τ point conversions for UWB short-range radars. IEEE Trans Geosci Remote Sens 58(4):2430–2443
69. Yang X, Chen P, Wang M, Guo S, Jia C, Cui G (2020) Human motion serialization recognition with through-the-wall radar. IEEE Access 8:186879–186889

Chapter 10
Emerging Technologies and Future Aspects

10.1 Introduction

The Internet of Things (IoT) is a new technology that connects a variety of physical objects, machines, sensors, servers, other devices, and people through communication networks [1, 2]. There are a plethora of applications and services offered by the IoT ranging from critical infrastructure to agriculture, retail, transportation, manufacturing, military, home appliances, and personal healthcare [3]. The huge scale of IoT networks brings new challenges related to management, processing of information, decision making and security/privacy. IoT is considered as one of the key technologies globally to make human life more productive, improve quality of life, economic growth, environment, safe, healthy, and comfortable by solving challenges and providing innovative solutions. It can be deployed in many fields including smart cities, homes, healthcare, industry, transportation, and agriculture [3–5]. Various domains in which IoT technology will prove beneficial and enhance service quality are presented in Fig. 10.1 [6].

The impact of IoT in conjunction with wireless and sensing technologies is significant and has various advantages. In the IoT, it is possible to collect, record and analyse new data streams faster and more accurately by making devices gather and share information directly with each other and the cloud. To perform IoT, a massive data amount having different content and formats has to be processed professionally, rapidly and intelligently using advanced techniques, algorithms, tools and models. This innovative paradigm is supported by the development of various technologies, such as the wireless communication, internet, cloud computing, machine learning algorithms and big data analysis [4].

The architecture of IoT consists of several technologies and layers to support the system and processes. The sensor layer which is the physical and the lowest layer consists of integrated smart objects along with the sensors which can also be wearable. The main purpose of the sensing layer is for sensing and gathering information about the environment. It senses some physical parameters or identifies other smart

S. K. Koul and R. Bharadwaj, *Wearable Antennas and Body Centric Communication*, Lecture Notes in Electrical Engineering 787,
https://doi.org/10.1007/978-981-16-3973-9_10

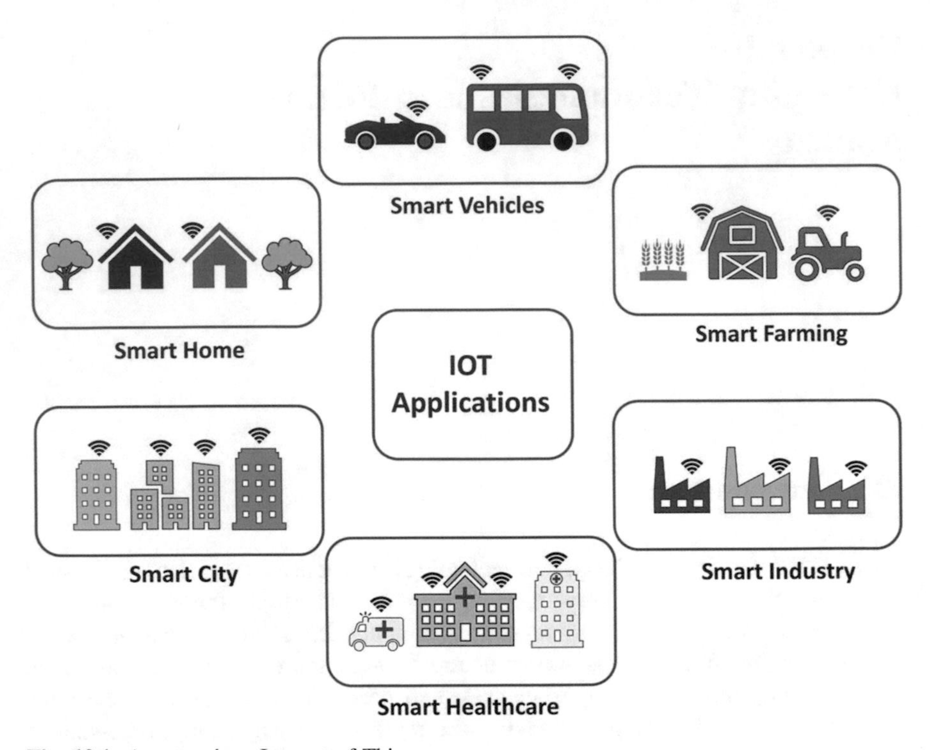

Fig. 10.1 An overview: Internet of Things

objects in the environment. The sensors can measure air quality, temperature, electricity, movement, and health related parameters. Connectivity is formed through wireless sensor networks (WSNs) and personal area network (PAN), through technologies such as Bluetooth, ZigBee, and ultra-wideband (UWB) or a local area network (LAN) and including WiFi and Ethernet connections.

The network layer is responsible for connecting to other smart things, network devices, and servers. Gateways and networks layer, where huge data volume is produced by the sensors, requires a high performance and robust and wired/wireless network infrastructure. These networks can be public, private or hybrid models to support the communication requests for bandwidth, latency, or security. Data processing layer deals with various analytics methods to extract applicable information from huge amount of raw data for processing in faster rate. During data processing, the information can be accessed, controlled, and integrated. Management service layer includes information processing through security controls, analytics, and devices management. The application layer is responsible for delivering application specific services to the user. It defines various applications in which the Internet of Things can be deployed, for example, smart homes, smart cities, and smart health care. Its features are also used for transmitting and processing sensor data. Finally, security check is performed across the whole IoT architecture dimensions [7–9]. An overview of the IoT architecture is shown in Fig. 10.2.

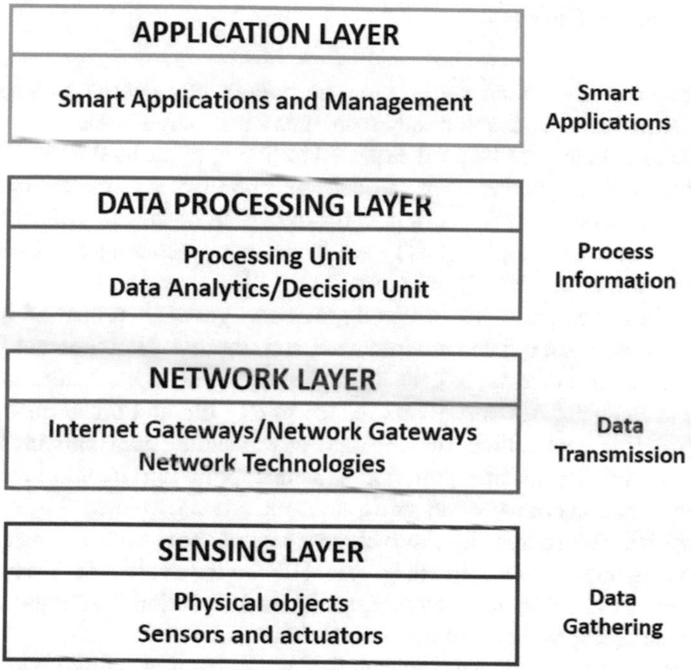

APPLICATION LAYER
Smart Applications and Management
Smart Applications

DATA PROCESSING LAYER
Processing Unit
Data Analytics/Decision Unit
Process Information

NETWORK LAYER
Internet Gateways/Network Gateways
Network Technologies
Data Transmission

SENSING LAYER
Physical objects
Sensors and actuators
Data Gathering

Fig. 10.2 Architecture of Internet of Things (IoT)

This chapter gives an overview of the IoT structure and discusses various applications and potentials of the IoT technology in various domains such as smart cities, smart home, and smart healthcare systems. IoT for personalized healthcare, smart hospital and remote monitoring services, intelligent diagnosis and decision-making aspects have been discussed in detail. With the advent of IoT ushers a new dimension to the development of practical, affordable, and easy-to-use healthcare solutions. The emergence of wearable devices with the IoT technology is a promising solution to the future health monitoring requirement in which antenna design and performance is of key importance. This chapter reports various technologies and state-of-the art techniques related to compact antenna design for body-centric communication for IoT applications. This chapter also explores the role of artificial intelligence (AI) and machine learning (ML) for healthcare IoT applications which would aid in accurate diagnosis, maximize the early detection of diseases and enhance the over-all healthcare system.

10.2 IoT Applications

IoT technology has played a prominent role in the upliftment and upgradation the of society. It converts cities into smart cities, electrical grids into smart grids, houses into smart home, tourisms into smart tourism and health into smart health system.

10.2.1 Smart Cities

An expansion of the urban ecosystems, as populations continue to sustain the transition from rural and some suburban areas into large urban areas, seventy percent of the human population is expected to live in cities by the year 2050 [10]. New technological solutions are needed to optimize the increasingly scarce infrastructure resources. When cities deploy state-of-the-art Information and Communication Technologies (ICT) on a large-scale, including IoT technologies, they are referred to as being "smart cities" [10–12].

Smart cities are aimed to efficiently manage growing urbanization, energy consumption, maintain a green environment, and improve the economic and living standards of their citizens. IoT technologies offer the opportunity to improve resource management of many assets related to city life and urban quality of life. Some of the key applications of the smart city include smart buildings, lighting management, intelligent transportation systems, pollution monitoring, resource monitoring, smart services, smart grids, unmanned aerial vehicle (UAV)s-assisted next-generation communication, surveillance/intelligence, crowd sensing, infrastructure management and asset management [13]. In the evolving IoT environment, the endpoint devices include environmental and situational sensors, actuators, vehicles, wearables, drones and robots [10–12].

There are various domains, such as the car and traffic monitoring and management, the city environment (streetlight, waste; pollution management), or the end-users themselves and their mobile appliances. Monitoring of vehicle traffic to detect traffic jam or damaged roadways and propose users for rerouting. Smart streetlights equipped with sensors that can detect human and car movement, which can dim with change in daylight or the amount of traffic in the zone under consideration. This is suitable for saving energy and ensuring security in the neighbourhood. Smart sensors can also detect pollution level in the vicinity and alert users and people living in the neighbourhood to take necessary action. City and urban planning can be based upon actual collected sensor data about how the city is used in terms of quantifying the inhabitant's mobility, and infrastructure needs. Monitoring and maintenance of structures such as roadways, buildings and bridges. IoT enables laying sensors on common roads for assistance with autonomous vehicles for public transportation, delivery, carpooling and shared transportation, etc. [10–12, 14].

10.2.2 Smart Home

Smart home is one of the most popular smart environments. Smart homes accommodate a variety of smart applications which include, smart energy metering/consumption, smart multimedia, and smart home healthcare [14, 15]. A typical smart home architecture divides the communication network into multiple components: a home network, with distributed sensors throughout the home; a gateway which collects the information from the sensor; a cloud hosted platform that receives the processed information from the gateway to store and analyse; and the mobile devices of the home occupants, who can connect to the gateway or the cloud server to receive information and notification about the home while away [15, 16].

Typical services of a smart home are presented in [17]. Centralized management can make electronic decisions such as monitoring, improving comfort, convenience, controlling surrounding conditions, and delivering required information [14]. Using machine learning and artificial intelligence methods from sensor data can track and detect changes in individuals' behavioural pattern and lifestyle [16]. The smart home domain includes all home equipment that can be connected together or to the Internet. Connected home appliances, such as fridges which can order new food products or beverages when it detects that a low threshold has been reached, pantry which can suggest recipes based upon available ingredients in the kitchen. The home can be equipped with small cameras and sensors for security purposes and detection of fire or any other hazard. Smart lightning and remote automation in which lights, electronic items could be turned off or on. Energy management to set the temperature and light in a room as a function of the number of people in the room, the time of day, the external conditions, the cost of the utility [15, 16].

10.2.3 Smart Vehicles

There have been significant investments in IoT devices and systems aimed at improving transportation systems, including traffic monitoring, live location streaming, and vehicle performance monitoring. Smart vehicles have sensors like radar, camera, light detection and range, petrol level indication along with actuator and controllers. These devices are driven by software, which is function specific software running on Electronic Control unit (ECU). Thus, machine learning software is also part of this and helps the vehicles to understand the surrounding environment and at the same time monitor its operations in real-time. Vehicle path predication and clustering which is based on road geometry and kinematic mechanism using vehicle trajectory-based data can be enhanced using machine learning algorithms which will aid to identify the destined direction of vehicle at different intersection [18]. The significance of vehicle trajectory in the task of collision avoidance, road hazard informing, scheduling, routing, and efficient method of movement in smart vehicle identification.

By placing smart antennas/sensors on the vehicles, sensing and detection of neighbouring vehicles is possible, monitoring distance between vehicles, estimation of the lane in which the vehicle is moving can prevent accidents and provide estimate of the proximity of vehicles. Monitoring the speed of a vehicle and detect presence of infrastructure through intelligent sensors is also another area which is being explored [19, 20].

Intelligent sensors and built-in cameras along with machine learning algorithms are proposed to monitor the driver's behaviour and generate alert if the driver is exhausted or drunk [21]. The data of the passengers in the vehicle can be transmitted to the emergency response team and infer important information.

10.2.4 Smart Industry

The explosion of the Internet of Things may lead to a breakthrough in industries: the capillary deployment of sensors, coupled with advanced analytics capabilities, can enable automatized and flexible processes that can be monitored in real time, reducing production and maintenance costs and unproductive downtime. Manufacturing considers data analytics, machine learning, cloud computing, robotics, and artificial intelligence to enhance over productivity [22–24]. Manufacturing systems go beyond simple connectivity and use the collected information to drive further intelligent actions and meet the demands for higher productivity, smart, green production, higher market share and flexibility.

New communication technologies could enable pervasive and continuous feedback inside the industry environments, allowing to achieve new industrial automation capabilities, such as intelligent logistics, real-time fault detection, asset tracking, remote visual monitoring, and remote robot control. With the advancement of computing and communication technologies, ML enables the analysis of massive quantities of data such as those produced by an IoT-based system, and can use the extracted knowledge (e.g., trained models) to aid real-time decision making in complex situations. Fault detection and isolation in industrial processes, real-time quality monitoring in additive manufacturing, and automatic fruit classification are some of the recent examples of deep learning (DL) in IoT-based Industry systems [22–25].

10.2.5 IoT for Healthcare

As the proportion of the world's population in older ages continues to increase, the need for improved strategies in provision of healthcare arises. Healthcare is considered one of the most important application areas of IoT, offering the potential for enhanced health care and management systems such as real-time monitoring, patient information administration, medical emergency, and population management [26, 27]. The dependence of healthcare on IoT is increasing by the day to improve access to personalized healthcare, increase the quality of care, patient support, improving efficiency of the healthcare system, reduce cost and general wellbeing. IoT benefits the healthcare system by real time remote monitoring, prevention of health issues, early pathology detection and diagnosis. Various IoT applications in healthcare are listed in Fig. 10.3.

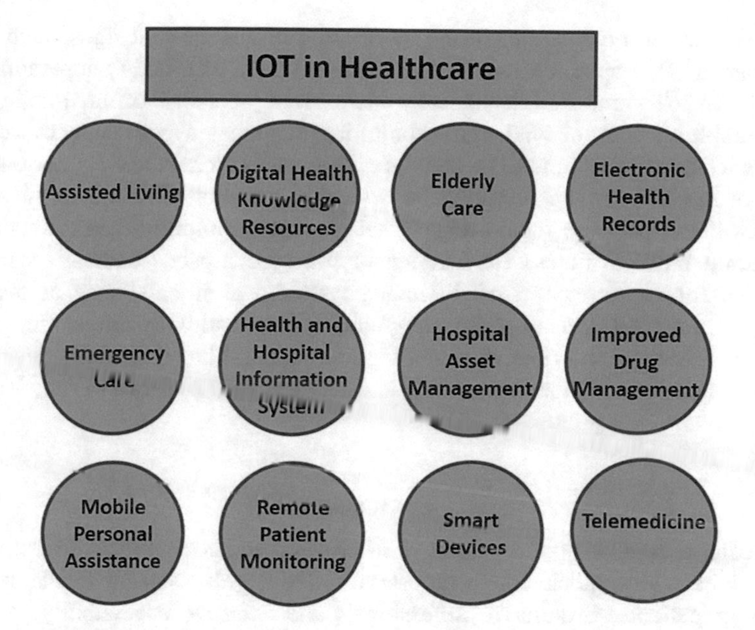

Fig. 10.3 IoT applications in healthcare

Healthcare applications supported by the IoT, connected things anyplace, anytime, with anyone perfectly using any network and any service, which lead to smart health care. For the IoT connected healthcare applications, the wireless body area network (WBAN) is gaining popularity as with the advancement of technology has led to miniaturised wearable devices. Multiple sensor nodes can be deployed on different positions of the body for e-health applications such as monitoring vital parameters, rehabilitation, early detection of medical issues and emergency notification. Portable devices such as smart phones can be employed as the gateway between the WBAN and the IoT cloud to transfer necessary information [28].

Healthcare remote monitoring through wireless sensors capable of sending information to relevant persons such as doctor, hospital, and relatives so that appropriate action should be taken. Several wearable sensors can be deployed in clothes, watches, wearable accessories, even jewels to continuously monitor the blood pressure, heartbeat, blood glucose level, blood oxygen level, standing position, etc. [27, 29]. Such wearable devices can also remind people to take their medication, and provide insightful suggestions to increase or reduce their prescribed quantities according to monitored metrics like blood glucose levels for diabetes, blood pressure etc. Various applications for in-house treatment and rehabilitation for elderly and injured people can be provided through IoT and still be under remote supervision of doctors who will be continuously obtaining health related parameters/status of the patients through wireless sensors and connectivity [28].

A smart health care system is presented in Fig. 10.4. The early diagnosis and treatment of chronic diseases can improve people's health conditions, which requires long-term continuous health monitoring of human vital signs, such as the heart rate (HR), respiration rate (RR), blood pressure (BP), body temperature, etc. [5, 30]. An IoT-connected healthcare system usually consists of three main parts: (1) wearable sensors for vital signs monitoring; (2) a gateway to connect wearable devices to the Internet; and (3) a cloud server for data storage and further analysis. Some of the IoT healthcare applications used to monitor different health aspects includes blood pressure monitoring, blood glucose monitoring, heart functioning, monitoring physical fitness etc. Different types of compact wearable sensors are presented for measurements of different physiological signals, such as the electrocardiogram (ECG), photoplethysmography (PPG), and body temperature which can provide critical information regarding the patients' health to individuals, doctors, healthcare professionals hospitals and medical data centres [14, 28, 30, 31].

10.2.5.1 Telehealth

Telemedicine is made possible through information and communication technology (ICT). It provides quick healthcare service delivery by storing health records, managing patients, healthcare professionals and improve accessibility, assistive care, and remote monitoring. Telehealth system offers teleconsultation through healthcare centres for analysing patients at home or hospital. This saves time and

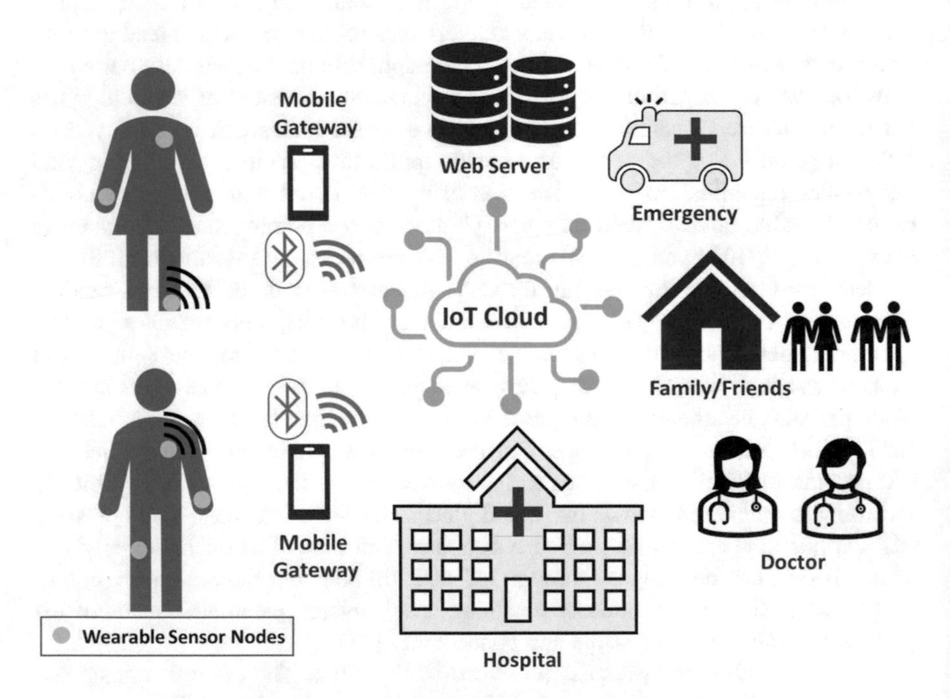

Fig. 10.4 Wireless body area network with IoT connected healthcare platform

provides secure access of health records by patients and doctors [32, 33]. Telehealth supports digital healthcare and multiple healthcare services are handled uniquely while servicing patients and in emergency scenarios. A wide range of research areas are needed to enhance telemedicine architecture such as network communications, artificial intelligence methods and techniques, IoT wearable sensors and hardware devices, smartphones, and cloud computing.

Telemedicine helps the patients and healthcare professionals through remote monitoring as seen in Fig. 10.4, [30, 34]. An important application of telemedicine is requirement of an immediate consultation in emergency situations, which can be tackled by solutions provided through teleconsultation. The doctors and medical professionals in hospitals can detect an emergency by monitoring real-time vital parameters such as blood viscosity, blood pressure, blood sugar of the patient through wearable devices. Hence, they can provide necessary treatment or alert nearest health center. The medical records of patients are stored securely in the cloud and the doctors create a network who can respond to emergency conditions of patients. The health data are stored in cloud, so that whenever the healthcare professionals need them, they can be easily retrieved and shared making cloud storage reliable. The cloud gives the comfort of security, data collection, data management, data accessibility and data analysis.

10.3 Antenna Design Requirements for IoT Body-Centric Communication Applications

Wearable compact antennas constitute a significant part of body-worn or handheld IoT devices to provide continuous communication between devices in various domain such as healthcare, telemedicine, and smart homes. Wearable antennas are an integral part for on- and off-body communication for remote monitoring, activity recognition, healthcare, sports, and leisure applications. The antennas should be very compact, low profile lightweight, durable, high radiation efficiency, energy efficient, preferably flexible, and conformal. Antenna is one of the key hardware element required to support the operation of the body-centric networks (BCNs) and its performance dictates working of the overall system. Antenna design is very challenging because of the strict requirements of low profile, small size, light weight, flexible shape, broad bandwidth, high gain, etc. [3]. Frequency domains such as ISM narrowband (2.4, 5.8 GHz), UWB (3.6–10 GHz) and mmWave frequency (24–30 GHz), (57–64 GHz), are some of the preferred frequency bands for operating various devices for body-centric IoT applications [35, 36].

The UWB technology is being used extensively in wearable communication due to its key features, such as low cost, low-power consumption, high data rate, fine time resolution, integration with various technologies. Apart from this, power spectral density over the entire UWB band is very low, which provides less interference [37]. It can be used for short-distance communication with high-speed data and low power consumption for continuous real-time monitoring of psychological patients' data.

With the emergence of advanced technologies of Internet of Things, 5G communications, and telehealth systems etc., importance of the BCNs is on the rise. The spectrum scarcity at 2.4 and 5 GHz and the resulting congestion have pushed a growing interest in the high frequency bands, where the large amount of bandwidth available greatly increases the system capacity and flexibility. In the development of BCN technologies, the mmWave band from 57 to 64 GHz has garnered much attention due to the scarcity of the frequency spectrum and increasingly high requirement of the network capacity and data rates [38]. The mmWave band is also becoming highly in demand because of license free operation, low interference, and confidentiality thanks to high atmospheric attenuation [38]. mmWave antenna yields high gain and bandwidth performance to achieve seamless wireless transmission. mmWaves suffer from higher path-loss resulting in fragile link, due to weak diffractions at these frequency bands. To overcome these issues, high gain and directive antenna, Multiple-Input Multiple-Output (MIMO) technology or phased array antennas, beamforming techniques antenna arrays using several elements are considered. mmWave Yagi-Uda antennas combine high gain with low cost and reduced size, and might result in compact and efficient antennas to be used in IoT sensors.

Several antennas are presented in open literature ranging from microstrip patch antennas, printed inverted-F antennas, loop antennas and slot antennas. Apart from this, novel antenna designs such as use of fractals, meta-materials, EGB structures, high dielectric constant substrates to improve the performance of the antenna play an important role in miniaturizing the antenna. Flexible and textile-based antennas also have vast applications in telemedicine and IoT healthcare aspects.

10.3.1 Band-Notch Antennas

For portable IoT sensors and WBANs, the compact-sized antenna has received much interest, and can be easily embedded in IoT devices. However, owing to the extremely broadband operation for UWB systems, there exists an inevitable overlap between UWB communication systems and many narrowband wireless communication systems. A compact-sized antenna for IoT applications which can achieve multiple-notched bands to avoid the potential interferences caused by these narrow-band communication systems within the UWB frequency band [39–41].

A microstrip UWB antenna with quintuple rejection bands were realized at the frequencies of 3.5, 4.5, 5.25, 5.7, and 8.2 GHz utilizing four rectangular complementary split ring resonators (R-CSRRs) on the radiating patch and placing two RSRRs near the feedline-patch junction of the conventional UWB antenna as shown in Fig. 10.5 [41]. A combination of the rectangular complementary split ring resonator (RCSRR) and the rectangular split-ring resonator (RSRR) for rejecting the WiMAX, INSAT, lower WLAN, upper WLAN, and ITU 8 GHz frequency

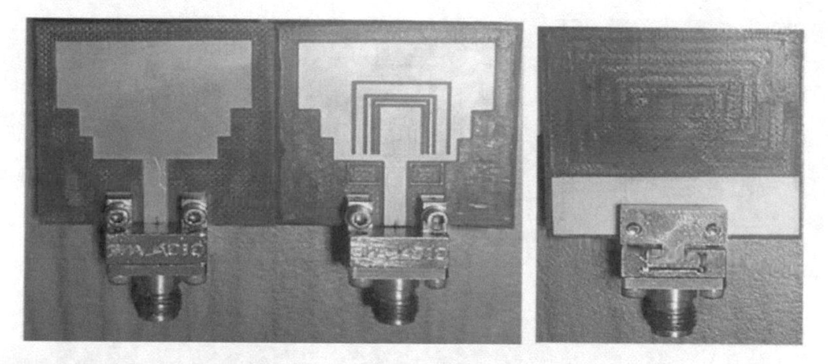

Fig. 10.5 Prototype of the reference and the band notched antenna (After [41], Sensors)

bands. A conventional stair-cased structured reference UWB antenna is designed and fabricated to operate within the desired UWB frequency band. This antenna was designed on Rogers RO5880 substrate with a thickness of 31 mils, a relative dielectric constant of 2.2, and a loss tangent of 0.0009.

10.3.2 Graphene and Nano-Particle Based Antennas

The possibility of printing electronic circuits will further promote the spread of the Internet of Things applications. Inks based on graphene have a chance to dominate this technology, as they potentially can be low cost and applied directly on materials like textile and paper. The printed patterns have excellent mechanical flexibility and has a great potential in wearable, deformable IoT applications [42].

A graphene-assembled film (GAF)-based compact and low-profile ultra-wide bandwidth (UWB) antenna is presented and tested for wearable applications [43] (Fig. 10.6a). The highly conductive GAFs together with the flexible ceramic substrate ensure the flexibility and robustness of the antenna. Two H-shaped slots are introduced on a coplanar-waveguide (CPW) feeding structure to adjust the current distribution and thus improve the antenna bandwidth. The compact GAF antenna with dimensions of $32 \times 52 \times 0.28$ mm^3 provides an impedance bandwidth of 60% (4.3–8.0 GHz) in simulation. The GAF with superb flexibility, light weight, and high conductivity of $\sim 10^6$ S/m enhances the performance of the antenna and is a good alternative material for the wearable antennas.

An inkjet-printed circular-shaped monopole UWB antenna with an inside-cut feed structure operating in 3.04–10.70 GHz in the UWB spectrum and 15.2–18 GHz is presented in [44] for wearable and Internet of Things applications (Fig. 10.6b). A commercially available PET substrate with dielectric constant of 3.2, a loss tangent of 0.022, and a thickness of 135 μm was used for designing the proposed antenna. The dimension of the antenna are $47 \times 25 \times 0.140$ mm^3.

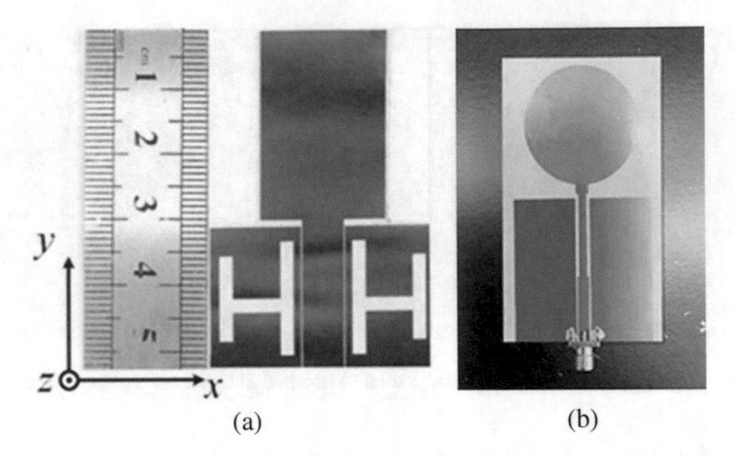

Fig. 10.6 **a** The fabricated GAF antenna prototype (After [43], Sensors). **b** Inkjet printed antenna on the paper substrate with silver nanoparticles ink (After [44], Microsystem Technologies)

10.3.3 3D Printing Based Antennas

Additive manufacturing (AM) or 3D Printing is a cluster of emerging technologies which is gaining a lot of interest in the electronic industry as it enables creation of objects bottom-up through layer-by-layer addition of versatile materials. AM use computer-aided design (CAD) virtual 3D models that are then translated into physical objects. This type of fabrication enables customized substrate structures, printed metallic and dielectric layers, electrical properties, and mechanical properties [45–47].

A 3D printed rectangular patch antennas on a removable fingernail for on-body commination operating at 28 GHz is presented in [46]. The 5G millimetre wave antenna has been embedded into a medallion using a 3D printing technique which combines fused deposition modelling (FDM) for the substrate and syringe dispensing for the metallic layers. The fake fingernail substrate is made of an Acrylonitrile butadiene styrene (ABS) of 0.5 mm thickness with a relative permittivity (ε_r) of about 2.7 and loss tangent of 0.005.

A rectangular UWB cavity-backed slot antenna manufactured by combining 3D printing and inkjet printing technology is introduced in [47]. The antenna substrate is made of polylactic acid (PLA) and is fabricated by using FDM and is shown in Fig. 10.7. PLA is chosen for the design due to its widespread use, low cost, low dielectric loss and nontoxicity. The measured permittivity for the infill was 2.72 with a loss tangent of 0.008. Using 3D printing technique, a cavity with slant sides providing high gain, has been designed. This shape is aimed at increasing the antenna impedance bandwidth and directivity. The rectangular-like proximity-coupled feeding line of the antenna is fabricated by depositing silver nano-particle ink on a layer of SU8. The antenna operates throughout the whole UWB (3.1–12 GHz) frequency band and is suitable for IoT applications.

Fig. 10.7 3D printed dielectric substrate with multiple infills and the additively manufactured cavity-backed UWB slot antenna (Reprinted with permission from IEEE [47])

10.3.4 Novel Electro-Textile and Materials

Smart textile systems represent new concept of garments which offer additional functionality such as sensing and communication, realized by wearable devices/antennas that are integrated into the "smart" garment. Fully textile antennas are usually flexible and lightweight solutions that can be seamlessly embedded in clothes without compromising wearability and user comfort suitable for several IoT applications [48–50]. These printed antennas are fully integrated, as its dielectric is the textile material composing the clothing itself. Typical e-textile materials used for these antennas are fabric coated with conductive material, and flexible waterproof foam sheets which were used for nonconductive spacer layers. E-textile are manufactured through integration of conductive and non-conductive yarns. Conductive yarns are constructed first by twisting copper threads (0.14 mm in diameter) with polyester threads by using a hollow-spindle machine after which the process is continued with the production of e-textile fabric by using the 'plain' weaving technique [48, 51]. Three antennas are fabricated on jeans substrate in the ISM band frequency range using three electro-textile materials: Cobaltex, Copper Polyester Taffeta and ShieldiT as shown in Fig. 10.8 [48].

10.3.5 Flexible Antennas

There is an unmet need for integrated, inexpensive, and conformal devices with a smaller footprint in the era of the IoT [52–54]. Wireless devices on flexible substrates play a critical role in numerous IoT applications including wearable's, healthcare, smart skins, functional clothing, and mobile network/internet devices. Flexible substrates have become essential for wearable IoT applications to provide increased flexibility in wearable sensors. Commonly used flexible materials for antennas in IoT applications, are Polyimides (PI), Polyethylene Terephthalate (PET), Polydimethylsiloxane (PDMS), Polytetrafluoroethylene (PTFE), polyethylene-naphtholate (PEN), and PDMS-coated silica nanoparticles, Rogers RT/Duroid and Liquid Crystal Polymer (LCP), Kapton,

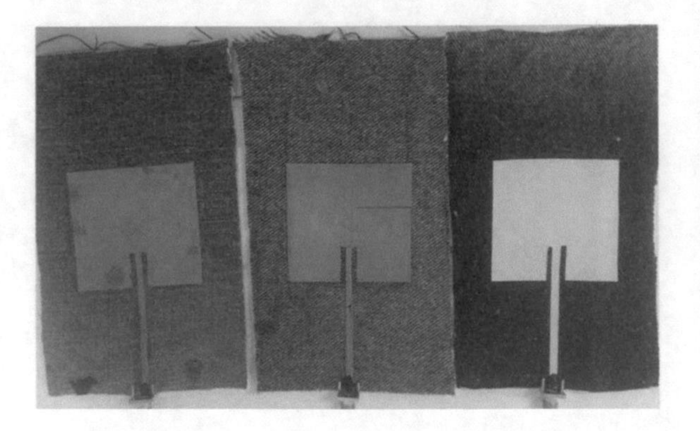

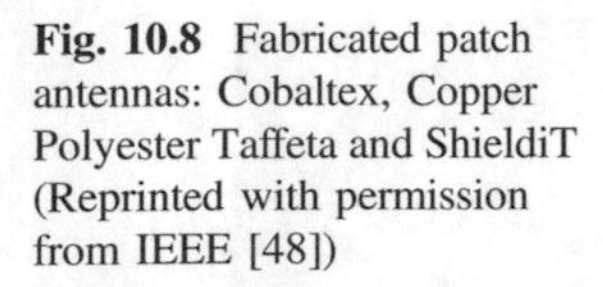

Fig. 10.8 Fabricated patch antennas: Cobaltex, Copper Polyester Taffeta and ShieldiT (Reprinted with permission from IEEE [48])

paper, textiles and fabrics. Flexible substrates such as Kapton Polyimide or PEN, is used due to its electrical and mechanical properties compared to other substrates [53, 54]. Various antenna performance characteristics are evaluated such as radiation pattern, gain, S-parameters, impedance mismatch for free space and on-body scenarios. Flexibility tests such as bending, crumpling, stretching and folding effects is carried out to study the variation in the antenna characteristics. Body-centric measurements related to specific absorption rate (SAR) analysis, humidity, bending effect, thermal and robustness test for wearable applications is reported [53, 54].

The UWB antennas, designed with fabric like jeans and cotton or perforated plastic as a substrate, can be used as a wearable device on the human body due to its low effect on the human body. A compact textile ultra-wideband (UWB) antenna with an electrical dimension of $0.24\ \lambda \times 0.24\ \lambda \times 0.009\ \lambda$ with microstrip line feed at lower edge and a frequency of operation of 2.96-11.6 GHz is proposed for UWB application as shown in Fig. 10.9, [52]. A wearable antenna using a flexible substrate working at 60 GHz is fabricated and has a slotted patch type structure. The structure of the proposed antenna is reported in [55]. The flexible printed circuit board has a thickness of 0.15 mm, $\varepsilon_r = 2.7$ and $\tan\delta = 0.005$.

10.3.6 Epidermal and Implantable Antennas

Advances in material science and the use of flexible material in skin mounted devices gave way to epidermal devices and sensors for medical applications. Epidermal antennas should be miniaturised, biocompatible, stretchable, adhesive, and fabricated on thin, conformal, and flexible substrates with minimal effect on the skin natural behaviour. Flexible substrates considered are medical grade adhesive dressings with an intermediate medical gauze, polyvinyl xyloglucan-poly (vinyl alcohol) based hydrogel films, bio silicon membrane, Dupont's Kapton polyimide, the silicon-based elastomer poly-dimethylsyloxane (PDMS), and the FDA-approved skin adhesive Tegaderm [56–58]. For the conductive antenna material, adhesive thin copper films or silver ink are generally used. A major advantage of

Fig. 10.9 Fabricated textile UWB antenna using jean cloth as substrate (After [52], Micromachines)

using polymer-based substrates lies in their flexibility and strength which arise from the strong intermolecular forces present in their structure. The epidermal Quadruple Loop (QL) antennas for Global System for Mobile Communications (GSM)-900 MHz, GSM-1800, and Bluetooth low energy (BLE) were fabricated using 75 microns thick Kapton substrates. For conductive layer, adhesive copper sheets were used. The QL antenna proposed is a very thin single-layer groundless structure which makes installation of neighbouring electronics and sensors feasible providing a compact structure to be placed on the epidermis layer [58]. An epidermal wideband antenna for medical radiometry is presented in [56]. The double asymmetric H-shaped slot antenna was designed to be matched to different parts of the body without fat layers (Fig. 10.10). The designed prototype was manufactured on a Rogers3003^{TM} flexible substrate with $\varepsilon_r = 3$, $\tan\delta = 0.0013$ at 10 GHz and thickness of 30 mil. The antenna is compact with 25 mm diameter and is relatively thin which can be conveniently placed on the human skin. The double H-shaped antenna shows good wideband matching results from around 1.5 GHz up to 5 GHz, in different body locations such as the neck, foot instep and foot sole.

The implantable medical devices also help in treatment and prevention of chronic health issues and monitor bio physical parameters. Implantable devices are made up of many components including antenna, battery, and sensors [59]. For implantable antennas some of the design requirements which should be met are miniaturization in size, wider bandwidth, biocompatibility, allowable SAR (specific absorption rate) and flexibility. Maity et al. have proposed a microstrip patch antenna with meander structure fractal geometry and operating within the ISM band with bandwidth of 1.5% in [60]. Fractal antennas are very compact, multiband, and wideband, and may be used in wearable communication systems. The effective area of a fractal antenna is significantly higher than the effective area of a regular printed antenna. The antenna has a volume of $11.44 \times 11.44 \times 0.275$ mm^3 and Silicon with $\varepsilon_r = 11.7$ is used as the substrate [60].

An indigestible capsule consists of an integrated communication system consisting of sensors, microcontroller and antenna. Ingestible capsule antenna design should be insensitive to surrounding lossy environment and electronics, sufficient bandwidth, miniaturization, low profile, sufficient gain, omnidirectional pattern, and

Fig. 10.10 Epidermal wideband antenna for medical radiometry (After [56], Sensors)

radiated safety [61]. The ingestible antenna resembles a patch antenna, which can achieve a directional pattern in the flat form, and omnidirectional patterns in the conformal form. Polyimide ($\varepsilon_r = 3.5$ and $\sigma = 0.008$) with a thickness of 0.15 mm is used as the flexible substrate so that the proposed antenna can spare a great deal of space. The total size of the capsule antenna in the flat form measures $34.5 \times 5.8 \times 0.15\ \text{mm}^3$ and is depicted in Fig. 10.11a. In the conformal form, the ingestible antenna fits into an $11 \times 22\ \text{mm}^2$ endoscope capsule, which occupies a volume of $\pi \times 5.8 \times (5.5^2 - 5.35^2)\ \text{mm}^3$. The antenna integrated into the capsule is shown in Fig. 10.11b.

10.3.7 Meta-Materials and Electromagnetic Band Gap (EBG) Structures

The performance of microstrip antennas has been enhanced with the introduction of artificial materials to suit multidisciplinary emerging technologies. Metamaterials (MTMs) possess several attractive features such as surface wave reduction and unique stopbands which are used to improve the bandwidth efficiency of small antennas for wearable medical and 5G communication IoT systems. Periodic split-ring resonators and metallic posts structures may be used to design materials with dielectric constant and permeability less than 1 [62].

Electromagnetic band gap (EBG) structures are widely used at microwave frequencies, as it improves the gain, bandwidth, tunability, and compactness. EBG structure with patch antenna is used for an increment of bandwidth and decrement of losses in the transmission line. An on-body antenna in practice needs not only to be compact and lightweight for better integration, it should also be efficient and induce minimal power absorption inside the human body. To overcome this problem, a band-gap material (EBG) structure can be integrated into the design.

Metamaterials have garnered significant attention in designing multiband antennas due to the possibility of tailoring ε and μ makes. A compact five band metamaterial slot antenna for WLAN/WiMAX/X-band system is proposed in [63].

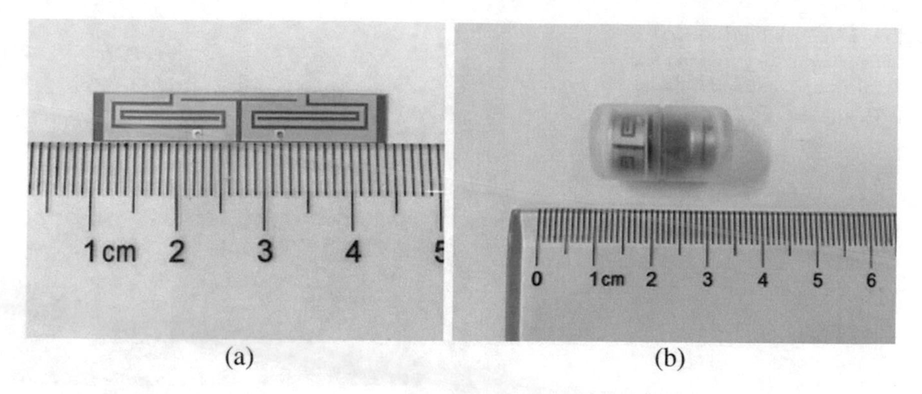

(a) (b)

Fig. 10.11 **a** Capsule antenna and **b** the integrated capsule system (Reprinted with permission from IEEE [61])

The proposed design consists of a ring monopole and metamaterial rectangular complementary split ring resonator (RCSRR) as the radiating part, two L and one T-shaped slot as the ground plane, respectively. The antenna has a dimension of $30.0 \times 24.8 \times 1.6\ \text{mm}^3$ and is printed on low cost, widely available FR4 substrate. The miniaturization process leads to about 46.8% reduction in volume of the proposed design, as compared to the conventional antenna.

A compact, metamaterial (MTM)-inspired UWB microstrip antenna is presented for microwave imaging system (MIS) application [64]. Two layers of left-handed metamaterial array (2×4) of the unit cell constructed using a modified a square split ring resonator (SRR) are placed on the radiating patch and the ground plane, respectively (Fig. 10.12). Each left-handed metamaterial (LHM) unit cell was constructed by modifying a square split ring resonator (SRR), resulting in negative permeability and permittivity with a stable negative refractive index. The results shows that it has a significant impact on the performance of the conventional patch antenna in terms of transmission co-efficient, efficiency and low loss. The proposed MTM antenna structure is fabricated on commercially-available, FR4 substrate having dielectric constant 4.6. The dimensions of the antennas are 26×22 $1.6\ \text{mm}^3$ and it achieves a bandwidth of 7.61 GHz from 3.1 to 10.71 GHz.

A millimetre-wave (mmWave) textile antenna with an EGB structure operating at 26 GHz band for 5G and IoT applications is proposed in [65]. The antenna is fabricated on a polyester fabric substrate along with the EBG structure which improves the performance of the antenna (Fig. 10.13). The EBG unit cells are used to surround the patch radiator, which offers the advantage of reducing the surface wave influence. Surface waves suppression can reduce the quantity of power wasted, and back radiation leads to improved antenna performance aspects such as enhancing the gain, side-lobe, and back-lobe reduction, and improving the energy efficiency. The gain and energy efficiency at 26 GHz were 8.65 dBi and 61%, respectively (an increase of 2.52 dB and 7% compared to a conventional antenna), and the specific absorption rate (SAR) was reduced by more than 69.9%. Design of

Fig. 10.12 A miniaturized metamaterial slot antenna for wireless applications (After [64], Applied Sciences)

Fig. 10.13 A textile EBG-based antenna (After [65], Electronics)

flexible wearable circular patch metamaterial antennas with high efficiency for 5G, IoT and biomedical applications are presented in [66]. Periodic split ring resonators (SRRs) and metallic posts structures may be used to design materials with specific dielectric constant and permeability. The circular patch metamaterial wearable antennas are compact and flexible. The directivity and gain of the antennas with Circular Split-Ring Resonators (CSRR) is higher by 2.5–3 dB than the antennas without CSRR.

10.3.8 MIMO Antennas

MIMO technology provides higher data throughput along with an increase in range and reliability by exploiting multipath which is difficult to tackle with single antenna systems. The antenna diversity techniques such as polarization, spatial, and pattern diversity is exploited by MIMO to enhance the power of the transmitted signals, and thereby it enhances the signal to noise ratio (SNR) and transmission capacity.

A practical challenge of MIMO design for space limited applications is to achieve small size of the physical structure while maintaining low mutual coupling

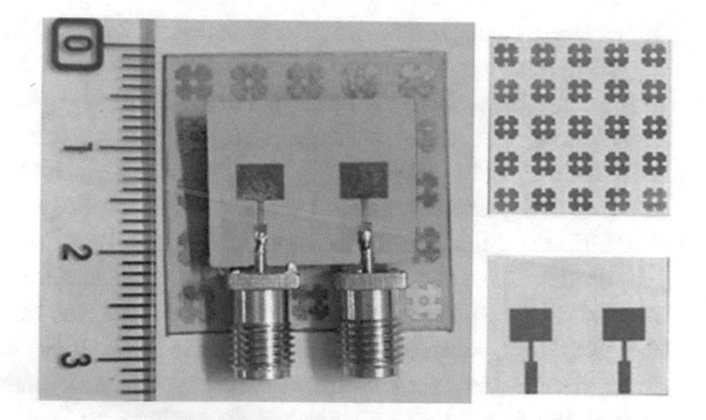

Fig. 10.14 Compact planar MIMO antenna for IoT applications (After [68], IEEE)

between individual antennas. Typical realisations exploit orthogonally allocated antennas with a common ground plane and various decoupling structures to improve isolation [67–69].

A two port, two element MIMO antenna operating at mmWave (24 GHz) range for wearable applications is proposed in [67]. The antenna is backed by 5 × 5 EBG, structure which reduces the back radiations from the antenna and increases the gain of the antenna [68]. The antenna structure is shown in Fig. 10.14. The antenna has good performance in on-body worn scenarios in term of conformability and spatial power density. The antenna is fabricated on a flexible Rogers 6002 material (ε_r = 2.94, tanδ = 0.0012, thickness = 0.254 mm). The EBG in this design reduce the near-field magnetic coupling and hence reduce the mutual coupling of the antenna.

A compact design of MIMO antenna (18 × 34 mm^2) with dual sharply rejected notch bands for portable wireless UWB applications is presented in [69]. The tapered microstrip fed slot antenna acts as a single radiating element with inverted L-shaped slits to introduce notches at wireless local area network and the IEEE INSAT/Super-Extended C-bands. The mutual coupling of less than −22 dB is achieved over the entire operating band (2.93–20 GHz).

10.4 Machine Learning for Improved Well-Being and Healthcare Applications

IoT is an emerging technology that is significantly improving the quality of services in the medical and health domains through innovative methods, miniaturised wearable devices and resources. With the rapid expansion of sensors and connected devices, intelligent IoT systems generate big data and provide smart solutions for remote monitoring IoT, disease detection, diagnostics, and assistance. ML mechanism provides proper learning and training to E-health automated machines/devices through which it can perform data classification, interpretation, analysis, and visualization. ML is a mathematical and statistical approach which is interrelated through data mining techniques with statistical, supervised, and unsupervised learning methods [70, 71]. Using machine learning and artificial intelligence methods from wearable devices/sensor data can enhance connecting, monitoring, decision making, assessment and diagnosis capabilities. Activity recognition, data processing, decision-making, image recognition, prediction-making, and voice recognition are few of the applications in which the classification capabilities of the ML algorithms can be explored in the healthcare domain.

The data is collected from several health wearable devices worn by patients in various areas such as hospital, homes or workstations which are processed to E-Health care monitoring system. These wearable sensor devices are connected through wireless sensor networks, routing antennas, and base stations. ML approaches is applied on raw data to analyse the health and obtain accurate diagnosis of the chronic diseases. Healthcare monitoring system can be utilized to help the patient in regular heart rate, temperature sensor, heartbeat sensor and blood pressure sensors monitoring. Collected patient's information is used for prediction, review analysis, decision making, and data visualization which is shared by the doctor, patient, and caretaker [70, 71] (Fig. 10.15). E-Health monitoring system can provide qualitative and security services like regular monitoring, valuable collection of data, proper diagnosis analysis, and in-time patient services [72]. Machine learning (ML) and deep learning (DL) techniques, which can provide embedded intelligence in the IoT devices and networks, can also be leveraged to cope with different security problems. Intelligent classification of various disease using IoT based data has been carried out by employing machine learning techniques. Commonly used machine learning-based algorithms and classifiers are Support vector machine (SVM), Decision tree, neural networks, k-Nearest Neighbour (k-NN) and Regression algorithms, Random Forest, and Naive Bayes [72, 73].

Personalized healthcare (PH) is a new patient-oriented healthcare approach which expects to improve the traditional healthcare system [74]. Modern healthcare systems now rely on advanced computing methods and technologies, such as Internet of Things devices, analytical ML models and clouds, to collect and analyse personal health data. It aids in remotely monitor patients, detect abnormal changes in one's health, early-diagnose of diseases, disease prediction, and patient self-management and find personalized treatments and medications. These models and data are integrated into different healthcare service applications and clinical decision support systems [74, 75].

10.4.1 Electronic Health Records Maintenance and Data Mining

The digitized healthcare system stores and analyses the collected data to build analytic models that provide a myriad of health services to the patients, such as real-time monitoring for identifying health anomalies. Patients' digitized health

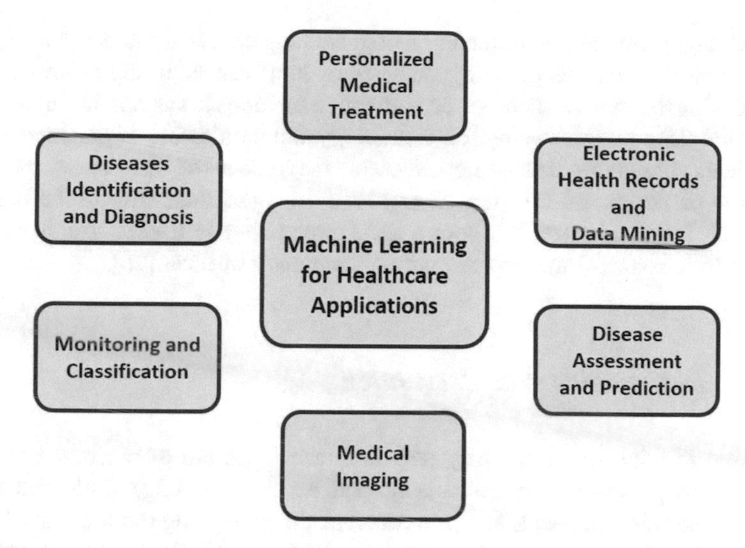

Fig. 10.15 Machine learning in IoT healthcare applications

records, clinical findings, doctor's prescription are stored securely as electronic health records (EHR) [72, 73]. IoT envisions the interconnection of healthcare devices and sensors to enable a new class of applications relying on medical data processing and storage. It is related to real-time retrieval, availability, and safe storage of health records in privacy preserved form. Vast amount of biomedical literature, journals, clinical notes are available which can aid in anomaly detection, enhance medical practitioner's knowledge, and optimise decisions. Deploying of machine-learning and other data-mining models, to analyse and identify the status of a patient's condition and to send cautions or alarms indicating the need for immediate medical care [74, 75].

10.4.2 Monitoring and Classification

Wearable devices can provide useful information related to internal anatomy, heart, lung, and other organ's function. Robust analytics would provide good first level health condition or anomaly detection which would enable early and timely detection of critical diseases. Data is collected in the form of signals or images from the sensor devices and other sources to identify activity, behavioural patterns, and clinical conditions of the patient. Activity monitoring, classification and recognition is possible using ML models to identify the persons movement, behaviour, daily routine, state, or posture [70–73]. The ML-based method will analyse the situation according to the trained dataset and generate the results. Based on such activity patterns the healthcare applications and the clinical decision support systems recommend lifestyle advice, special treatment, and care plans for the patient.

A ML algorithm can be trained to recognise lung cancer tissue from a variety of large samples which are pathology images of lung cancer tissue inferring results from previously stored data. A cost-effective inventive system to monitor and analyse the sleep patterns using IoT technology and monitoring of parameters based on patients bio-status including physical movement of the body, heartbeat, Saturation of Peripheral Oxygen (SPO_2) level (oxygen saturation in the blood for the proper functioning of the body). Data processing is carried using statistical methods to measure sleep patterns with an accuracy of 95% [76].

10.4.3 Diagnostics and Prevention

Supervised learning-based anomaly detection techniques are of practical importance owing to the presence of noted datasets and expert knowledge. This will lead to accurate diagnosis, treatment, and prevention of the underlying disease. Health monitoring and imaging devices such as Electrocardiogram (ECG), Photoplethysmogram (PPG), Electroencephalogram (EEG), MRI, CT scan can provide important information related to heart, brain etc. by emitting electrical and other sensory signals which will aid in diagnosis of various health related issues [71, 72]. Signal processing methods, image feature extraction and classification, 3D image processing, followed by robust anomaly detection would result in intelligent and smart biomedical signal/image classification, analysis, and disease diagnosis through intelligent ML algorithms.

10.4.4 Assessment and Prediction

Predictive modelling for diagnosis and providing medications is an emerging requirement of near future healthcare analytics. Assessment of an individual's health can lead to early diagnosis of diseases even with presence of few symptoms. Sophisticated ML algorithms and computational skills will improve the accuracy and precision of clinical predictions [73, 74]. To detect the stress beforehand, heartbeat rate has been used as one of the parameters. IoT along with ML can be used to alert the situation when the person is in stress by monitoring the heartbeat rate, thus predicting stress which is communicated to the person regarding his condition [77].

10.5 Conclusion

The fields of computer science and electronics have merged to result into one of the most notable technological advances in the form of realization of the Internet of Things. The Internet of Things will redesign the society, and will bring seamless 'anytime, anywhere' services such as remote monitoring, sensing over fast reliable and secure networks. IoT empowered devices simultaneously enhance the quality of care with regular monitoring and reduce the cost of care and actively engage in data collection and analysis of the same. Miniaturized antennas, novel materials, design methodologies and various technologies such as ISM band, UWB, mmWave communication will lead to high communication speeds and enhanced performance of IoT based wearable devices/systems. The popularity of miniature wearable devices, embracement of artificial intelligence and machine learning as well as the rapid growth of big data analytics are creating promising opportunities and unique prospects in customized and futuristic IoT-healthcare services.

References

1. Chris Y, AbouGhaly MA, Kerim B, Harb HAM (2019) An extended review on internet of things (IoT) and its promising applications communications on applied electronics (CAE). Found Comput Sci FCS, New York, USA 9(26):8–22
2. Bhatt C, Dey N, Ashour AS (2017) Internet of things and big data technologies for next generation healthcare. Springer, Cham, Switzerland
3. John Dian F, Vahidnia R, Rahmati A (2020) Wearables and the internet of things (IoT), applications, opportunities, and challenges: a survey. IEEE Access 8:69200–69211
4. Čõlakovic A, Hadialic M (2018) Internet of things (IoT): a review of enabling technologies, challenges, and open research issues. Comput Netw 144:17–39
5. Ali ZH, Ali HA (2020) Towards sustainable smart IoT applications architectural elements and design: opportunities, challenges, and open directions. J Supercomput
6. Nuruddeen AU (2019) An overview: internet of things, 5G communication system and cloud computing
7. Minoli D, Occhiogrosso B (2018) Ultrawideband (UWB) technology for smart cities IoT applications. In: IEEE international smart cities conference (ISC2), Kansas City, MO, USA, pp1–8
8. Sethi P, Sarangi SR (2017) Internet of things: architectures protocols and applications. J Electr Comput Eng 2017:1–25
9. Architecture of Internet of Things (IoT). https://www.geeksforgeeksorg/architecture-of-internet-of-things-iot/
10. Kim T, Ramos C, Mohammed S (2017) Smart city and IoT. Futur Gener Comput Syst 76:159–162
11. Minoli D, Occhiogrosso B (2018) Internet of things applications for smart cities. In: Hassan Q (ed) Internet of things A to Z: technologies and applications. Chapter 12, IEEE Press/Wiley
12. Ullah Z, Al-Turjman F, Mostarda L, Gagliardi R (2020) Applications of artificial intelligence and machine learning in smart cities. Comput Commun 154:313–323
13. Kumar S, Tiwari P, Zymbler M (2019) Internet of things is a revolutionary approach for future technology enhancement: a review. J Big Data 6(1):111
14. Mavromoustakis CX, Mastorakis G, Batalla JM (2016) Internet of things (IoT) in 5G mobile technologies. Springer, Cham, Switzerland, p 8
15. Jabbar WA, Alsibai MH, Amran NSS and Mahayadin SK (2018) Design and implementation of IoT-based automation system for smart home. In: International symposium on networks, computers and communications (ISNCC), Rome, Italy, pp 1–6
16. Zaidan AA, Zaidan BB (2020) A review on intelligent process for smart home applications based on IoT: coherent taxonomy, motivation, open challenges, and recommendations. Artif Intell Rev 53:141–165

17. Smart city—What is smart home? http://www.infiniteinformationtechnology.com/iot-smart-city-what-is-smart-home
18. Askari H, Khajepour A, Khamesee MB, Wang ZL (2019) Embedded self-powered sensing systems for smart vehicles and intelligent transportation. Nano Energy 66
19. Celesti AG, Carnevale L, Fazio M, Ĺay-Ekuakille A, Villari M (2018) An IoT cloud system for traffic monitoring and vehicular accidents prevention based on mobile sensor data processing. IEEE Sens J 18(12):4795–4802
20. Kamble SJ, Kounte MR (2019) On road intelligent vehicle path predication and clustering using machine learning approach. Third international conference on I-SMAC (IoT in Social, Mobile, Analytics and Cloud) (I-SMAC), Palladam, India, 2019, pp 501–505
21. Hansen JHL, Busso C, Zheng Y, Sathyanarayana A (2017) Driver modeling for detection and assessment of driver distraction: examples from the UTDrive test bed. IEEE Signal Process Mag 34(4):130–142
22. Breivold HP (2017) Internet-of-things and cloud computing for smart industry: a systematic mapping study. In: 2017 5th International conference on enterprise systems (ES), Beijing, pp 299–304
23. Li L, Ota K, Dong M (2018) Deep learning for smart industry: efficient manufacture inspection system with fog computing. IEEE Trans Industr Inf 14(10):4665–4673
24. Arachchige PCM, Bertok P, Khalil I, Liu D, Camtepe S, Atiquzzaman M (2020) A trustworthy privacy preserving framework for machine learning in industrial IoT systems. IEEE Trans Industr Inf 16(9):6092–6102
25. IIC: Industrial IoT Reference Architecture. https://iiot-world.com/connected-industry/iic-industrial-iot-reference-architecture/
26. Philip V, Suman VK, Menon VG, Dhanya KA (2017) A review on latest internet of things based healthcare applications. Int J Comput Sci Inf Secur 15(1):248
27. Selvaraj S, Sundaravaradhan S (2020) Challenges and opportunities in IoT healthcare systems: a systematic review. SN Appl Sci 2:139
28. Wu F, Wu T, Yuce M (2018) An internet-of-things (IoT) network system for connected safety and health monitoring applications. Sensors 19(1):21
29. Laplante PA, Laplante N (2016) The internet of things in healthcare: potential applications and challenges. IT Prof 18(3)
30. Wu T, Wu F, Redouté J, Yuce MR (2017) An autonomous wireless body area network implementation towards IoT connected healthcare applications. IEEE Access 5:11413–11422
31. Wu T, Wu F, Qiu C, Redouté JM, Yuce MR (2020) A rigid-flex wearable health monitoring sensor patch for IoT-connected healthcare applications. IEEE Internet Things J 7(8):6932–6945
32. Albahri AS et al (2021) IoT-based telemedicine for disease prevention and health promotion: state-of-the-art. J Netw Comput Appl 173
33. Latha R, Vetrivelan P, Geetha S (2019) Telemedicine setup using wireless body area network over cloud. Procedia Compu Sci 165:285–291
34. Albalawi U, Joshi S (2018) Secure and trusted telemedicine in internet of things IoT. IEEE 4th World Forum on Internet of Things (WF-IoT) Singapore 2018:30–34
35. Varma S, Sharma S, John M, Bharadwaj R, Dhawan A, Koul SK (2021) Design and performance analysis of compact wearable textile antennas for IoT and body-centric communication applications. Int J Antennas Propag
36. Sabban A (2020) Wearable systems and antennas technologies for 5G, IoT and medical systems. CRC Press
37. Mustaqim M, Khawaja BA, Chattha HT, Shafique K, Zafar MJ, Jamil M (2019) Ultra-wideband antenna for wearable internet of things devices and wireless body area network applications. Int J Numer Model 32:2590
38. Ramos T, Varum T, Matos J (2018) Compact multilayer Yagi-Uda based antenna for IoT/5G sensors. Sensors 18(9):2914
39. Shome PP, Khan T, Laskar RH (2019) A state-of-art review on band-notch characteristics in UWB antennas, Int J RF Microwave Comput-Aided Eng

40. Khan MM, Sultana A (2020) Novel and compact ultra-wideband wearable band-notch antenna design for body sensor networks and mobile healthcare system. Eng Proc 3(1):1
41. Rahman M, Park JD (2018) The smallest form factor UWB antenna with quintuple rejection bands for IoT applications utilizing RSRR and RCSRR. Sensors 18(3):911
42. Pan K, Fan Y, Leng T et al (2018) Sustainable production of highly conductive multilayer graphene ink for wireless connectivity and IoT applications. Nat Commun 9:5197
43. Fang R, Song R, Zhao X, Wang Z, Qian W, He D (2020) Compact and low-profile UWB antenna based on graphene-assembled films for wearable applications. Sensors 20(9):2552
44. Saha TK, Knaus TN, Khosla A, Sekhar PK (2018) A CPW-fed flexible UWB antenna for IoT applications. Microsyst Technol 1–7
45. Xia ZX, Leung KW, Lu K (2019) 3-D-Printed wideband multi-ring dielectric resonator antenna. IEEE Antennas Wirel Propag Lett 18(10):2110–2114
46. Njogu P, Sanz-Izquierdo B, Elibiary A, Jun SY, Chen Z, Bird D (2020) 3D printed fingernail antennas for 5G applications. IEEE Access 8:228711–228719
47. Palazzi V et al (2017) A novel additive-manufactured multiple-infill ultra-lightweight cavity-backed slot antenna for UWB applications.In: 47th European microwave conference (EuMC), Nuremberg, Germany, pp 252–255
48. Agbor I, Biswas DK and Mahbub I (2018) A comprehensive analysis of various electro-textile materials for wearable antenna applications. In: Texas symposium on wireless and microwave circuits and systems (WMCS), Waco, TX, pp 1–4
49. Loss C, Gonçalves R, Lopes C, Pinho P, Salvado R (2016) Smart coat with a fully-embedded textile antenna for IoT applications. Sensors 16(6):938
50. Abd Rahman NH, Yamada Y, Amin Nordin MS (2019) Analysis on the effects of the human body on the performance of electro-textile antennas for wearable monitoring and tracking application. Materials 12(10):1636
51. Corchia L, Monti G, Tarricone L (2019) Wearable antennas: nontextile versus fully textile solutions. IEEE Antennas Propag Mag 61(2):71–83
52. Yadav A, Kumar Singh V, Kumar Bhoi A, Marques G, Garcia-Zapirain B, de la Torre Díez I (2020) Wireless body area networks: uwb wearable textile antenna for telemedicine and mobile health systems. Micromachines 11(6):558
53. Paracha KN, Abdul Rahim SK, Soh PJ, Khalily M (2019) Wearable antennas: a review of materials, structures, and innovative features for autonomous communication and sensing. IEEE Access 7:56694–56712
54. Ali Khan MU, Raad R, Tubbal F, Theoharis PI, Liu S, Foroughi J (2021) Bending analysis of polymer-based flexible antennas for wearable general IoT applications: a review. Polymers 13 (3):357
55. Ur-Rehman M et al (2018) A wearable antenna for mmWave IoT applications. In: IEEE international symposium on antennas and propagation and USNC/URSI national radio science meeting, Boston, MA, pp 1211–1212
56. León G, Herrán LF, Mateos I, Villa E, Ruiz-Alzola JB (2020) Wideband epidermal antenna for medical radiometry. Sensors 20(7):1987
57. Occhiuzzi C, Ajovalasit A, Sabatino MA, Dispenza C, and Marrocco G (2015) Rfid epidermal sensor including hydrogel membranes for wound monitoring and healing. In: IEEE International Conference RFID (RFID), pp 182–188
58. Damis HA, Khalid N, Mirzavand R, Chung H, Mousavi P (2018) Investigation of epidermal loop antennas for biotelemetry IoT applications. IEEE Access 6:15806–15815
59. Malik NA, Sant P, Ajmal T, Ur-Rehman M (2021) Implantable antennas for bio-medical applications. IEEE J Electromagn RF Microwaves Med Biol 5(1):84–96
60. Maity S, Roy Barman K, Bhattacharjee S (2018) Silicon-based technology: circularly polarized microstrip patch antenna at ism band with miniature structure using fractal geometry for biomedical application. Microwave Optical Technol Lett 60(1):93–101
61. Zhang K et al (2018) A conformal differentially fed antenna for ingestible capsule system. IEEE Trans Antennas Propag 66(4):1695–1703

62. Zhang K, Soh PJ, Yan S (2020) Meta-wearable antennas—a review of metamaterial based antennas in wireless body area networks. Materials 14(1):149
63. Ali T, Saadh AWM, Biradar RC, Anguera J, Andújar A (2017) A miniaturized metamaterial slot antenna for wireless applications. AEU-Int J Electron Commun 82:368–382
64. Mahmud M, Islam M, Misran N, Singh M, Mat K (2017) A negative index metamaterial to enhance the performance of miniaturized UWB antenna for microwave imaging applications. Appl Sci 7(11):1149
65. Wissem EM, Sfar I, Osman L, Ribero JM (2021) A textile EBG-based antenna for future 5G-IoT millimeter-wave applications. Electronics 10(2):154
66. Sabban A (2020) New compact wearable metamaterials circular patch antennas for IoT, medical and 5g applications. Appl Syst Innovation 3(4):42
67. Jha KR, Bukhari B, Singh C, Mishra G, Sharma SK (2018) Compact Planar multistandard MIMO antenna for IoT applications. IEEE Trans Antennas Propag 66(7):3327–3336
68. Iqbal A et al (2019) Electromagnetic bandgap backed millimeter-wave MIMO antenna for wearable applications. IEEE Access 7:111135–111144
69. Chandel R, Gautam AK, Rambabu K (2018) Tapered fed compact UWB MIMO-diversity antenna with dual band-notched characteristics. IEEE Trans Antennas Propag 66(4):1677–1684
70. Jagannath J, Polosky N, Jagannath A, Restuccia F, Melodia T (2019) Machine learning for wireless communications in the internet of things: a comprehensive survey. Ad Hoc Networks 93
71. Machine Learning in Healthcare to Enhance Your Solution. https://innovecs.com/blog/machine-learning-in-healthcare/
72. Godi B, Viswanadham S, Muttipati AS, Samantray OP and Gadiraju SR (2020) E-healthcare monitoring system using IoT with machine learning approaches. In: international conference computer science engineering and applications (ICCSEA), Gunupur, India, pp 1–5
73. Panda S, Panda G (2020), Intelligent Classification of IoT Traffic in healthcare using machine learning techniques. In: 6th International conference on control, automation and robotics (ICCAR), Singapore, pp 581–585
74. Ahamed F, Farid F (2018) Applying Internet of things and machine-learning for personalized healthcare: issues and challenges. International conference on machine learning and data engineering (iCMLDE), Sydney, Australia, pp 19–21
75. Jindal M, Gupta J, Bhushan B (2019) Machine learning methods for IoT and their future applications. In: 2019 International conference on computing, communication, and intelligent systems (ICCCIS), Greater Noida, India, pp 430–434
76. Saleem K, Sarwar Bajwa I, Sarwar N, Anwar W, Ashraf A (2020) IoT healthcare: design of smart and cost-effective sleep quality monitoring system. J Sens 17
77. Pandey PS (2017) Machine learning and IoT for prediction and detection of stress. In: 17th International conference on computational science and its applications (ICCSA), Trieste, pp1–5

Printed in the United States
by Baker & Taylor Publisher Services